Quality Engineering Series
Volume 4

TAGUCHI METHODS™ : DESIGN OF EXPERIMENTS

Taguchi Methods™ Design of Experiments is part of a seven-volume Quality Engineering Series that the American Supplier Institute (ASI), Inc., Dearborn, Michigan, is publishing in conjunction with the Japanese Standards Association (JSA), Tokyo, Japan. ASI is publishing the English editions of the series and JSA is publishing the Japanese editions. The other volumes in the series are:

Volume 1:
Taguchi Methods: Research and Development

Volume 2:
Taguchi Methods: On-Line Production

Volume 3:
Taguchi Methods: Signal-to-Noise Ratio for Quality Evaluation

Volume 5:
Taguchi Methods: Case Studies from Japan

Volume 6:
Taguchi Methods Case Studies from the U.S. and Europe

Volume 7:
Taguchi Methods: Case Studies in Measurement from Japan

Quality Engineering Series
Volume 4

TAGUCHI METHODS

Design of Experiments

DR. GENICHI TAGUCHI
Publication Committee Chairman

YOSHIKO YOKOYAMA
Chief Editor

YUIN WU
Translator for the English Edition

Japanese Standards Association

ASI Press

Published by the ASI Press, an ASI company, in conjunction with the Japanese Standards Association, Tokyo, Japan. Printed in the United States of America.

Library of Congress Cataloging-in-Publication Data

Taguchi, Gen'ichi, 1924-
Taguchi methods : design of experiments / Genichi Taguchi, Toshiko Yokoyama, chief editor, Yuin Wu, technical editor for the English edition.
p. cm. — (Quality engineering series ; v. 4)
Translation from Japanese.
Includes bibliographical references and index.
ISBN 0-941243-18-4 : $45.00
1. Quality control—Statistical methods. 2. Experimental design. 3. Taguchi methods (Quality control) I. Yokoyama, Yoshiko, 1941- . II. Title. III. Series.
TS156.T3415 1993
658.5'62'015195—dc20 93-16742
CIP

Contents

About the Series...

"The Taguchi Method," currently the focus of attention in North America and European countries, is an engineering approach to quality control or what is called the quality engineering advocated by Dr. Genichi Taguchi. Called Taguchi Methods in the U.S., it has a long history in Japan where related research has been accumulated. In order to systematically present Taguchi Methods in the form of a series of publications, the Japanese Standards Association, the major promoter of Taguchi Methods for many years, initiated the Publication Committee in August 1987 with the full support of the American Supplier Institute, Inc., the promoter of Taguchi Methods in the U.S.

During the Publication Committee's discussion examination about what structure this series should take, it was suggested that actual application examples be published as soon as possible because such examples help readers understand and master the methods of quality engineering and have been requested by many users of Taguchi Methods. Aside from the four volumes that summarize the basic methods and approaches of quality engineering, it was planned that many application examples from different fields be published. This publication series is a global endeavor, in that both Japanese and English editions are being published simultaneously, where application examples from Japan and the U.S. are used.

Although the series title "Quality Engineering" is not necessarily new, this name was selected so as to separate it from traditional quality control and design techniques and to emphasize that this is new quality design engineering. Quality engineering offers a new and efficient engineering technology of quality design for improved productivity.

We sincerely hope this series will help readers' understanding and will promote application of quality engineering. Your questions and comments will be greatly appreciated.

QUALITY ENGINEERING SERIES
PUBLICATION COMMITTEE

EDITORS AND AUTHORS

(title is omitted and the order is arbitrary)

Chief Editor

YOSHIKO YOKOYAMA
Japanese Standards Association

Executive Secretary

GENICHI TAGUCHI
Japanese Standards Association

IKU-O BABA
Baba Institute of Technology

HIROSHI YANO
National Research Institute of Metrology
Ministry of International Trade and Industry

MASTATAKA YOSHIZAWA
NX Activity Promotion Room
Fuji-Xerox, Ltd.

MASANOBU KAWAMURA
Director, Japanese Standards Association

Quality Engineering Series Publication Committee

(title is omitted and the order is arbitrary)

Committee Chairman

Genichi Taguchi
Consultant, Japanese Standards Association

Executive Secretary

Hiroshi Yano
National Research Institute of Metrology
Ministry of International Trade and Industry

Committee Members

Seiso Konishi
Semiconductor Division
Hitachi Co., Ltd.

Shogo Yamamoto
Executive Director, TQC Promotion Division
Director, Hamanakako Denso Co., Ltd.

Yoshiko Yokoyama
Affiliate, Japanese Standards Association

Masataka Yoshizawa
NX Activity Promotion Room,
Fuji-Xerox, Ltd.

Masanobu Kawamura
Director, Japanese Standards Association

Yuin Wu
Vice Chairman, American Supplier Institute Inc.

Shin Taguchi
Vice President, American Supplier Institute Inc.

Editorial Advisors

PREFACE

This book illustrates experimental design from the standpoint of quality engineering. Before the name Taguchi Methods or quality engineering was popularly used, stability design, which is the core of quality engineering, was allocated as one of the applications in experimental design. It was considered that the learning of the philosophy and approaches for stability design follows after becoming acquainted with the use of experimental design approaches. It is a fact, however, that it is not necessary to study before applying quality engineering approaches. In this sense, this book is not arranged in a traditional way.

Quality engineering differs from traditional experimental design with respect to the following:

1. The analysis of variance mainly for the calculation of Signal-to-Noise Ratios (S/N Ratio).
2. There is a clear distinction between the use of inner and the outer layouts, which is an imperative in quality design.
3. There is more importance on the comparison of S/N Ratios, which are the indices to evaluate variability, rather than the comparison of the means of directly measured characteristics.
4. The layouts of orthogonal arrays are used from the very beginning, and instead of starting from one-way, two-way layouts and then gradually increasing the number of factors.
5. Mixed-type orthogonal arrays are used in principle. The orthogonal arrays with linear graphs are used only for the cases when mixed-type arrays cannot be used.

The above new approaches might appear to be incongruous to those who already are knowledgeable in traditional experimental design. However, that does not mean that this book ignores the areas which are included in traditional experimental design such as:

1. The issue of quality characteristic selection
2. Methods of layout
3. Data analysis

(1) is included in Chapters 2, 3 and part of Chapter 1 through the introduction of S/N Ratio. (2) is included in Chapters 5 and 10, and (3) is included in Chapter 4, and is also explained after Chapter 5 when necessary. The data analysis of non-continuous variables is explained in Chapter 11. The case of missing data is illustrated in Chapter 12. An additional illustration is provided in the Appendix. Since the explanation was made with quality engineering in mind, there might be some

redundancy with Volume 1 of this series: *Taguchi Methods: Research and Development*.

This book should be useful for quality design engineers to learn experimental design by themselves. For this reason, some frequently appeared questions are explained and controversial issues are discussed. The book may be difficult for some readers to get the points; therefore, repeat readings are suggested. It is my hope that this book will aid in the application of quality engineering.

Case examples are provided for illustration purposes. Those examples were provided by many supporters, and I express sincere gratitude to them. I also express many thanks to editorial members and the staff of the Japanese Standards Association for publication and proofreading.

September 1988
Genichi Taguchi
Publication Committee Chairman

PREFACE TO ENGLISH VERSION

This unique and concise book is quite different in content than similar books on the subject of "Design of Experiments (DOE)" or "Experimental Design."

Experimental Design is applicable to all areas where there is a necessity to acquire information efficiently, such as in scientific and non-scientific research, as well as in engineering. However, this book is written specifically for Quality Engineering applications.

One of this book's major differences is in its simplicity. Analysis of variance of raw data, referred to as "response analysis," is substituted by the comparison of the means of Signal-to-Noise Ratios (S/N Ratios). In this book, decomposition of data is made only for the determination of S/N Ratios and for tolerance design.

Many people are under the misconception that interactions are not considered in Taguchi Methods; however, the opposite is true. In fact, Dr. Taguchi considers interactions one of the most important issues in his approach.

Since a S/N Ratio, which is an index for robustness of quality, shows the magnitude of the interaction between a control factor and noise factors, control factors and noise factors must be separately assigned to different groups for the study of robustness. This is significantly different from the traditional DOE approach, where there is no distinction between control and noise factors. Instead of individually calculating those interactions between control and noise factors, as performed in Japan forty years ago, S/N Ratios are used as indices for comparing the differences in S/N Ratios between control factor levels. This enables engineers to simplify their calculations, saving tremendous amounts of time in data analysis.

In this book, orthogonal arrays are used in all case examples except for the explanation of S/N Ratio calculations in the early chapters. The reason for doing so is to test the robustness of conclusions. The most important and most difficult step in the quality engineering approach is to secure the robustness of conclusions by the selection of quality characteristics: the characteristics that are additive, reproducible, or having no interactions between control factors. When there are insignificant interactions between control factors, it is likely that conclusions are robust and reproducible. Robustness of conclusions is tested by the use of orthogonal arrays, followed by confirmatory experiments. This is one of the reasons why mixed-type orthogonal arrays are recommended. An additional "plus" is that this has the effect of discouraging engineers from assigning interactions between control factors.

Unlike other DOE books, the test of significance by the F test is not used to compare the impacts of factors. The F test is just lightly mentioned in the appendix. Its theory is based on the distribution of "response data," the raw data obtained from experimentation. Since the distribution of response data is useless in robust design, optimization is conducted through the comparison of S/N Ratios. The use of dynamic S/N Ratios is indicative of the farewell to the distribution of response data. The use of dynamic S/N Ratios leads towards "Robust Technology Development." With the methods described in this book, we are able to achieve robustness and reduce research and development time cycles to a small fraction of what they once were.

May 11, 1993
Yuin Wu
Translator for the English Version

CHAPTER 1

WHAT IS EXPERIMENTAL DESIGN FOR QUALITY?

1.1 FUNCTIONAL VARIATION AND ITS ASSOCIATED QUALITY LOSS

In any component part, raw material, or final assembly, tolerance limits and a target value are assigned for each characteristic, which in turn, are used to determine whether or not the units will pass or fail inspection procedures. For example, consider a film with thickness specified as 50±2μm, or electric coils with resistance specified as 8.2±0.2Ω. The manufacturers of these products will be content as long as the units are within specification. Many manufacturing companies actually use fraction non-defective, or fraction defective, as their index for quality evaluation. Any out-of-specification production will be either discarded or repaired, thus imparting additional costs.

Based upon a Loss Function concept, we advocate a new approach, which represents a repudiation of the mind set of "just trying to produce products to specification." We emphasize the importance of producing product as close to the mid-value of the specification limits as possible. To understand this concept, it is necessary to develop an understanding of the Loss Function, and its relationship with tolerance limits. The following notation will be introduced.

The mid-value of the specification limits, $\pm\Delta$, is denoted by y_0, with loss, A_0 yen, of scrapping or repairing an out-of-specification unit. Let y denote the quality level of a product. In traditional practice, as long as y is within $y_0 \pm \Delta$, y is treated as if it is at y_0. This can lead to quality problems at subsequent processing steps, or final assembly. As illustrated in Figure 1.1, with the traditional approach, a loss of magnitude, A_0, occurs only when the quality level is outside of $y_0 \pm \Delta$; however, with a modern quality engineering approach, a loss occurs continuously with any deviation from y_0.

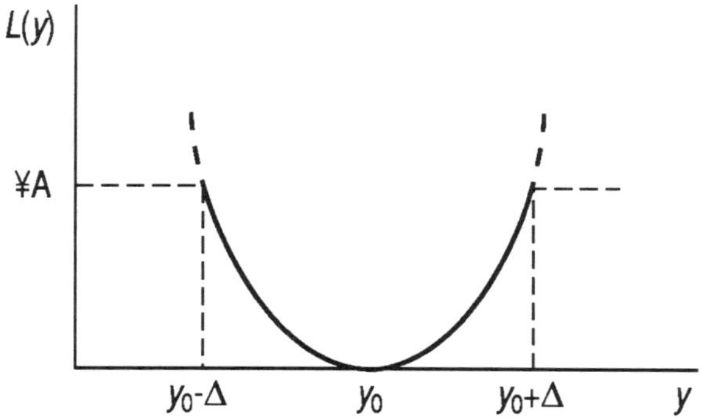

Figure 1.1. The Loss In and Out of Specification

The loss caused by deviations of y being deviated from y_0, denoted by $L(y)$, is given by the Taylor series expansion of $L(y)$ about y_0:

$$L(y) = L(y_0 + y - y_0)$$
$$= L(y_0) + \frac{L'(y_0)}{1!}(y - y_0) + \frac{L''(y_0)}{2!}(y - y_0)^2 + \cdots \tag{1.1}$$

$L(y)$ is at its minimum when y is equal to y_0. Without loss of generality, let $L(y_0)=0$. Since the differential coefficient is equal to zero, $L'(y_0)=0$, and the loss can be approximated by the quadratic:

$$L(y) \approx \frac{L''(y_0)}{2!}(y - y_0)^2 = k(y - y_0)^2 = k\sigma^2 \tag{1.2}$$

Assuming that the loss $L(y)$ is equal to A, when y deviates from y_0 by Δ,

$$k = \frac{A}{\Delta^2} \tag{1.3}$$

σ^2 in equation (1.2) is just the average of $(y\text{-}y_0)^2$.

Equation (1.2) is used to justify that the loss, $L(y)$, is proportional to the average of the squared deviations of y from y_0, and that the quality level must be as close to the center of the specification limits as possible, in order to minimize quality loss—instead of just being within the specification limits.

There is a direct relationship between drawing tolerance and quality loss once the product is shipped to the customer. To illustrate this, consider a color TV power circuit, whose function is to transform 100V AC into 115V DC. The receiver circuit, or the "brown tube," are designed in such a way that optimum performance is achieved when the output voltage is 115V. When the output voltage, y, deviates from 115V, a quality loss occurs, which can be estimated with the use of equation (1.2):

$$L(y) = k(y-115)^2 \tag{1.4}$$

To calculate the proportionality constant, the voltage level at which the TV ceases to perform its intended purpose must be determined. For example, suppose that at a voltage deviation of (y-115 volts) = 25 volts, the TV screen becomes too dark, and the consumer either must be forced to buy a new set or have it repaired. Assuming an average cost of buying or repairing of 20,000 yen, the proportionality constant is calculated by:

$$k = \frac{20{,}000}{25^2} = 32 \tag{1.5}$$

The Loss Function, $L(y)$, is:

$$L(y) = 32 \times (y - 115)^2 \text{ (¥)} \tag{1.6}$$

In general, in terms of y_0, the target value of an objective characteristic, y, and Δ_0, the tolerance limit where upon actual trouble occurs at the customer's or client's site, the Loss Function is approximately given by:

$$L(y) = \frac{A_0}{\Delta_0{}^2}(y - y_0)^2 \tag{1.7}$$

Once the Loss Function is established, it can be used to establish drawing tolerances. For example, suppose that the output voltage produced in the factory is 112 volts. If this unit were shipped, a loss would occur of magnitude:

$$L(112) = 32 \times (112 - 115)^2 = 288 \text{ (¥)} \tag{1.8}$$

In the factory, an unacceptable output voltage deviation can be remedied by simply replacing a resistor element in the circuit. The labor cost for this adjustment to the product is just ¥90, and the total repair cost amounts to ¥100. By using the ¥100 figure on the left side of the equation, we can solve this in terms of y to obtain a tolerance which is based upon the Loss Function:

$$y = 115 \pm \sqrt{\frac{100}{32}} = 115 \pm 1.8 \text{ (V)} \tag{1.9}$$

This result shows that although trouble-free performance in the customer's home can be obtained if the output voltage is in the range of 115±25V, the drawing tolerance should be only 115±1.8V. Thus, a circuit having 112V output should not be considered for shipment to the customer. If this unit were shipped, it would result in a ¥288 loss to society, compared to the ¥100 cost of remedying the problem. The objective in quality engineering is to discover the countermeasure which results in an overall reduction in the total quality loss to both the company and society. It is therefore important to evaluate the two kinds of losses—the loss due to function and the one due to harmful effects. (For harmful effects, see Quality Engineering Series, Volume 1, *Taguchi Methods: Research and Development*, Chapter 1.)

1.2 Types of Quality Characteristics, the Loss Function and S/N Ratio

(1) **Smaller-the-better characteristics**

A non-negative characteristic which must be reduced to as low a level as possible is called a "smaller-the-better" characteristic. As examples of smaller-the-better characteristics, consider the following: (i) the level of pollutants in exhaust gas; (ii) the number of pinhole defects; and (iii) the amount of wear on a mechanical component. Those characteristics whose ideal values are zero, but are allowed to be negative, do not belong to the case of smaller-the-better characteristics. Instead, they are referred to as nominal-the-best characteristics having target values of zero.

Since the target value of smaller-the-better characteristics is zero, the square of the deviations of characteristic y from its target is y^2. Denoting the average squared deviations from the target, as σ^2, the loss due to variation is:

$$L = k\sigma^2 \tag{1.10}$$

For a sample of n observations, $y_1, y_2, \ldots, y_n$ of a smaller-the-better characteristic, σ^2 if given by:

$$\sigma^2 = \frac{1}{n}(y_1{}^2 + y_2{}^2 + \cdots + y_n{}^2) \tag{1.11}$$

In quality engineering, the reciprocal of σ^2 is referred to as the Signal-to-Noise (S/N) Ratio. The logarithmic of the S/N Ratio multiplied by 10 gives the S/N Ratio in decibel units, which is denoted by η.

$$\eta = 10 \log \frac{1}{\sigma^2} = -10 \log \sigma^2 \tag{1.12}$$

Quite often, η is also used to denote the reciprocal of σ^2, or the anti-log value of S/N Ratio before the logarithmic transformation. The reason of using decibel scale will be described later.

The S/N Ratio is a terminology which is used in communications engineering to describe the ratio of the signal and the noise of a circuit. In quality engineering, it is used to measure quality in terms of the reciprocal of variability per unit. The S/N Ratio for smaller-the-better characteristics is just the magnitude of the variation (error) per unit (the square of the unit).

(2) **Larger-the-better characteristics**

A non-negative characteristic which must be increased to as high a level as possible is defined to be a "larger-the-better" characteristic. Tensile or binding strength are examples of larger-the-better characteristics, wherein it is desirable that these characteristics be as large as possible. For a larger-the-better characteristic, trouble occurs when the quality level is lower than Δ_0; its loss is A_0. The loss for a larger-the-better characteristic decreases when the quality level increases as illustrated by Figure 1.2.

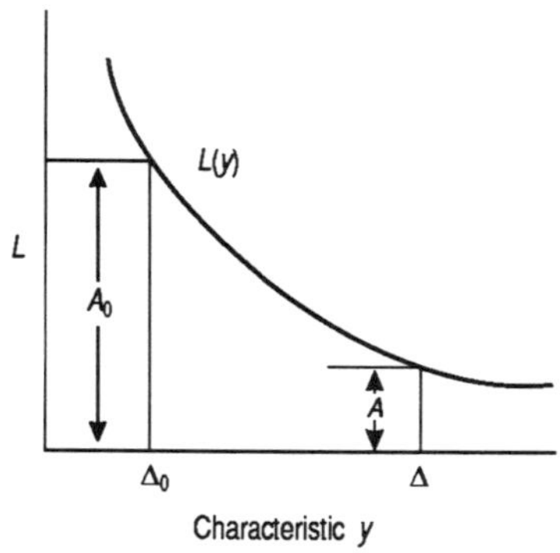

Figure 1.2. The Loss Function of Larger-the-Better Characteristics

In the figure, Δ is the manufacturing tolerance, while A denotes the loss due to producing product which is out-of-tolerance.

$$L(\infty) = 0 \tag{1.13}$$

$$L'(\infty) = 0 \tag{1.14}$$

If a Taylor series expansion is made of the Loss Function, $L(y)$, about the reciprocal of y,

$$L(y) = L(\infty) + \frac{L'(\infty)}{1!} \times \frac{1}{y} + \frac{L''(\infty)}{2!} \times \frac{1}{y^2} + \cdots \tag{1.15}$$

and if terms of cubic order or higher are omitted, then

$$L(y) = k \times \frac{1}{y^2} \tag{1.16}$$

Since the loss, $L(y)$, is A_0 when $y=\Delta_0$,

$$A_0 = k \times \frac{1}{\Delta_0^{\ 2}} \tag{1.17}$$

Therefore,

$$k = A_0 \Delta_0^{\ 2} \tag{1.18}$$

The Loss Function is given by:

$$L(y) = A_0 \Delta_0^{\ 2} \times \frac{1}{y^2} \tag{1.19}$$

Letting σ^2 denote the average of the squared values from a sample of n pieces of data, $y_1, y_2, ..., y_n$, of a larger-the-better characteristic,

$$L = k\sigma^2 \tag{1.20}$$

where σ^2 is:

$$\sigma^2 = \frac{1}{n}\left(\frac{1}{y_1^{\ 2}} + \frac{1}{y_2^{\ 2}} + \cdots + \frac{1}{y_n^{\ 2}}\right) \tag{1.21}$$

As in the case of the smaller-the-better characteristic, the reciprocal of σ^2 is referred to as the S/N Ratio, and it is converted to decibel units by multiplying its logarithmic value by 10.

$$\eta = 10 \log \frac{1}{\sigma^2} = -10 \log \sigma^2 \tag{1.22}$$

(3) **Nominal-the-best characteristics**

A characteristic wherein it is desirable not to deviate from either side of a target value is referred to as a nominal-the-best characteristic. When a certain quality characteristic, y, has a target, y_0, its Loss Function is:

$$L = k(y - y_0)^2 \tag{1.23}$$

as described in Section 1.1.

To improve quality, it is important to minimize $\sigma_0^{\ 2}$, which is the mean of $(y\text{-}y_0)^2$. However, it would be inefficient to define -10 log σ^2 as the S/ N Ratio, and attempt to maximize it as is the case for a smaller-the-better or larger-the-better characteristic. To understand this, keep in mind that the S/N Ratio is a characteristic which is used to determine whether or not quality improvements have been made. In order to optimize the process, it is inefficient to try to minimize $\sigma_0^{\ 2}$ or the mean of the squares of deviations from the target. It is important in the process of determining the optimum design or manufacturing conditions, to begin the research without being required to adhere to a particular target value. If we are free from the restriction of the target, y_0, we might be able to find the condition with minimum variability. Then, if the mean deviates from the target, it will be necessary to adjust the process to reduce the deviation from a prescribed target. Generally, such an adjustment can be made by using one variable for adjustment called the adjusting factor. It is therefore better to discuss the magnitude of variation after the deviation is adjusted.

Let $y_1, y_2, ..., y_n$ denote sample realizations for a quality characteristic, y, having a target y_0. Assuming that an adjustment is made to the process to move it to its target, then after adjustment, the individual value, y_i', is given by:

$$y_i' = y_i \frac{y_0}{m} \tag{1.24}$$

m in equation (1.24) is the population mean of the average, $\bar{y}$, which is estimated by $\bar{y}$. In a routine manufacturing setting, where design and manufacturing conditions have been set, the adjustment which is made, such as changing the cutting tool position, is made in small increments.

Therefore, the value of y_i' after adjustment is:

$$y_i' = y_i + (y_0 - m) \tag{1.25}$$

In equation (1.24), letting σ^2 be the average of $(y_i\text{-}\bar{y})^2$ before correction, $\sigma^{2\prime}$, the average of $(y_i'-y')^2$ after correction, is given by:

$$\sigma^{2\prime} = \sigma^2 \left(\frac{y_0}{m}\right)^2 = y_0{}^2 \frac{\sigma^2}{m^2} \tag{1.26}$$

The Loss Function after correction, denoted by L_1 is:

$$L_1 = k\sigma^{2\prime} = ky_0{}^2 \frac{\sigma^2}{m^2} \tag{1.27}$$

At the product design stage, the Loss Function, L_1, is used to compare different design condition. However, it is important to use:

$$\frac{\sigma^2}{m^2} \tag{1.28}$$

as the base of comparison. The reciprocal of equation (1.28) is called the S/N Ratio, and its logarithmic value multiplied by 10 is called the S/N Ratio in decibel units.

$$\eta = 10 \log \frac{m^2}{\sigma^2} \tag{1.29}$$

Procedures for the estimation of m^2 and σ^2 will be discussed in Chapter 2.

For the case of Equation (1.25), however, σ^2, the average of $(y_i-\bar{y})^2$ before correction is equal to $\sigma^{2\prime}$, the average of $(y_i - \bar{y}')^2$ after correction. The Loss Function after correction, L_2, is:

$$\begin{aligned} L_2 &= \text{average of } k\sigma^{2\prime} = k(y_i - \bar{y})^2 \\ &= k\sigma^2 \end{aligned} \tag{1.30}$$

The S/N Ratio in this case is the reciprocal of σ^2, and in decibel units,

$$\eta = -10 \log \sigma^2 \tag{1.31}$$

[Note: σ^2 is the average of $(y_i - \bar{y})^2$ before correction]

In the case when the target differs from the average by only a small percentage, Equation (1.25) may be used for the Loss Function after correction (although Equation (1.24) will always apply). In the case of a nominal-the-best characteristic with zero as a target, Equation (1.31) should be used for formulating the S/N Ratio. (Footnote: See Chapter 7 of Quality Engineering Series, Volume 1, *Taguchi Methods: Research and Development*).

In the case when the deviation cannot be corrected, its loss, denoted by L_0 is:

$$L_0 = k(\text{average of } y_i - y_0)^2 \tag{1.32}$$

The S/N Ratio, is the reciprocal of the average of $(y_i-y_0)^2$. In decibel units,

$$\eta = -10 \log \sigma_0{}^2 \tag{1.33}$$

[Note: $\sigma_0{}^2$ = average of $(y_i-y_0)^2$]

In nominal-the-best characteristics, there is a type called dynamic characteristics. Driveability of cars, dyeability, function of ski or golf balls, rollability of material or press machine, controllability of control devices are examples of dynamic characteristics. In such cases, an operation is required in order to bring the characteristic into its target. In dynamic characteristics, target values change from time to time. In the case of a vehicle accelerator pedal, the pedal is pushed to increase the speed, where the depth of pushing is the input and the speed of the vehicle is the output.

The input factor, M, is referred to as the signal factor. Ideally, the relationship between the input and the output must be maintained constant, but there are many disturbances, which are caused by noise factors or error factors. In the accelerator pedal example, it is desirable that the relationship between the pushing depth of the accelerator and the car speed not be affected by noise factors such as road conditions or tire concentration characteristic, and let M, the signal factor, denote the amount of dye. The relationship between y and M could be of the form:

$$y = \beta M \tag{1.34}$$

The symbol, β, in Equation (1.34) is the amount of color change per unit of dye, called the sensitivity. That is, color concentration increases by β when the amount of dye increases by one unit. Such a relationship, however, is not perfect since there exists an error variance, σ^2. (The steps for determination of the error variance are explained in Chapter 3.) In the case of a dyeing process, a good dye implies that Equation (1.34) always holds regardless of yarn size or cloth type used, etc.; however, these factors vary by various dyeing conditions. σ^2 is the average of variation caused by noise factors under different conditions of M.

When the sensitivity β is too large, it is difficult to obtain the right amount of color, since a small amount of dye results in significant color changes. There must be an appropriate sensitivity, β_0. If the dye concentration can be diluted by adding filler, the σ^2, in the following equation,

$$y = \beta M \tag{1.35}$$

is given by:

$$\sigma_0{}^2 = \left(\frac{\beta_0}{\beta}\right)^2 \sigma^2 \tag{1.36}$$

The Loss Function is then:

$$L = k\sigma_0{}^2 = \left(\frac{\beta_0}{\beta}\right)^2 \sigma^2 = k\beta_0{}^2\left(\frac{\sigma}{\beta}\right)^2 \tag{1.37}$$

Since the value of β_0 has nothing to do with the quality of design, comparison of various designs is made using $(\sigma/\beta)^2$ as a base. $(\sigma/\beta)^2$ is the S/N Ratio of dynamic characteristics, and 10 times its logarithmic value is the S/N Ratio in decibel unit.

$$\eta = 10 \log \frac{\beta^2}{\sigma^2} \tag{1.38}$$

The S/N Ratio described above is often applied for the evaluation of measurement methods. The readings of a good measurement system should be linearly proportional to the true values of the input, and not be affected by various disturbances. When applied to measurement systems, the factor which are associated with changing the true values of the input is the signal factor, and the factors which disturb the measurement system are noise factors. In actual measurement systems, calibration is made so that β_0 is equal to 1. Therefore, the S/N Ratio given by Equation (1.38) is the one after calibration; the reciprocal of its anti-log value is the error variance after calibration. Measurement system applications are described in more detail in Quality Engineering, Volume 3, *Taguchi Methods: S/N Ratios for the Evaluation of Quality.*

1.3 Brief Description of Quality Improvement

In the previous section, the loss caused by functional variation can be evaluated using the Loss Function. There are various reasons for functional variation, or deviation of the output characteristic from the target. They are classified into three types:

1. Outer noise: environmental conditions such as temperature, humidity, dust, input voltage, etc.
2. Inner noise: variation during the initial period and variation caused by deterioration of components or elements.
3. Between product variation: variation caused by the manufacturing process.

In quality engineering, there are two countermeasures—off-line and on-line quality control. For on-line countermeasures, the reader should refer to Quality Engineering Series, Volume 2, *Taguchi Methods: On-Line Production*. Off-line quality control techniques include parameter design and **tolerance design**, wherein experimental design techniques are applied. Parameter design is also called robust design, and the S/N Ratio is the characteristic used for the evaluation of robustness.

Parameter design is used to achieve robust quality by reducing the effects of the outer and inner noise terms with the selection of different design alternatives, or by varying the levels of the design parameters for component parts or system elements. S/N Ratios, as described in the previous section, are used to achieve robust quality (insensitivity to noise conditions). The Loss Function will be used to evaluate the overall improvements in the design.

Even the best parameter design effort, sometimes the effects caused by noise cannot be completely eliminated. If these effects remain significant, then countermeasures must be taken to remove these causes. However, the methods which are used to eliminate these causes will normally result in an overall cost increase. It is important, therefore, to carefully examine the cost versus quality tradeoffs. This is called tolerance design. Of course, the Loss Function will still be used for tolerance design.

1.4 CLASSIFICATION OF FACTORS

To which type of factor should be classified depends on the purpose of experiment. The factors which are used to improve stability are classified into two types: factors which are used to determine the S/N Ratio, the index for stability, and the group of factors for which comparisons at different levels are made with respect to the S/N Ratio.

In order to determine the S/N Ratio of smaller-the-better and larger-the-better characteristics, noise factors must be introduced. In the case of nominal-the-best characteristics, a signal factor and noise factors are introduced. Sometimes a signal factor is not introduced, and in such a case, only noise factors are considered. Control factors and indicative factors belong to the second group of factors—those which are varied, and comparisons are made based upon the S/N Ratio.

A brief description of each of these factors follows:

(1) Control factors

Control factors are used to select the best conditions for stability in design or manufacturing processes. For example, in a study to improve bonding strength, the type of bonding material or the number of coatings of the bond material are examples of control factors. In another study to improve the stability of the DC output voltage converted from 100V AC power, all elements in the circuit are control factors. It is important to study as many control factors as deemed possible. It is nice to have an "attitude of trying to study all possible factors which can be changed."

(2) Indicative factors

Indicative factors behave like control factors in terms of their technical significance, but are not used to select their best levels. They are used to select alternatives. For example, condition of usage, product variety types or size. High values of the S/N Ratio signifies that the product function is not easily affected by product usage conditions. In this sense, indicator factors behave like noise factors. For example, in an experiment for which the driveability of a car is evaluated, the effects by inner or outer noise at a lower speed and a higher speed are obviously different. A characteristic such as turning radius will effect driveability differently at different speeds. Of course it would be nice to have the same driveability at any speed, but there is a significant difference from a technical viewpoint. In such a case, it is important to be aware of differences in S/N Ratio at different speeds. A factor which is treated this way is called an indicative factor. It is not a control factor since we cannot say "please do not drive at a high speed because the S/N Ratio is lower."

(3) **Signal factors**

A signal factor is the factor used to evaluate the input signal and is denoted by *S*. A signal factor for dynamic characteristics is used to realize the requirement of the output and sometimes it is also called the active signal factor. In the case of a dyeing process, the amount of dye is the signal factor if it is used to adjust color concentration to the target.

The characteristics used in measurement studies are referred to as passive type characteristics. These are the signal factors; the factors which change the true value.

(4) **Noise factors**

Noise factors are all factors which cause variation such as inner noise, outer noise factors, etc. Inner noise factors are those factors which result in the deviation of components from the target, or deterioration during usage. Outer noise factors include environmental conditions where products are used.

For smaller-the-better characteristics, noise factors are the causes which increase the level of the characteristic. For larger-the-better characteristics, noise factors cause a decrease in the level of the characteristic. For example, for tensile strength of steel members, a larger-the-better characteristic, high temperatures weakens the strength of the structural member. Thus, temperature is a noise factor.

In the case of a nominal-the-best dynamic characteristics, a system is designed to perform to target conditions. Those causes which disturb system performance are noise factors. In the car driveability study, road conditions, load position, and the air pressure of tires are examples of noise factors.

For smaller-the-better characteristics, it is desirable to have a small value and small variability caused by various noise factors. In larger-the-better characteristics, it is desirable to have a large value and small variation. In nominal-the-best characteristics, it is desirable to operate a system to meet a certain output without being affected various noise.

Factors are classified into the above four categories. What factor should be classified and treated as which category is determined by the objective of an experiment.

1.5 EXPERIMENTAL DESIGN FOR QUALITY IMPROVEMENT

Experimental design or layout refers to the determination of the experimental conditions run in order to determine design parameters or manufacturing conditions for stable quality. Traditional experimental design is mainly used to improve the average level of a process. In modern quality engineering, experimental design is used to come up with robust designs for quality improvement.

However, it does not mean that experimental design has never been used to reduce functional variation. For example, suppose one were to find the best of three additive types, A_1, A_2 and A_3 (factor A has three levels) which are used in a plastic to minimize the effect of temperature. Elongation of the plastic is measured at four different temperatures, B_1, B_2, B_3 and B_4 (that is, factor B has four levels) for each additive, A_1, A_2 and A_3. Since the S/N Ratio is the characteristic for the evaluation of stability, it is rational to use S/N Ratio to discover the best condition. The use of S/N Ratio also directly relates to the Loss Function.

To determine whether or not elongation increases linearly with temperature, analysis is conducted using a technique called analysis of variance (ANOVA), which is introduced in Chapter 2, to determine if elongation increase is different with A_1, A_2, or A_3 (the effect of B being different at different levels of A is called the interaction A×B). If, as a result, the tendency of elongation increase is different at different levels of A, the best level of A is selected from the graph of temperature and elongation increase.

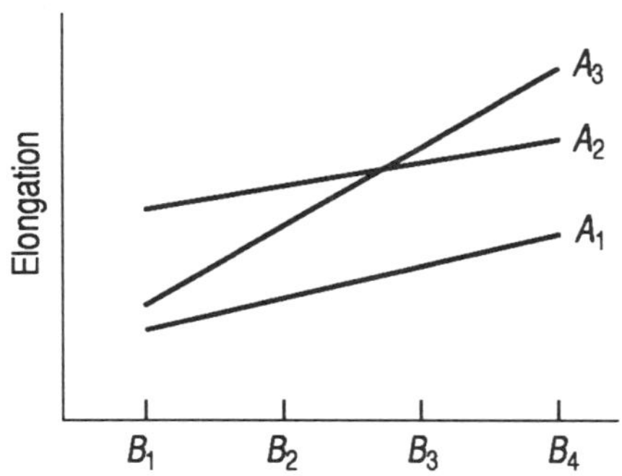

Figure 1.3.

In the study of functional stability, some factors are affected by environmental conditions, while others are not. For such a study, the interaction between control factor A and noise factor B, would be examined to determine whether or not a particular choice of A results in a design which is less sensitive to noise. However, it would be quite tedious to investigate all interactions when many control factors are studied at the same time.

In order to discover the stable condition, two groups of factors are separately laid out: control factors and indicator factors are assigned into one group (called the inner layout), signal factor and noise factors are assigned into another group (the outer layout). It is important to remember the words "inner" and "outer"; the thought that control factors which are used to find the best combination for stability and the factors used to calculate S/N Ratios must be separately assigned.

Sometimes the factors assigned to the inner layout are called s, and those assigned to the outer layout are called outer factors.

As mentioned earlier, it is important to investigate as many control factors as possible. It is desirable that the interval between levels to be as wide as possible. Sometimes a narrow interval has shown a significant effect in the traditional approach for the improvement of average. We have also seen the type of experiment where level interval was greatly widened just to obtain a significant effect or a large degree of contribution. But such an approach was not well justified. In order to reduce the variation of current condition or existing design, it is important to consider a fairly large level interval, since it is believed that a narrow level interval does not really change the situation of the effects if the inner or the outer noise factors. It is especially effective in the development stage as described in Chapter 9 where simulation or the analysis by computers are used instead of physically produce a hardware. Signal factor and noise factor, will be discussed starting from Chapter 5.

Control factors are assigned as the s to set up the inner experimental conditions. The outer factors are assigned to determine the way to collect results.

The techniques of the layout for quality improvement will be explained from Chapter 4. When there are k kinds of set up for the inner layout and n kinds of set up for the outer layout, data are collected for all of the combinations. It is called "direct product" of the inner and the outer layouts.

	1	2	....	n
1				
2				
:				
:				
k				

From each of the inner condition, a S/N Ratio is determined using n pieces of outside data, then from k pieces of S/N Ratios, the condition for stability will be investigated. See Chapters 4 and later for actual procedures.

Indicative factors are generally assigned as the s, but sometimes they are separated from both the inner and outer layouts and collect $k \times n$ pieces of data for each level of the indicative factor.

EXERCISES

Question (1.1). In an experiment for an air pump, two designs:

A_1: current design A_2: new design

were compared. For each design, 7 specimens of a sliding part in a pump were tested for durability. Data was collected on the amount of wear on the spring side, yielding the following results:

A_1: 0.09 0.13 0.05 0.04 0.08 0.07 0.05

A_2: 0.03 0.05 0.05 0.04 0.04 0.02 0.04

a) Calculate the variance (average squared error) on the amount of wear for designs A_1 and A_2 respectively. How much could the variance be improved if the new design were used?

b) When the amount of wear at any spot exceeds 0.2 mm, trouble occurs in the market causing a loss of ¥80,000. It is estimated that the current average squared error of wear is $(0.1\text{mm})^2$. Estimate the loss of the current design using the Loss Function caused by wear.

c) It is assumed that the amount of wear in the market is proportional to the variance of the durability test results. Estimate the loss of the new design caused by wear, and compare with the current design.

Question (1.2). Pick a quality improvement project in your working area, and identify control and noise factors for improving quality.

CHAPTER 2

DECOMPOSITION OF SUM OF SQUARES (1)—THE CASE WITH A FIXED TARGET

In the evaluation of quality based on loss due to variability, S/N Ratio is an important measure. In the cases of smaller-the-better or larger-the-better, the S/N Ratio is obtained from the average of squares or the average of the squares of reciprocal. For the S/N Ratio of nominal-the-best characteristic, knowledge of the decomposition of sum of squares is required. In this chapter, determination of S/N Ratio with no signal factor is explained. Also, the preliminary part of analysis of variance is explained.

For algebraic treatment in this and the next chapters, see Chapter 15 of Quality Engineering Series, Volume 1, *Taguchi Methods: Research and Development*.

2.1 DEVIATION FROM THE TARGET, INDIVIDUAL DIFFERENCES AND THEIR VARIATION

The specification of the thickness of a component part is 100±5μm and the loss for rejection, A, is 300 yen. 10 pieces of component parts were made under a certain condition and deviations from the target were measured to obtain the following results:

5 5 2 4 3 8 5 4 3 6

All deviations from the target have positive values, showing they are thicker than the target. Such "deviation" is obtained from the average of deviations from the target. Letting the deviation be $\bar{y}$,

$$\bar{y} = \frac{5+5+\cdots+6}{10} = \frac{45}{10} = 4.5 \tag{2.1}$$

Table 2.1. Decomposition of Data

$y_i = \bar{y} + (y_i - \bar{y})$
5 = 4.5 + (0.5)
5 = 4.5 + (0.5)
2 = 4.5 + (-2.5)
4 = 4.5 + (-0.5)
3 = 4.5 + (-1.5)
8 = 4.5 + (3.5)
5 = 4.5 + (0.5)
4 = 4.5 + (-0.5)
3 = 4.5 + (-1.5)
6 = 4.5 + (1.5)

Deviation from the target, denoted by y_i, includes average deviation, $\bar{y}$ and deviations from the average, y_i-$\bar{y}$. The deviations from the average, y_i-$\bar{y}$, are individual deviations, also called individual differences.

In Table 2.1, the total of the squares of data on the left side is the sum of the squares of deviations from the target, showing the variability from the target. For reasoning behind using squares to express the magnitude of variability, see the Appendix.

The total of variability is called total variation represented or denoted by the symbol S_T.

$$S_T = 5^2 + 5^2 + \cdots + 6^2 = 229 \qquad (f=10) \quad (2.2)$$

The first term on the right side in Table 2.1 is deviation; the total of their squares is the total amount of deviation, or the magnitude of the average of data. It is called general mean, represented by the symbol S_m.

$$S_m = \bar{y}^2 \times 10 = \frac{(\text{Total})^2}{10} = \frac{45^2}{10} = 202.5 \qquad (f=1) \quad (2.3)$$

The second term is individual deviation from the mean deviation. The total of the squares of the second term is called error variation or the variation of individual deviation, represented by the symbol S_e.

$$S_e = 0.5^2 + 0.5^2 + (-2.5)^2 + \cdots + 1.5^2 = 26.5 \qquad (f=9) \quad (2.4)$$

It can be seen that the total of S_m and S_e is equal to S_T.

$$\begin{aligned} S_e &= (y_1 - \bar{y})^2 + (y_2 - \bar{y})^2 + \cdots + (y_{10} - \bar{y})^2 \\ &= y_1{}^2 - 2y_1\bar{y} + (\bar{y})^2 + y_2{}^2 - 2y_2\bar{y} + (\bar{y})^2 + \cdots + y_{10}{}^2 - 2y_{10}\bar{y} + (\bar{y})^2 \\ &= y_1{}^2 + y_2{}^2 + \cdots + y_{10}{}^2 - 2\bar{y}(y_1 + y_2 + \cdots + y_{10}) + 10(\bar{y})^2 \\ &= y_1{}^2 + y_2{}^2 + \cdots + y_{10}{}^2 - 2\frac{(y_1 + y_2 + \cdots + y_n)^2}{10} + 10\left(\frac{y_1 + y_2 + \cdots + y_n}{10}\right)^2 \end{aligned}$$

$$= y_1^{\,2} + y_2^{\,2} + \cdots + y_{10}^{\,2} - \frac{(y_1 + y_2 + \cdots + y_n)^2}{10}$$

$$= S_T - S_m \qquad (f{=}10\text{-}1) \quad (2.5)$$

Therefore,

$$S_e = 229 - 202.5 = 26.5 \qquad (2.6)$$

This agrees with the results in Equation (2.4).

Since variation is the total variation, we need to know the average variation. Degrees of freedom are used to indicate the average. Those in the parentheses of Equations (2.2), (2.3) and (2.4) are degrees of freedom. The number of degrees of freedom is the number of unknowns in the calculation of variation. It is the independent number of squares. Generally, symbol f is used to denote the number of degrees of freedom, and quite often, subscripts such as T, m or e are attached to symbol f. For further reference on degrees of freedom, see the Appendix.

The total sum of squares is calculated using all of the data. Therefore, its number of degrees of freedom is the number of data points. It is expressed as f=10.

The general mean is calculated by dividing the square of the total by the number of data points. It means the square of the average times the number of data points. Since it can be calculated from one figure,namely, the average, its number of degrees of freedom is equal to 1, or f_m=1.

The error variation is the total of the squares of 10 individual points of data. But there is a restriction that the total of the individual difference is equal to zero. Therefore, the number of unknowns of individual differences equal to 9. The number of degrees of freedom, or f_e, is 9.

Variation and degrees of freedom are summarized as follows:

$$S_T = S_m + S_e \qquad (2.7)$$

$$f_T = f_m + f_e \qquad (2.8)$$

Equation (2.7) is called decomposition of the sum of squares.

2.2 EXPECTATION OF VARIATION AND PURE VARIATION

In the previous section, the average is used for deviation. This is an estimate from data. The true deviation, although it is never known, is considered to be the average of the infinite number of deviations from the target. It is denoted by the symbol m. This is expressed by:

$$E(y) = m \qquad (2.9)$$

Thus, the expectation of y is the true mean(deviation), m. σ^2, the variance of infinite number of y's, is denoted by $V(y)$.

It is known that the expectation of y^2 is not equal to m^2. y^2 and σ2 have the following relationship:

$$E(y^2) = m^2 + \sigma^2 \tag{2.10}$$

The above equation may be understood by conducting an experiment using a group of random numbers which have a certain average and variance. Observe the change in the average of squares and variance by increasing the size of random numbers and comparing with the true values.

From Equation (2.10), the total of the squares of k pieces of data, S_T,includes k units of $m^2+\sigma^2$.

$$E(S_T) = k(m^2 + \sigma^2) \tag{2.11}$$

It is known that the expectation of the total and the variance of k pieces of y's are:

$$E(\Sigma y) = km \tag{2.12}$$

$$V(\Sigma y) = k\sigma^2 \tag{2.13}$$

The square of the total of y's is given by:

$$E[(\Sigma y)^2] = (km)^2 + k\sigma^2 \tag{2.14}$$

Since the general mean, S_m, is k times the square of the average, its expectation, $E(S_m)$, is:

$$E(S_m) = E\ [k\left(\frac{\Sigma y}{k}\right)^2] = E\left[\frac{(\Sigma y)^2}{k}\right] = \frac{(km)^2 + k\sigma^2}{k}$$

$$= km^2 + \sigma^2 \tag{2.15}$$

The first term of the right side of the equation is the total of the squares of true deviation; it is the real deviation. km^2 is called the pure variation of the deviation, denoted by adding a prime (′) to the variation.

$$E(S_m) = S_m' + \sigma^2 \tag{2.16}$$

where $S_m' = km^2$

The expectation of the error variation is:

$$E(S_e) = E(S_T - S_m) = k(m^2 + \sigma^2) - (km^2 + \sigma^2)$$
$$= k\sigma^2 - \sigma^2 \qquad (2.17)$$

The first item of the right side of the equation is called the pure variation of the error, denoted by S_e'.

$$E(S_e) = S_e' - \sigma^2 \qquad (2.18)$$

From the above, S_T is also decomposed into the following pure variation.

$$S_T = S_m' + S_e' \qquad (2.19)$$

2.3 VARIANCE AND THE EXPECTATION OF VARIANCE

Variation per unit degree of freedom is called variance, denoted by the symbol V. The square root of variance indicates the spread of data which has the same unit as the original data and is called standard deviation.

The number of degrees of freedom for the sum of squares is equal to 10; its variance is:

$$V_T = \frac{S_T}{f_T} = \frac{229}{10} = 22.9 \qquad (2.20)$$

This shows the average variability from the target. Since the number of degrees of freedom for the general mean, S_m, is equal to 1, its variance, V_m, is equal to S_m. The error variance is the error variation divided by the degrees of freedom.

$$V_e = \frac{S_e}{f_e} = \frac{26.5}{9} = 2.94 \qquad (2.21)$$

From the expectation of variance, we can see what the variance is estimating. It is calculated from the expectation of variation divided by its degrees of freedom. Since the variation and variance of the general mean are equal, their expectations are also equal.

$$E(V_m) = km^2 + \sigma^2 \qquad (2.22)$$

The expectation of the error variance is obtained by dividing Equation (2.17) by the degrees of freedom, (k-1).

$$E(V_e) = \frac{k\sigma^2 - \sigma^2}{k-1} = \frac{(k-1)\sigma^2}{k-1} = \sigma^2 \qquad (2.23)$$

Thus, the calculated error variance estimates, σ^2, the variance of infinite number of y's. Therefore, the pure variation of the general mean, S_m', is estimated by:

$$S_m' = km^2 = S_m - \sigma^2 = S_m - V_e = 202.5 - 2.94 = 199.56 \tag{2.24}$$

From Equation (2.18), the pure variation of the error is given by:

$$S_e' = S_e + \sigma^2 = S_e + V_e = 26.5 + 2.94 = 29.44 \tag{2.25}$$

The rate of pure variation in the total variation is called degrees of contribution using symbol ρ.

$$\rho_m = \frac{S_m'}{S_T} \times 100 = \frac{199.56}{229.0} \times 100 = 87.1 \ \ (\%) \tag{2.26}$$

$$\rho_e = \frac{S_e'}{S_T} \times 100 = \frac{29.44}{229.0} \times 100 = 12.9 \ \ (\%) \tag{2.27}$$

The following decomposition also holds:

$$\rho_m + \rho_e = 100 \ \ (\%) \tag{2.28}$$

2.4 Preparation of Analysis of Variance Table

For easy observation, analysis of variance table (ANOVA table) is prepared as shown in Table 2.2.

Table 2.2. ANOVA Table

Factor	f	S	V	S'	$\rho(\%)$
m	1	202.5	202.5	199.56	87.1
e	9	26.5	2.94	29.44	12.9
T	10	229.0	22.9	229.0	100.0

The column of variance, V, is variation divided by degrees of freedom. In many cases, the total variance is not shown in the table. The total variance must be the total variation divided by the total degrees of freedom. Notice that only in the column of variance, the total is not the variances added together.

The column of expectation is not shown in this table. If it is to be shown, put it to the right side of the column of variance. Expectation of variances were:

$$E(V_m) = km^2 + \sigma^2 \tag{2.29}$$

$$E(V_e) = \sigma^2 \tag{2.30}$$

If deviation, m, is non-existing or very small, V_m and V_e are about the same. In this example, V_m is about 100 times as large as V_e; probably with some deviation. If V_m is about the same as V_e (sometimes V_m is smaller than V_e), deviation is not recognized and its correction does not make sense. In such a case,there is no necessity to distinguish the error (individual difference) and deviation and S_m is considered as the error. This is called pooling with the error. The pooled error variance is normally denoted by $V_{(e)}$ and calculated by:

$$V_{(e)} = \frac{(\text{error variation, } S_e) + (\text{variance of pooled factors})}{f_e + (\text{degrees of freedom of pooled factors})} \tag{2.31}$$

When there are general mean and error variations only and the general mean is pooled with the error, the pooled variance is equal to the total variance.

2.5 CORRECTION OF DEVIATION AND QUALITY IMPROVEMENT

In routine manufacturing processes, deviation is normally small if it exists. In Chapter 1, it was described that if deviation is corrected, the variability after correction becomes the average of $(y\text{-}\bar{y})$. By the application of the decomposition of sum of squares, we can determine the variation without correction, V_T, and the variation after correction, V_e.

Without correcting deviation, the average of squares of deviations from the target is:

$$V_T = \frac{5^2 + 5^2 + \cdots + 6^2}{10} = \frac{229}{10} = 22.9 \tag{2.32}$$

The Loss Function is:

$$L = \frac{300}{5^2} \times 22.9 = 12 \times 22.9 = 274.8 \text{ (¥/unit)} \tag{2.33}$$

From the decomposition, it is known that the deviation is significant. From an engineering viewpoint, deviation can be corrected in many cases.

When deviation is corrected, the variability after correction is the error variance, which is S_e divided by degrees of freedom. Its Loss Function is:

$$L = \frac{300}{5^2} \times 2.94 = 35.3 \text{ (¥/unit)} \tag{2.34}$$

There is a 239.5 yen quality improvement per unit product compared with the case without correction.

2.6 S/N Ratio for Nominal-the-Best Characteristics—The Case When a Constant Value Is Desirable

In the previous section, determination of deviation and evaluation of quality improvement by correcting deviation were discussed. One of the common approaches in quality engineering is to find a combination of control factors to reduce variability after correcting the deviation. S/N Ratio is an effective characteristic to use for this purpose. To use the S/N Ratio with a target, we must consider the cases when deviation can and cannot be corrected, as discussed in Chapter 1.

Let $y_1, y_2, \ldots, y_k$ be the data with target y_0. When the average deviates from the target but cannot be corrected, it is not necessary to decompose the sum of squares. If the average of $(y_i - y_0)^2$ is V_T, the Loss Function is given by:

$$L = \text{average of } k(y_i - y_0)^2 = kV_T \tag{2.35}$$

The S/N Ratio is:

$$\eta = -10 \log V_T \tag{2.36}$$

In the example of the previous section, $V_T = 22.9$ from Equation (2.32),

$$\eta = -10 \log V_T = -13.6 \text{ (db)} \tag{2.37}$$

Next, the average will be adjusted to the target using an adjusting factor. As shown in Chapter 1, there are two cases for the value after correction, y_i', is:

$$y_i' = y_i \frac{y_0}{m} \tag{2.38}$$

$$y_i' = y_i + (y_0 - m) \tag{2.39}$$

The difference is based on the way of correction. The former case is to change the sensitivity of the output and the latter case is to change something such as the position of the cutting tool. In both cases, deviation can be corrected, but the variability of y' after correction will be different.

The S/N Ratio for the case of Equation (2.38) will be discussed in the next section.

In the case of Equation (2.39), let σ^2 be the average of the squares of $(y_i - \bar{y})^2$ before correction and σ'^2 be the average of the squares of $(y_i' - \bar{y})^2$ after correction. In this case, σ'^2 is the same as σ^2. Therefore, its Loss Function, L_2, is:

$$L_2 = k\sigma'^2 = k \text{ [mean of the squares of } (y_i - \bar{y}), \sigma^2] \tag{2.40}$$

The S/N Ratio in this case is the reciprocal of σ^2 and the S/N Ratio in decibels is:

$$\eta = -10 \log \sigma^2 \tag{2.41}$$

Since the estimate of V_e, the error variance after decomposing the sum of squares into deviation, m, and individual difference, e, is equal to σ^2, the S/N Ratio is calculated by:

$$\eta = -10 \log V_e \tag{2.42}$$

In the previous example, V_e=2.94 from the ANOVA table (Table 2.2), the S/N Ratio is:

$$\eta = -10 \log V_e = -4.7 \text{ (db)} \tag{2.43}$$

Of the two cases, Equation (2.38) is more general and is used often in parameter design. It will be discussed in the next section. But if there is only a few percent difference between the target, y_0, and the average,the case of (2.38) may be treated as the case of (2.39). For the case when the target is equal to zero, it is also treated as the case of(2.39).

2.7 DECOMPOSITION OF DATA INTO AVERAGE AND INDIVIDUAL DIFFERENCE

Up to Section 2.5, the data we discussed was the differences from the target, y_0. It was explained in Chapter 1 that when there is a large amount of correction or when the sensitivity is to be adjusted, it is better to determine σ'^2, the variability after correction, by:

$$\sigma^{2\prime} = \sigma^2 \left(\frac{y_0}{m}\right)^2 = y_0{}^2 \frac{\sigma^2}{m^2} \tag{2.44}$$

To calculate m^2 and σ^2 in the equation, the data before subtracting the target is used. The total sum of squares is decomposed into the variation of the mean, S_m, and the error variation, S_e. The procedures are exactly the same as the previous section.

Convert the data into the ones before subtracting the target.

$$S_T = 105^2 + 105^2 + \cdots + 106^2 = 109{,}229 \qquad (f{=}10) \tag{2.45}$$

$$S_m = \frac{(105 + 105 + \cdots + 106)^2}{10} = \frac{1{,}045^2}{10} = 109{,}202.5 \qquad (f{=}1) \tag{2.46}$$

$$S_e = S_T - S_m = 26.5 \qquad (f{=}9) \tag{2.47}$$

If it is tedious to use original data having more digits, the following calculation may used by letting the total of deviations from the target be T_0:

$$S_m = \frac{[T_0 + (\text{target value}) \times (\text{number of data points})]^2}{(\text{number of data points})}$$

$$= \frac{(45 + 100 \times 10)^2}{10} = 109{,}202.5 \qquad (f{=}1) \quad (2.48)$$

Letting the sum of squares of deviations from the target be S_{T0} and the magnitude of deviation be S_{m0}, the error variation, S_e, is obtained as follows:

$$S_e = S_{TO} - S_{m0} = 229 - 202.5 = 26.5 \qquad (f{=}9) \quad (2.49)$$

The total variation,.S_T, is:

$$S_T = S_m + S_e = 109{,}202.5 + 26.5 = 109{,}229 \qquad (f{=}10) \quad (2.50)$$

The same results can be obtained by subtracting a value which is close to the average instead of using the target.

Table 2.3. ANOVA Table

Factor	f	S	V	S'	ρ(%)
m	1	109,202.5	109,202.5	109,199.56	99.97
e	9	26.5	2.94	29.44	0.03
T	10	109,229.0	109,22.9	109,229.00	100.00

The above results are summarized in Table 2.3. The pure variation S_m'(10 times the square of the true mean, m) is calculated as follows:

$$S_m' = S_m - V_e = 109{,}202.5 - 2.94 = 109{,}199.56 \qquad (2.51)$$

The pure error variation is:

$$S_e' = S_e + V_e$$
$$= 26.5 + 2.94 = 29.44 \qquad (2.52)$$

Although the same symbol m is used, it becomes either the variation of deviation or variation of simple average; it depends on the nature of data calculated. The error e is the same either way.

The objective of this example is to evaluate the variation after correction in the case of Equation (2.44) and its reciprocal, S/N Ratio. For this calculation, it is necessary to evaluate the true mean, m, and the true individual difference, σ^2 in Equation (2.44).

Since the expectations of the general mean, S_m, and the true individual difference, σ^2, are:

$$E(S_m) = km^2 + \sigma^2 \tag{2.53}$$

$$E(V_e) = \sigma^2 \tag{2.54}$$

m^2 is calculated as:

$$m^2 = \frac{S_m - V_e}{k} \tag{2.55}$$

In decibel units,

$$\eta = 10 \log \frac{\frac{1}{k}(S_m - V_e)}{V_e} \tag{2.56}$$

Numerically,

$$\eta = 10 \log \frac{\frac{1}{10}(109{,}202.5 - 2.94)}{2.94} = 35.70 \text{ (db)} \tag{2.57}$$

If the pure variations have been calculated, the S/N Ratio may also be obtained as follows. S_m, the pure variation of m (true mean), is m^2 multiplied by the number of data points.

$$S_m' = 10m^2 \tag{2.58}$$

S_e', the pure variation of e, is σ^2, the true individual difference multiplied by the number of data points.

$$S_e' = 10\sigma^2 \tag{2.59}$$

Therefore, m^2 and σ^2 are obtained from S_m' and S_e' divided by 10 respectively.

The variation after adjustment in Equation (2.44), or $\sigma^{2\prime}$ is:

$$\sigma^{2\prime} = y_0{}^2 \frac{\sigma^2}{m^2} = y_0{}^2 \frac{S_e'/10}{S_m'/10} \tag{2.60}$$

Since the target has nothing to do with quality, the index for quality evaluation, η, is given by:

$$\eta = 10 \log \frac{m^2}{\sigma^2} = 10 \log \frac{S_m'/10}{S_e'/10} \tag{2.61}$$

Numerically,

$$\eta = 10 \log \frac{S_m'/10}{S_e'/10} = 10 \log \frac{10{,}919.956}{2.944} = 35.69 \text{ (db)} \tag{2.62}$$

Improving the S/N Ratio determined from the above is to improve the quality after correcting the deviation. Assuming the average is adjusted to a target of 100, the variability after adjustment, $\sigma^{2\prime}$, is:

$$\sigma^{2\prime} = \frac{{y_0}^2}{\text{anti–log of } \eta} = \frac{100^2}{2{,}706.8} = 2.69 \tag{2.63}$$

Since the deviation is only a few percent, the variability after correction does not differ too much no matter which of Equations (2.38) or (2.39) is used.

In this example, only one S/N Ratio was determined. When the inner factors are assigned to an orthogonal array, S/N Ratio is calculated from each of the inner array conditions. Using decibel values to conduct analysis as described after Chapter 4, you will discover the factors which can improve S/N Ratio.

$$S = 10 \log m^2 = 10 \log \frac{1}{k} (S_m - V_e) \tag{2.64}$$

The numerator of the S/N Ratio is sometimes called sensitivity, denoted by the symbol S.

2.8 S/N Ratio for the Case of One Way Layout

2.8.1 The Error After Adjustment

The variability of characteristics with a target is evaluated after the deviation from the target is adjusted. In the previous sections, the data obtained from repetitions is used. In some cases, data is collected batch to batch or day to day, and quite frequently, adjustments are made every batch or everyday. In such cases, the between batch or between day difference is not the error after adjustment.

Table 2.4. Deviation from the Target

Unit: μm

Coil							Total
K_1	-10	-8	-4	2	-6	-4	-30
K_2	3	-12	9	-7	4	-9	-12
K_3	12	2	12	8	10	4	48

The target value of a steel plate thickness is 760 μm. Three coils were produced and the thickness of six points both in longitudinal and cross directions were measured. Table 2.4 shows the results after subtracting the target value. No adjustment was made before measurement.

This is an example of collecting data with one factor; the coils, in this case. It is called one factor layout. There are three sets of data, so the number of levels is three.

The deviation on the whole data is:

$$\bar{y}=\frac{-30-12+48}{18}=0.333 \qquad (2.65)$$

It seems that there is very little deviation. But there is a deviation in each coil.

$$\begin{cases}\bar{K}_1=\dfrac{-30}{6}=-5\\[2mm] \bar{K}_2=\dfrac{-12}{6}=-2\\[2mm] \bar{K}_3=\dfrac{48}{6}=8\end{cases} \qquad (2.66)$$

The deviation of the "i"th coil from the target is:

$$y_{ij}=\bar{y}+(\bar{K}_i-\bar{y})+(y_{ij}-\bar{K}_i) \qquad (2.67)$$

The deviation on the whole is:

$$y_{ij}=\bar{y}+(y_{ij}-\bar{y}) \qquad (2.68)$$

The variation between coils is included in the difference between the total variation, S_T, and the variation due to the whole deviation, S_m. Such variation between coils disappears after adjustment is made. Therefore, the variation after adjustment is expressed in the second term of Equation (2.67). This will be discussed in the next section.

2.8.2 Decomposition of Sum of Squares and Analysis of Variance

First, the total sum of squares and the whole deviation, S_m, are calculated.

$$S_T=(-10)^2+(-8)^2+\cdots+4^2=1{,}088 \qquad (f{=}18) \quad (2.69)$$

$$S_m=\frac{(-30-12+48)^2}{18}=\frac{6^2}{18}=2 \qquad (f{=}1) \quad (2.70)$$

The variation of $(K\text{-}y)$ in the second term of Equation (2.67) is called the variation due to K, denoted by S_K.

$$S_K=6\times[(\bar{K}_1-\bar{y})^2+(\bar{K}_2-\bar{y})^2+(\bar{K}_3-\bar{y})^2] \qquad (2.71)$$

Since there are six $(\bar{K}_i\text{-}\bar{y})$s, six is multiplied to the bracket [ ].

$$S_K=\frac{K_1{}^2+K_2{}^2+K_3{}^2}{\text{No. of replications of } K_i\ (6)}-S_m \qquad (f{=}2) \quad (2.72)$$

In Equation (2.71), $(\overline{K}_1 - \overline{y}) + (\overline{K}_2 - \overline{y}) + (\overline{K}_3 - \overline{y}) = 0$

Therefore, its degree of freedom is equal to 2. It is expressed as f=2. In general, the degrees of freedom of the variation for k levels is equal to k-1. The variation is calculated as:

$$S_K = \frac{(-30)^2 + (-12)^2 + 48^2}{6} - 2 = 556 \qquad (f=2) \quad (2.73)$$

The error variation after removing the differences between coils is given by the total of the squares of $(y_{ij}\text{-}\overline{K}_i)$.

$$\begin{aligned} S_e = &(y_{11} - \overline{K}_1)^2 + (y_{12} - \overline{K}_1)^2 + \cdots + (y_{16} - \overline{K}_1)^2 \\ &+ (y_{21} - \overline{K}_2)^2 + (y_{22} - \overline{K}_2)^2 + \cdots + (y_{26} - \overline{K}_2)^2 \\ &+ (y_{31} - \overline{K}_3)^2 + (y_{32} - \overline{K}_3)^2 + \cdots + (y_{36} - \overline{K}_3)^2 \end{aligned} \qquad (2.74)$$

It is actually calculated by:

$$S_e = S_T - S_m - S_K \qquad (f=15) \quad (2.75)$$

In the 18 terms of Equation (2.74), the following restrictions exist:

$$\begin{cases} (y_{11} - \overline{K}_1) + (y_{12} - \overline{K}_1) + \cdots + (y_{16} - \overline{K}_1) = 0 \\ (y_{21} - \overline{K}_2) + (y_{22} - \overline{K}_2) + \cdots + (y_{26} - \overline{K}_2) = 0 \\ (y_{31} - \overline{K}_3) + (y_{32} - \overline{K}_3) + \cdots + (y_{36} - \overline{K}_3) = 0 \end{cases} \qquad (2.76)$$

Therefore, the number of unknowns is 18-3=15, i.e., f=15. For actual calculation, the degrees of freedom of the total variation (18) is subtracted by the degree of freedom of the deviation(1) and the degrees of freedom of K(2) to get 15.

The error variation is calculated as:

$$S_e = S_T - S_m - S_K = 1{,}088 - 2 - 556 = 530 \qquad (f=15) \quad (2.77)$$

Table 2.5 shows the ANOVA table.

Table 2.5. ANOVA Table

Factor	f	S	V	S'	ρ(%)
m	1	2	2		
K	2	556	278	489.5	45.0
e	15	530	35.33		
(e)	(16)	(532)	(33.25)	(598.5)	(55.0)
T	18	1,088	60.44	1,088.0	100.0

In the ANOVA table, the deviation is small and is pooled with the error. The expectation of S_K is:

$$E[S_K] = S_K' + (\text{degrees of freedom of } K)\ (\sigma^2, \text{ true error variance}) \qquad (2.78)$$

$$E[V_K] = (\text{number of replicationsof } K) \times \sigma_k^{\ 2} + \sigma^2 \ (\text{true error variance}) \tag{2.79}$$

$$\sigma_K^{\ 2} = \frac{S_K' \,/\, (\text{number of replications of } K)}{\text{degrees of freedom of } K} \tag{2.80}$$

where

$$S_K' = S_K - (\text{degreesof freedom of } K) \times V_e = 556 - 2 \times 33.25 = 489.5 \tag{2.81}$$

For the expectation of variance, see the Appendix.

The pure variation of K, or S_K', is therefore,

$$S_e' = S_e + (\text{degrees of freedom of } K) \times V_e = 532 + 2 \times 33.25 = 598.5 \tag{2.82}$$

where the pooled S_e was used.

From the above calculation, the variation becomes 33.25 if the deviation of the individual coil is adjusted.

2.8.3 Calculation of S/N Ratio

When the error after adjustment is proportional to the mean, the S/N Ratio is:

$$\eta = 10 \log \frac{m^2}{\sigma^2} = 10 \log \frac{\frac{1}{k}(S_m - V_e)}{V_e} \tag{2.83}$$

The average in the above equation must be calculated from S_m, which is calculated from the average of original thickness, instead of using the deviations from the target.

$$S_m = \frac{(18 \times 760 + 6)^2}{18} = 10{,}405{,}922 \qquad (f{=}1) \tag{2.84}$$

For V_e, the error variance before pooling m is used.

$$\eta = 10 \log \frac{\frac{1}{6}(10{,}405{,}922 - 35.33)}{35.33} = 46.91 \ \text{(db)} \tag{2.85}$$

This is the S/N Ratio of the adjusted individual coil.

EXERCISE

Question (2.1). The specification of diameter for a microprocessor window is 3.00±0.25(μm), and the loss of becoming defective is 300 yen. One wafer each was produced from two processing conditions and the diameter of 10 positions was measured. Deviations from the target value were as follows:

A_1 0.5 0.0 -0.5 -1.0 0.0 -1.0 -0.5 0.5 1.5 1.0

A_2 1.5 1.0 0.5 0.5 1.0 0.5 0.5 1.0 0.9 0.5

(a) Make an ANOVA table by decomposing the total variation into the portions of deviation and individual difference.

(b) What is the loss due to the variation in terms of the mean square of the deviation from the target?

(c) Compare the above with the loss after adjusting the deviations.

Question (2.2). The h_{FE} values were measured from 8 transistors in a wafer produced by a certain method.

183 166 167 175 175 177 170 177

(a) Make an ANOVA table by decomposing the total variation into the variation of the mean and the variation of the individual difference.

(b) Determine the S/N Ratio of nominal-the-best characteristic.

CHAPTER 3

DECOMPOSITION OF SUM OF SQUARES—PROPORTIONAL AND LINEAR EQUATIONS

In Chapter 2, the S/N Ratio calculation for the nominal-the-best characteristic was explained. The characteristic for the case when the input directly affects the output is referred to as a "dynamic characteristic." It is a member of the family of nominal-the-best characteristics. There are two types of dynamic characteristics: (1) active type dynamic characteristics which is used to obtain the objective output by adjusting the input, and (2) passive type dynamic characteristics when the input is given to one direction such as measuring systems or receivers. In both cases, it is desirable that the relationship between the input and the output not be affected by error. The factor which is used for the input is called the signal factor. In this chapter, the calculation of S/N Ratios using signal factors in quality engineering is explained. The decomposition of the sum of squares into proportional or linear effects and the error variation is described.

3.1 DECOMPOSITION OF VARIATION AND THE S/N RATIO FOR DYNAMIC CHARACTERISTICS

3.1.1 The Proportional Equation

In a dyeing process, the color after dyeing is adjusted by the amount of dye. Two types of dyes, denoted by A_1 and A_2, were compared to evaluate the performance or quality, referred to as "dyeability". The amount of dye was varied to obtain a different color concentrations. Therefore, the signal factor is the amount of dye. Letting the amount of dye be denoted by M, three levels of M were set as follows:

$M_1 = 2,\ \ M_2 = 4,\ \ M_3 = 6$ owf (% of fiber weight)

Four observations were made at each level of M as shown in Table 3.1.

Table 3.1. Amount of Dye and Concentration

Repetition	A_1			Total	A_2			Total
	M_1	M_2	M_3		M_1	M_2	M_3	
1	12.1	24.5	36.6		9.8	20.0	29.3	
2	11.8	23.8	36.4		9.5	19.8	30.0	
3	11.7	23.8	36.7		9.7	19.4	29.6	
4	12.3	24.6	36.5		9.4	19.5	29.8	
	47.9	96.7	146.2	290.8	38.4	78.7	118.7	235.8

Color concentration increases when the amount of dye increases. Letting β denote the increment of concentration per unit amount of dye, the relationship between concentration, denoted by y, and amount of dye, M has the following relationship:

$$y = \beta M \tag{3.1}$$

The above equation indicates that when the amount of dye is equal to zero, the color concentration must be equal to zero. Such a relationship is referred to as a proportional equation or proportional regression equation. The procedures to estimate the proportional coefficient, β, will be explained first.

When the signal factor has k levels with r_0 repetitions each, the individual observation, y_{ij}, is written as:

$$y_{ij} = \beta M_i \ (i = 1, \cdots, k, \ j = 1, \cdots, r_0) \tag{3.2}$$

In (3.2), there are kr_0 simultaneous equations with one unknown. These are called observational equations. When the number of equations is larger than the number of unknowns, there is no solution to satisfy the equations. In such a case, the value of β which minimized the squares of differences between the right and the left hand sides is determined as the solution. This is called the least square method. In this case, β is obtained by minimizing the following residual sum of squares:

$$S_e = (y_{11} - \beta M_1)^2 + \cdots + (y_{kr0} - \beta M_k)^2 \tag{3.3}$$

β is given by:

$$\beta = \frac{y_1 M_1 + y_2 M_2 + \cdots + y_k M_k}{r_0(M_1{}^2 + M_2{}^2 + \cdots + M_k{}^2)} \tag{3.4}$$

where y_i is the total of the observations under signal factor level M_i

Using the data in Table 3.1, β is calculated as:

$$\beta(A_1) = \frac{47.9 \times 2 + 96.7 \times 4 + 146.2 \times 6}{4(2^2 + 4^2 + 6^2)} = 6.071 \tag{3.5}$$

$$\beta(A_2) = \frac{38.4 \times 2 + 78.7 \times 4 + 118.7 \times 6}{4(2^2 + 4^2 + 6^2)} = 4.928 \tag{3.6}$$

The relationship between amount of dye and concentration is given by:

$$\begin{cases} y(A_1) = 6.071M \\ y(A_2) = 4.928M \end{cases} \tag{3.7}$$

From the residual sum of squares in equation (3.3), the error is calculated by:

$$\begin{aligned} S_e &= \Sigma(y_{ij} - \frac{\Sigma y_i M_i}{r_0 \Sigma M_i^{\,2}} M_i)^2 \\ &= \Sigma y_{ij}^{\,2} - 2\frac{(\Sigma y_i M_i)^2}{r_0 \Sigma M_i^{\,2}} + \frac{(\Sigma y_i M_i)^2}{(r_0 \Sigma M_i^{\,2})^2} r_0 \Sigma M_i^{\,2} \\ &= \Sigma y_{ij}^{\,2} - \frac{(\Sigma y_i M_i)^2}{r_0 \Sigma M_i^{\,2}} \end{aligned} \tag{3.8}$$

The first term on the right side of equation (3.8) is the total of the squares of all data points. From the equation of β in equation (3.4), it is known that the second term is the total of $(\beta M)^2$. Thus,

$$\begin{aligned} r_0[(\beta M_1)^2 + (\beta M_2)^2 + \cdots + (\beta M_k)^2] &= r_0 \Sigma \left(\frac{\Sigma y_i M_i}{r_0 \Sigma M_i^{\,2}} M_i \right)^2 \\ &= \frac{(\Sigma y_i M_i)^2}{r_0 \Sigma M_i^{\,2}} \end{aligned} \tag{3.9}$$

This is the magnitude of the proportional term, and is referred to as variation of proportional term and denoted by S_β. Letting the total variation, or the total of the squares of all data points, be denoted by S_T, the residual sum of squares is given by:

$$S_e = S_T - S_\beta \tag{3.10}$$

Thus, the total variation is decomposed into proportional and residual terms:

$$S_T = S_\beta + S_e \tag{3.11}$$

The degrees of freedom for each variation term is as follows:

S_T: kr_0
S_β: 1
S_e: kr_0-1

The total variation of A_1 is decomposed as follows:

$$S_T(A_1) = 12.1^2 + 11.8^2 + \cdots + 36.5^2 = 8{,}255.78 \qquad (f=12) \quad (3.12)$$

$$S_\beta(A_1) = \frac{(47.9 \times 2 + 96.7 \times 4 + 146.2 \times 6)^2}{4(2^2 + 4^2 + 6^2)}$$

$$= \frac{1{,}359.8^2}{224} = 8{,}254.71 \qquad (f=1) \quad (3.13)$$

$$S_e(A_1) = S_T(A_1) - S_\beta(A_1) = 1.07 \qquad (f=11) \quad (3.14)$$

$$V_e(A_1) = \frac{S_e(A_1)}{11} = 0.097\,3 \qquad (3.15)$$

The above results are summarized in the form of an analysis of variance table (ANOVA Table) as shown in Table 3.2 (although it is not important).

Table 3.2. ANOVA Table for Proportional Equation

Factor	f	S	V
β	1	S_β	S_β
e	kr_0-1	S_e	V_e
T	kr_0	S_T	

Similarly, the total variation of A_2 is decomposed:

$$S_T(A_2) = 5{,}440.08 \qquad (f=12) \quad (3.16)$$

$$S_\beta(A_2) = 5{,}439.17 \qquad (f=1) \quad (3.17)$$

$$S_e(A_2) = 0.83 \qquad (f=11) \quad (3.18)$$

$$V_e(A_2) = 0.075\,5 \qquad (3.19)$$

The error variance is the average of the squares of residuals when the amount of dye, M, and color concentration, y, are expressed as a proportional relationship. The smaller the error variance, the better the fit of the proportional relationship. Comparing A_1 and A_2, it is seen that the variance of A_2 is smaller, and it seems that A_2 is better. However, this is not necessarily true. As explained in Chapter 1, large sensitivity in β means that there will be significant color changes with only a small amount of dye. It is difficult to adjust color. In actual processing, there must be an appropriate or ideal sensitivity, which is denoted by β_0. If sensitivity is adjusted to the ideal level, β_0, such as diluted by filler, the error variance σ^2, of the following equation:

$$y = \beta M \tag{3.20}$$

would be:

$$\sigma_0^{\,2} = \left(\frac{\beta_0}{\beta}\right)^2 \sigma^2 \tag{3.21}$$

The level of the ideal sensitivity, β_0, has nothing to do with the quality of the dyeing process. Therefore, β_0 is not studied where the signal factor exists. Instead, the reciprocal of $(\sigma/\beta)^2$ in decibel units is considered as the S/N Ratio, which is denoted by η. This S/N Ratio is used to evaluate the error after considering the differences of the slope, and is written as:

$$\eta = 10 \log \frac{\beta^2}{\sigma^2} \quad \text{(db)} \tag{3.22}$$

It seems that the S/N Ratios of A_1 and A_2 might be calculated from $\beta(A_1)$, $\beta(A_2)$, $V_e(A_1)$ and $V_e(A_2)$. But strictly, it is not exactly so since there is an error included in the estimate of β. Therefore, the square of β cannot exactly be equal to its true value, β^2. This is the same situation as is the case of estimating the mean in the previous chapter.

As explained before, S_β, the variation of β, is the total of $(\beta M)^2$ in equation (3.9). Since the true value of β is unknown, let the total of $(\beta M)^2$ be called the net variation or pure variation, denoted by S_β'. The expectation of S_β, denoted by $E[S_\beta]$, is given by:

$$\begin{aligned} E[S_\beta] &= S_\beta' + \sigma^2 \\ &= r_0[(\beta M_1)^2 + (\beta M_2)^2 + \cdots + (\beta M_k)^2] + \sigma^2 \\ &= r\beta^2 + \sigma^2 \end{aligned} \tag{3.23}$$

where

$$r = r_0(M_1^{\,2} + M_2^{\,2} + \cdots + M_k^{\,2}) \tag{3.24}$$

(See Appendix 5 for more details.) Since net variation, S_e', is the error variance, σ^2 multiplied by kr_0, the expectation of S_e, denoted by $E[S_e]$, is given by:

$$E[S_e] = S_e' - \sigma^2 = kr_0\sigma^2 - \sigma^2 = (kr_0 - 1)\sigma^2 \tag{3.25}$$

The expectation of the error variance is obtained after dividing the above result by its degrees of freedom.

$$E[V_e] = \sigma^2 \tag{3.26}$$

Letting

$$\sigma^2 = V_e \tag{3.27}$$

β_2 is estimated by:

$$\beta^2 = \frac{1}{r}(S_\beta - V_e) \tag{3.28}$$

From the above, the S/N Ratio in decibel units is given by:

$$\eta = 10\log\frac{\beta^2}{\sigma^2} = 10\log\frac{\frac{1}{r}(S_\beta - V_e)}{V_e} \tag{3.29}$$

where r is called the "effective divisor." In the dyeing example, r is calculated as:

$$r = r_0(M_1{}^2 + M_2{}^2 + \cdots + M_k{}^2) = 4 \times (2^2 + 4^2 + 6^2) = 224 \tag{3.30}$$

The S/N Ratios of A_1 and A_2 are calculated as:

$$\begin{aligned}\eta(A_1) &= 10\log\frac{\frac{1}{224}(8{,}254.71 - 0.0973)}{0.0973}\\ &= 10\log 378.735 = 25.78 \quad \text{(db)}\end{aligned} \tag{3.31}$$

$$\begin{aligned}\eta(A_2) &= 10\log\frac{\frac{1}{224}(5{,}439.17 - 0.0755)}{0.0755}\\ &= 10\log 321.612 = 25.07 \quad \text{(db)}\end{aligned} \tag{3.32}$$

The error variances of the proportional equation were compared using equations (3.15) and (3.19), revealing that A_2 is smaller. By comparing S/N Ratios, however, there is no difference between the two. This means that the error would be the same if the sensitivity, β, were adjusted to the same level.

Assuming that the most convenient level of sensitivity to use in the operation is $\beta_0=5$, and an adjustment can be made, the variance after the adjustment, $\sigma_0{}^2$, would be:

$$\sigma_0{}^2 = \left(\frac{\beta_0}{\beta}\right)^2 \sigma^2 = \frac{\beta_0{}^2}{\text{anti–log of } \eta} \tag{3.33}$$

$$\sigma_0{}^2(A_1) = \frac{5^2}{378.735} = 0.066 \tag{3.34}$$

$$\sigma_0{}^2(A_2) = \frac{5^2}{321.612} = 0.078 \tag{3.35}$$

The above error variances are used to calculate the Loss Function.

3.1.2 The Case of the Linear Equation

When the signal factor level is equal to zero, the output characteristic may not be equal to zero all the time. However, the proportional equation may often be applied in many cases by considering a certain level of the signal factor as a base, and the output at that base is set equal to zero. This is called a reference point proportional equation. Once the coordinate system of measuring the signal is shifted, the procedures are exactly identical to the case of the proportional equation. Therefore, the case of the reference proportional equation is not explained here.

In a study of the driveability of a car, for example, the signal factor is the factor which changes direction of the vehicle. It is the steering angle in this case. When the steering angle is fixed, the car runs along a constant circle. Good driveability means that the steering angle and turning radius have a linear relationship. When the steering angle is fixed, the turning radius must be constant without being affected by various types of noise. In normal street driving, the steering angle is about 250 degrees.

For the study, signal factor levels were set at

$M_1 = 200 \quad M_2 = 250 \quad M_3 = 300$ (degree)

The results of the experiment are presented in 6 measurements at each steering angle.

Table 3.3. Steering Angle and Turning Radius

Unit: *m*

Replication	M_1	M_2	M_3	Total
	44.0	34.2	27.4	
	43.7	31.9	31.8	
	46.6	35.5	27.3	
	37.2	42.8	25.5	
	38.2	33.7	27.6	
	39.5	28.1	34.6	
Total	249.2	206.2	174.2	629.6

It would be desirable that the relationship between turning radius, y, and steering angle, M, be proportional. In this case, it is more appropriate to consider a linear equation (of a linear regression equation) instead of a proportional equation. A linear equation can be written as

$$y = a + bM \tag{3.36}$$

However, it is better to use the following equation:

$$y = m + \beta(M - \overline{M}) \tag{3.37}$$

where

$$\overline{M} = \frac{M_1 + M_2 + M_3}{3} \tag{3.38}$$

Equation 3.37 is called "Chevychev's orthogonal polynomial" equation (see Appendix 12). In this equation, only terms up to the linear term are shown. The difference between equations (3.36) and (3.37) is in the orthogonality between the terms in the equation. In the previous sections, variation of characteristics is decomposed into the variations of individual factors. It is possible because the property of the orthogonality of the factorial effects is utilized. It is important to maintain orthogonality for the factors and terms in the inner array as well as in the outer array. In equation (3.36), the first and the second terms are not orthogonal, but the terms in equation (3.37) are.

Suppose there are r_0 pieces of observations at each of the k signal factor levels. An observation, denoted by y_{ij}, is put into equation (3.37):

$$y_{ij} = m + \beta(M_i - \overline{M}) \quad (i=1, \cdots, k,\ j=1, \cdots, r_0) \tag{3.39}$$

In order to find the two unknowns, m and β, from kr_0 observational equations, the least square method is used, as was the case of the proportional equation to determine values of m and β which minimize the following residual sum of squares:

$$S_e = [y_{11} - m - \beta(M_1 - \overline{M})]^2 + \cdots + [y_{kr0} - m - \beta(M_k - \overline{M})]^2 \tag{3.40}$$

The equations used to determine m and β are given by

$$m = \frac{y_{11} + y_{12} + \cdots + y_{kr0}}{kr_0} \tag{3.41}$$

$$\beta = \frac{y_1(M_1 - \overline{M}) + y_2(M_2 - \overline{M}) + \cdots + yk(M_k - \overline{M})}{r_0[(M_1 - \overline{M})^2 + (M_2 - \overline{M})^2 + \cdots + (M_k - \overline{M})^2]} \tag{3.42}$$

where y_i is the total of r_0 pieces of data under signal factor level, M_i. Residual sum of squares, S_e, is given by

$$S_e = [y_{11} - m - \beta(M_1 - \overline{M})]^2 + \cdots + [y_{kr0} - m - \beta(M_k - \overline{M})]^2$$
$$= \Sigma y_{ij}{}^2 - \frac{(\Sigma y_{ij})^2}{kr_0} - \frac{[\Sigma(M_i - \overline{M})y_i]^2}{r_0\,\Sigma(M_i - \overline{M})^2} \qquad (3.43)$$

The first term of the right side in the equation is the total of the squares of individual data and is referred to as S_T. The second term is called the general mean, denoted by S_m, which is equal to the square of the average, m, multiplied by kr_0.

The third term

$$r_0\beta^2[(M_1 - \overline{M})^2 + (M_2 - \overline{M})^2 + \cdots + (M_k - \overline{M})^2] \qquad (3.44)$$

is called the linear term variation, denoted by S_β. That is the magnitude of the linear term, $\beta(M - \overline{M})$. S_e is calculated by

$$S_e = S_T - S_m - S_\beta \qquad (f{=}kr_0\text{-}2) \quad (3.45)$$

where

$$S_T = \text{Total of squares} \qquad (f = k_{r0}) \quad (3.46)$$

$$S_m = \frac{(\Sigma y_{ij})^2}{kr_0} \qquad (f{=}1) \quad (3.47)$$

$$S_\beta = \frac{[\Sigma(M_i - \overline{M})y_i]^2}{r_0\,\Sigma(M_i - \overline{M})^2} \qquad (f{=}1) \quad (3.48)$$

The numerator and denominator of S_β can be written as

$$[\Sigma(M_i - \overline{M})y_i]^2 = [\Sigma M_i y_i - \frac{\Sigma M_i \Sigma y_i}{k}]^2 \qquad (3.49)$$

$$r_0\,\Sigma(M_i - \overline{M})^2 = r_0\,[\Sigma M_i{}^2 - \frac{(\Sigma M_i)^2}{k}] \qquad (3.50)$$

From equation (3.45), the following decomposition can be written:

$$S_T = S_m + S_\beta + S_e \qquad (3.51)$$

3.1.3 Table of Coefficients for the Orthogonal Polynomial Equation

When the signal factor levels are equally spaced, the coefficients of a linear equation or the variation of the linear term can be very easily calculated. Appendix Table 2 shows the coefficients of an orthogonal polynomial equation.

The table also includes the coefficients for higher order terms, but only the linear term is discussed here. Symbol b_1 in the table is referred to as the linear term, which is the same as β in the previous section. Also A is referred to as y, the total of r_0 pieces of data at each level. Substituting

b_1 and A to β and y respectively, the equations for the linear coefficient, β, and its variation, S_β, are given by

$$\beta = \frac{W_1y_1 + W_2y_2 + \cdots + W_ky_k}{r_0\lambda Sh} \tag{3.52}$$

$$S_\beta = \frac{(W_1y_1 + W_2y_2 + \cdots + W_ky_k)^2}{r_0\lambda^2 S} \tag{3.53}$$

The figures of $W_1, W_2, \ldots, W_k$, λS and $\lambda^2 S$ in the equations are found from the column of b_1 for the corresponding number of levels, k, where h is the signal factor level interval.

In the example of driveability of cars, with steering angle, M, having three levels, with an interval, h, of 50 degrees, the total sum of squares and S_m are calculated as follows:

$$S_T = 44.0^2 + 43.7^2 + \cdots + 34.6^2 = 22{,}741.68 \qquad (f{=}18) \tag{3.54}$$

$$S_m = \frac{629.6^2}{18} = 22{,}022.01 \qquad (f{=}1) \tag{3.55}$$

To calculate S_β, the values of W are found from the column of k=3 and b_1.

$W_1 = -1 \quad W_2 = 0 \quad W_3 = +1 \quad \lambda^2 S{=}2$

$$\begin{aligned} S_\beta &= \frac{(W_1y_1 + W_2y_2 + \cdots + W_ky_k)^2}{r_0\lambda^2 S} \\ &= \frac{(-1 \times 249.2 + 0 \times 206.2 + 1 \times 174.2)^2}{6 \times 2} \\ &= \frac{(-75)^2}{12} = 468.75 \qquad (f{=}1) \end{aligned} \tag{3.56}$$

$$S_e = S_T - S_m - S_\beta = 250.92 \qquad (f{=}16) \tag{3.57}$$

The above calculations are often presented in an ANOVA table:

Table 3.4. ANOVA Table

Factor	f	S	V
m	1	22,022.01	22,022.01
β	1	468.75	468.75
e	16	250.92	15.682
T	18	22,741.68	

The equation for the S/N Ratio is the same as the case of proportional equation.

$$\eta = 10 \log \frac{\beta^2}{\sigma^2} = 10 \log \frac{\frac{1}{r}(S_\beta - V_e)}{V_e} \tag{3.58}$$

where

$$r = r_0[(M_1 - \overline{M})^2 + (M_2 - \overline{M})^2 + \cdots + (M_k - \overline{M})^2] \tag{3.59}$$

In the case of equispaced levels, r is calculated by

$$r = r_0 S h^2 \tag{3.60}$$

When there are 3 levels, S=2 from the table and r is calculated as

$$r = r_0 S h^2 = 6 \times 2 \times 50^2 = 30{,}000 \tag{3.61}$$

The S/N Ratio in decibel units is calculated as

$$\begin{aligned}\eta &= 10 \log \frac{\frac{1}{r}(S_\beta - V_e)}{V_e} = 10 \log \frac{\frac{1}{30{,}000}(468.75 - 15.682)}{15.682} \\ &= 10 \log 0.000\,963 = -30.16 \text{ (db)}\end{aligned} \tag{3.62}$$

3.2 DECOMPOSITION OF VARIATION (WITH COMPOUNDING)

In order to find a design whose input/output relationship is not affected by noise, it is important to collect data under various noise factor conditions. For the purpose of improving quality, it is common that there are many inner conditions and therefore, it is desirable that the outer conditions be less to be efficient. For such purpose, only major noise factors are considered, and those noise factors are "compounded". In the nominal-the-best case, noise factor compounding is made by considering the following 3 levels of noise factors whose qualitative trend is known:

N_1 = Combination of noise factor conditions which makes the output result smaller. This is called "negative side worst condition".

N_2 = Standard condition

N_3 = Combination of noise factor conditions which makes the output result larger. This is called "positive side worst condition".

In many cases, only N_1 and N_3 are used. For the layout of signal and noise factors, see Chapter 7. Sometimes, the calculation of the S/N Ratio with a compounded noise factor is different from that which was described in the previous section. In this section, the S/N Ratio for the case with compounding is shown here. This section may be omitted for those readers who are not prepared at this time.

3.2.2 The Case of the Proportional Equation

In the following S/N Ratio equation,

$$\eta = 10 \log \frac{\frac{1}{r}(S_\beta - V_e)}{V_e} \tag{3.63}$$

the numerator is used to estimate β^2. When noise factors are compounded, the above equation cannot be used. S_e is calculated from S_T-S_β. It includes not only the deviation from linearity, but also the effect of the compounded noise factor, which is referred to as N. In the case of the proportional equation, the effect of N is the difference in the linear coefficient, β, caused by different levels of N. This is also called the interaction between N and β. When the slope, β, is different at different levels of N, it is not easy to control color concentration, the objective characteristic in a dyeing process. In that sense, it is the error, but sometimes, it is too large to consider to be the experimental error. To illustrate the calculations, Table 3.5 is prepared describing the results of the outer conditions with compounded factors.

Table 3.5. Results with Compounding

	M_1	M_2	⋯	M_k	
N_1	y_{11}	y_{21}	⋯	y_{k1}	N_1
·	·		·	·	·
·	·	y_{22}	·	·	·
·	·		·	·	·
N_3	y_{13}	y_{23}	⋯	y_{k3}	N_3
	y_1	y_2	⋯	y_k	T

Y_{ij} is the total of r_0 repetitions. Sometimes r_0=1.

Letting the coefficients of proportional equations at each level of N be $\beta_1, \beta_2, ..., \beta_n$, respectively, the output characteristics, y_j (j=1, 2, ..., n), at the "j"th level of N is given by

$$y_j = [\bar{\beta} + (\beta_j - \bar{\beta})]M = \bar{\beta}M + (\beta_j - \bar{\beta})M \tag{3.64}$$

where

$$\bar{\beta} = \frac{M_1 y_1 + M_2 y_2 + \cdots + M_k y_k}{n r_0 (M_1{}^2 + M_2{}^2 + \cdots + M_k{}^2)} = \frac{\beta_1 + \cdots + \beta_n}{n} \tag{3.65}$$

$\bar{\beta}$ is the same as the β in Section 3.1.1.

$$\beta_j = \frac{M_1 y_{1j} + M_2 y_{2j} + \cdots + M_k y_{kj}}{r_0 (M_1{}^2 + M_2{}^2 + \cdots + M_k{}^2)} \tag{3.66}$$

Residual variation, S_e, is the total of the squares of the differences between observations and the results calculated from equation (3.64). The total variation, S_T, is decomposed into the following:

$$S_T = S_\beta + S_{N\times\beta} + S_e \tag{3.67}$$

where

$$S_T = \text{Total of the squares of } nkr_0 \text{ pieces of data} \qquad (f = nkr_0) \tag{3.68}$$

S_β is the total of the squares of the first term in the right side of equation (3.64)

$$S_\beta = \left[\frac{M_1 y_1 + M_2 y_2 + \cdots + M_k y_k}{n r_0 (M_1{}^2 + M_2{}^2 + \cdots + M_k{}^2)}\right]^2 (M_1{}^2 + \cdots + M_k{}^2) n r_0$$
$$= \frac{(M_1 y_1 + M_2 y_2 + \cdots + M_k y_k)^2}{n r_0 (M_1{}^2 + M_2{}^2 + \cdots + M_k{}^2)} \qquad (f=1) \tag{3.69}$$

In the same equation, the second term in the right side is the magnitude of the difference between the proportional coefficients at different levels of N, of the interaction between N and β, denoted by S_N or $S_{N\times\beta}$.

$$S_N = S_{N\times\beta} = \text{total of the squares of } (\beta_j - \bar{\beta})M$$
$$= (\text{total of the squares of } \beta_j M)$$
$$- (\text{total of the squares of } \bar{\beta}M)$$
$$= \frac{L_1{}^2 + \cdots + L_n{}^2}{r_0(M_1{}^2 + M_2{}^2 + \cdots + M_k{}^2)} - S_\beta \qquad (f\!=\!2) \quad (3.70)$$

where

$$L_j = M_1 y_{1j} + M_2 y_{2j} + \cdots + M_k y_{kj} \qquad (3.71)$$

S_e is the total of the squares of the differences between right and left sides of equation (3.64). The following equation can be derived, whose derivation is not shown here:

$$S_e = S_T - S_\beta - S_N \qquad [(f\!=\!n(kr_0\text{-}1)] \quad (3.72)$$

The above results are summarized in Table 3.6, the ANOVA table.

Table 3.6. ANOVA Table for Proportional Equation With Compounding

Factor	f	S	V
β	1	S_β	S_β
N	n - 1	S_N	V_N
e	$n(kr_0$ - 1)	S_e	V_e
$(N$+$e)$	$(nkr_0$ - 1)	(S_N')	(V_N')
T	nkr_0	S_T	

The magnitude of the error to evaluate S/N Ratio is the total of S_N, the variation of the compounded noise factors, and S_e, the variation of the error. It is denoted by S_N' or $S_{(N+e)}$.

$$S_N' = S_T - S_\beta \qquad (f\!=\!nkr_0\text{-}1) \quad (3.73)$$

There are two cases for treating V_N and V_e. When V_N is several times greater than V_e, the pooled error variance, V_N', is calculated from N and e.

$$\eta = 10 \log \frac{\frac{1}{r}(S_\beta - V_e)}{V_N'} \qquad (3.74)$$

where

$$r = nr_0(M_1{}^2 + M_2{}^2 + \cdots + M_k{}^2) \qquad (3.75)$$

To estimate β^2, S_β is subtracted by the experimental error, V_e, instead of V_N', which includes the compounded error. But when the V_N is about the same as V_e, there is no necessity to distinguish and two and use the following equation:

$$\eta = 10 \log \frac{\frac{1}{r}(S_\beta - V_N')}{V_N'} \quad (3.75)$$

If it is too tedious to compare V_N and V_E, equation (3.74) may be used all the time. However, it should also be noted that sometimes the use of equation (3.75) is inappropriate.

3.2.3 A Simple Example

In a cutting process, the dimensions of products are controlled by the cutting depth of a bite. In this case, the cutting depth is the signal factor. To evaluate the processability under a certain condition, cutting depth was set at the following conditions:

$M_1 = 0.5 \quad M_2 = 1.0 \quad M_3 = 1.5$ (mm)

Noise factors such as feeding speed, cutting speed, and tool wear, etc. were compounded to get 3 levels: N_1 (negative side worst condition), N_2 (normal condition) and N_3 (positive side worst condition). The number of repetitions is one. The response, the amount of cut, was measured under each signal factor level and compounded noise factor level as shown in Table 3.7.

Table 3.7. Amount to Cut

	M_1	M_2	M_3	
N_1	0.477	1.001	1.489	2.967
N_2	0.486	1.019	1.581	3.086
N_3	0.499	1.038	1.581	3.118
	1.462	3.058	4.651	9.171

The proportional equation was used since the amount of cut is zero when cutting depth is zero.

$$S_T = 0.477^2 + 0.486^2 + \cdots + 1.581^2 = 11.046\,775 \quad (f=9) \quad (3.76)$$

$$S_\beta = \frac{(0.5 \times 1.462 + 1.0 \times 3.058 + 1.5 \times 4.651)^2}{3(0.5^2 + 1.0^2 + 1.5^2)}$$

$$= 11.037\,713 \quad (f=1) \quad (3.77)$$

$$S_N = \frac{L_1{}^2 + L_2{}^2 + L_3{}^2}{(0.5^2 + 1.0^2 + 1.5^2)} - S_\beta = 0.005\,810\,5 \quad (f=2) \quad (3.78)$$

$$\begin{cases} L_1 = 0.5 \times 0.477 + 1.0 \times 1.001 + 1.5 \times 1.489 = 3.473\,0 \\ L_2 = 0.5 \times 0.486 + 1.0 \times 1.019 + 1.5 \times 1.581 = 3.633\,5 \\ L_3 = 0.5 \times 0.499 + 1.0 \times 1.038 + 1.5 \times 1.581 = 3.659\,0 \end{cases} \tag{3.79}$$

$$S_e = S_T - S_\beta - S_N = 0.003\,252 \qquad (f=6) \tag{3.80}$$

Since V_N is several times greater than V_e, the S/N Ratio is calculated by Equation (3.74).

$$\begin{aligned} \eta &= 10 \log \frac{\frac{1}{r}(S_\beta - V_e)}{V_N'} \\ &= 10 \log \frac{\frac{1}{3(0.5^2 + 1.0^2 + 1.5^2)}(11.037713 - 0.000\,542)}{0.001\,133} \\ &= 10 \log 927.8 = 29.67 \quad \text{(db)} \end{aligned} \tag{3.81}$$

The ANOVA table is shown in Table 3.8.

Table 3.8. ANOVA Table for Proportional Equation With Compounding

Factor	f	S	V
β	1	11.037 713	11.037 713
N	2	0.005 810	0.002 905
e	6	0.003 252	0.000 542
$(N+e)$	(8)	(0.009 062)	(0.001 133)
T	9	11.046 775	

3.2.4 The Case of the Linear Equation

The way of treating the case of the linear equation with compounding is similar to the case of the proportional equation. The ANOVA table and variance of calculations are shown below.

$$S_m = \frac{(y_1 + y_2 + \cdots + y_k)^2}{nkr_0} \qquad (f=1) \tag{3.82}$$

$$S_T = \text{Total of } nkr_0 \text{ pieces of data} \qquad (f=nkr_0) \tag{3.83}$$

$$S_\beta = \frac{[(M_1 - \overline{M})y_1 + (M_2 - \overline{M})y_2 + \cdots + (M_k - \overline{M})y_k]^2}{nr_0[(M_1 - \overline{M})^2 + (M_2 - \overline{M})^2 + \cdots + (M_k - \overline{M})^2]} \qquad (f=1) \tag{3.84}$$

Table 3.9. ANOVA Table for Linear Equation with Compounding

Factor	f	S	V
m	1	S_m	V_m
β	1	S_β	V_β
N	n-1	S_N	V_N
e	$n(kr_0\text{-}1)\text{-}1$	S_e	V_e
$(N+e)$	$(nkr_0\text{-}2)$	(S_N')	(V_N')
T	nkr_0	S_T	

When the signal factor levels are equally spaced, coefficients of orthogonal polynomial equation are used as shown in equation (3.53).

$$S_N = \frac{N_1^{\,2} + \cdots + N_n^{\,2}}{kr_0} - S_m \qquad (f{=}2) \quad (3.85)$$

$$S_e = S_T - S_m - S_\beta - S_N \qquad [f{=}n(kr_0\text{-}1)\text{-}1] \quad (3.86)$$

$$S_N' = S_T - S_m - S_\beta \qquad (f{=}nkr_0\text{-}2) \quad (3.87)$$

In S_e, the interaction between N and β is included. In the case of the linear equation, S_N is the main effect of N, the compounded noise factors, since it is judged that the interaction is not significant.

The S/N Ratio calculation is similar to the case of the proportional equation. When V_N is several times greater than V_e, the pooled error variance is calculated from N and e to get V_N'.

$$\eta = 10 \log \frac{\frac{1}{r}(S_\beta - V_e)}{V_N'} \quad (3.88)$$

where

$$r = nr_0\,[(M_1 - \overline{M})^2 + (M_2 - \overline{M})^2 + \cdots + (M_k - \overline{M})^2]$$

If the levels of M are equally spaced, then one should use the coefficients b_1 and S at k level, as shown in the Appendix, Table 2,

$$r = nr_0 S h^2$$

If V_N and V_e are about the same, then there is no necessity to distinguish them.

$$\eta = 10 \log \frac{\frac{1}{R}(S_\beta - V_N')}{V_N'} \quad (3.89)$$

For the case when there is no signal factor, compounding is made with repetitions. The data is decomposed into the following:

$$S_T = S_m + S_N + S_e \quad (3.90)$$

Decomposition of the data with a 3-level compounded factor and r_0 repetitions is shown in Table 3.10.

Table 3.10. ANOVA Table with Compounding But Without Signal Factor

Factor	f	S	V
m	1	S_m	V_m
N	2	S_N	V_N
e	$3(r_0\text{-}1)$	S_e	V_e
$(N{+}e)$	$(3r_0\text{-}2)$	(S_N')	(V_N')
T	$3r_0$	S_T	

The S/N Ratio is calculated by

$$\eta = 10 \log \frac{\frac{1}{r}(S_m - V_e)}{V_N'} \tag{3.91}$$

Variation of N, or S_N, is calculated by

$$S_N = \frac{N_1{}^2 + \cdots + N_n{}^2}{r_0} - S_m \tag{3.92}$$

3.3 Other Cases

In some cases, a factor does not belong to either the class of signal or noise factors. An example is the calibration of deviation or sensitivity is made from batch to batch or from variety to variety, as shown in Chapter 2. It is important to understand the basics of the S/N Ratio and be able to calculate variations of all kinds. However, most of the calculations follow the applications which have been illustrated in this section.

Exercises

Question (3.1). In a machining process, cutting depth is used as the signal with the following three levels:

$M_1 = 0.5 \quad M_2 = 1.0 \quad M_3 = 1.5$ (mm)

Noise factors such as material types or tool wear were compounded to be the following two levels:

N_1 = negative side worst condition
N_2 = positive side worst condition
The amount of cut was shown in the following two-way layout:

	M_1	M_2	M_3
N_1	0.72	1.45	2.14
N_2	0.77	1.59	2.43

(a) Determine the S/N Ratio using the proportional equation.
(b) Determine the S/N Ratio using the linear equation.
(c) Compare the variability on the amount of cut when the sensitivity, β is equal to 1 for the cases of the proportional and linear equations.

Question (3.2). An illumination meter is used to measure the thickness of transparent color film. When the thickness becomes out of specification (+/-2μm), the loss is ¥300. In order to compare the effect of the distance between the light source and the sensor,

A_1 = current distance
A_2 = 10μm linger

were set. Thickness of the following films were measured:

M_1 = 47 μm
M_2 = 50 μm
M_3 = 53 μm

A_1	M_1	M_2	M_3
	156.15	145.15	134.13
	157.16	146.15	135.14

A_2	M_1	M_2	M_3
	161.16	151.15	137.14
	162.14	151.03	136.14

(a) The range of thickness measurement if 50±0.3 (μm). Use M_2 = 50 as a reference point to decompose the total variation, S_T, into the proportional term and residual portion based on proportional equation and construct an ANOVA table. (Since M_2 is the reference point, M_1 = -3, M_2 = 0, M_3 = 3. Subtract the mean of A_1M_2, or 165.65, from the results under A_1 and subtract the mean of A_2M_2, or 151.09, from the results under A_2. After decomposition, construct the following ANOVA table.)

ANOVA Table

Factor	f	S	V
β	1		
e	4		
T			

(b) Use the following equation to calculate the S/N Ratios of A_1 and A_2, then compare which is better. S/N Ratio is always larger-the-better.

$$\eta = \frac{\frac{1}{r}(S_\beta - V_e)}{V_e}$$

where

$$r = 2[(-3)^2 + 0^2 + 3^2] = 36$$

(c) The loss due to the thickness measurement error of illumination meter is given by:

$$L = \frac{A}{\Delta^2} \times \frac{1}{\eta} = \frac{300}{2^2} \times \frac{1}{\eta}$$

Compare in terms of annual loss. The annual production of this film is 15,000 m^2.

CHAPTER 4

INTRODUCTION TO ORTHOGONAL ARRAYS

In quality engineering, control factors used for the improvement of a S/N Ratio are referred to as "inner factors." The factors which are used to determine the S/N Ratio (The signal factor and noise factors) are called the "outer factors." The determination of the inner experimental conditions referred to as the inner layout; the determination of the outer conditions for the S/N Ratio is referred to as the outer layout. When many factors are studied, orthogonal arrays (O.A.) are used. (See Appendix for more information on their use.) O.A. can be used in either the inner and the outer layouts. In parameter design, however, it is usually applied in the inner layout alone. In this chapter, a simple explanation for the use of orthogonal arrays is provided, with analysis based upon the use of the S/N Ratios of the inner factors. To simplify discussion, only the explanation, L_4, an O.A. of four conditions, the smallest O.A. array, will be used. However, keep in mind that this array is seldom used in actual applications.

4.1 LAYOUT OF FACTORS

4.1.1 The Multi-factor Layout and Orthogonal Array

Many control factors are usually considered in making design improvements. The inner layout is used to determine the level combination of the inner design conditions to achieve robust design.

For example, consider a design improvement study with factors *A*, *B* and *C*, each evaluated at two conditions (levels). Therefore, since *A* has 2 levels, *B* has 2 levels and *C* has 2 levels, there are 8 possible run combinations:

1. $A_1B_1C_1$
2. $A_1B_1C_2$
3. $A_1B_2C_1$

. ...

. ...

8. $A_2B_2C_2$

When all run combinations of the three factors are to be studied, the combinations represent a three-way layout of the inner conditions. When all combinations of a limited number of factors are studied, the layout is referred to as a "multi-way layout."

In the determination of stable design conditions, however, a multi-way layout is not used when there are many control factors. Instead,orthogonal arrays are used. It is the characteristic of an orthogonal array layout that only a certain number of combinations are investigated, instead of all combinations.

There are 2-level and 3-level series in orthogonal arrays. Some typical arrays are illustrated in Appendix 4. Table 4.1 depicts the smallest array, $L_4(2^3)$

Table 4.1. Orthogonal Array $L_4(2^3)$

No. \ Column	1	2	3
1	1	1	1
2	1	2	2
3	2	1	2
4	2	2	1

"No." is the layout numbers—1 to 4 for the $L_4(2^3)$ array. Orthogonal array $L_4(2^3)$ consists of three columns and four rows and the numbers in rows and columns are 1 and 2. (1 and 2 in 2-level series, 1, 2 and 3 in 3-level series.) By corresponding 2-level factors to columns, each row indicates the level combination of factors.

When any of the two columns has numbers 1 and 2 in the four combinations, (11), (12), (21) and (22), occur the same number of times, the two columns are said to be "balanced" or "orthogonal." By counting the number of occurrences of combinations (11), (12), (21) and (22) in any two columns, it is seen that any pair of columns are orthogonal.

As described before, there are eight combinations when there are three 2-level factors. With orthogonal array $L_4(2^3)$, only four experiments will be conducted. Which combinations will be experimented depends on which factor is assigned to which column. If *A*, *B* and *C* correspond to columns 1, 2 and 3 respectively, the following combinations will be tested:

1. $A_1B_1C_1$
2. $A_1B_2C_2$
3. $A_2B_1C_2$
4. $A_2B_2C_1$

The motivation for using orthogonal arrays will be discussed later.

4.1.2 Layout of Using Mixed Series Orthogonal Arrays

In robust design, the recommended orthogonal arrays are the L_{12}, L_{18} and L_{36}, which are called mixed series orthogonal arrays. (See Chapter 5.)

For example, with an L_{12}, up to 11 factors can be assigned. If the majority of the factors are 3-level factors and they are:

(1) One 2-level and up to seven 3-level factors, or

(2) One 6-level and up to six 3-level factors, then it is recommended that an L_{18} array be used.

If there are too many 3-level factors for an L_{18}, then an L_{36} may be used. In L_{36}, up to 13 3-level factors can be assigned.

4.2 DECOMPOSITION OF VARIATION IN ORTHOGONAL ARRAY

4.2.1 Illustrated Example, $L_4(2^3)$

In order to reduce the wear of brushes in a pump, three 2-level factors were assigned to an L_4 orthogonal array. The amount of wear (mm) was measured at four locations after 1000 hours of operation. Since wear is a smaller-the-better characteristic, S/N Ratio for a smaller-the-better characteristic is used in the analysis.

Table 4.2. Orthogonal Array $L_4(2^3)$

Factor · Column / No.	*A* 1	*B* 2	*C* 3	η(db)	$\eta - \bar{y}$
1	1	1	1	$y_1 = 8.4$	-5.1
2	1	2	2	$y_2 = 24.6$	11.1
3	2	1	2	$y_3 = 6.8$	-6.7
4	2	2	1	$y_4 = 14.2$	0.7

The average of the decibel values is:

$$\overline{y} = \frac{y_1 + y_2 + y_3 + y_4}{4} = 13.5 \text{ (db)} \tag{4.1}$$

The decibel values at the four conditions are different, probably because of the different combinations of factors *A*, *B* and *C*. If factors *A*, *B* and C have no effect on wear, then the four decibel values should be the same. The effect caused by these factors is written as $y_i - \overline{y}$.

Let the effect of *A* at its first level, or $(y_i - \overline{y})$, be a_1 and the effect at its second level be a_2. Similarly, the effect of *B* at its first and second level be b_1 and b_2 respectively. And the effect of *C* at its first and second level be c_1 and c_2 respectively. Since the total of $y_i - \overline{y}$ is equal to zero, and there are two runs for each level of each factor, when a level of one factor is better by Δ, the other level of the same factor becomes worse by Δ. In other words, the following relationship holds.

$$a_1 + a_2 = 0 \quad b_1 + b_2 = 0 \quad c_1 + c_2 = 0 \tag{4.2}$$

The four experimental conditions are written as:

$$\begin{cases} y_1 - \overline{y} = a_1 + b_1 + c_1 = -5.1 \cdots\cdots \ (1) \\ y_2 - \overline{y} = a_1 + b_2 + c_2 = 11.1 \cdots\cdots \ (2) \\ y_3 - \overline{y} = a_2 + b_1 + c_2 = -6.7 \cdots\cdots \ (3) \\ y_4 - \overline{y} = a_2 + b_2 + c_1 = 0.7 \cdots\cdots \ (4) \end{cases} \tag{4.3}$$

Since there are conditions $a_1+a_2=0$, $b_1+b_2=0$ and $c_1+c_2=0$, the number of unknowns in Equation (4.3) is not 6 but 3 (1 for each factor). As explained before, the number of such unknowns is called the number of degrees of freedom. (Degree of freedom), *A*, *B* and *C* each have 1 degree of freedom, respectively. Normally, the degrees of freedom can be determined by:

(number of levels) - 1

The symbol *f* is used to denote degrees of freedom.

a_1 is calculated by adding (1) and (2). Since $b_1+b_2=0$ and $c_1+c_2=0$,

$$a_1 = \frac{(-5.1) + 11.1}{2} = 3.0 \tag{4.4}$$

a_2 is calculated by (3)+(4).

$$a_2 = \frac{(-6.7) + 0.7}{2} = -3.0 \tag{4.5}$$

Similarly, b_1=-5.9, b_2=5.9, c_1=-2.2 and c_2=2.2 are calculated.

From the above results, $(y_i - \overline{y})$ is decomposed to the portions of the effects of *A*, *B* and *C*.

As described in Chapters 2 and 3, the sum of squares of a change is called variation to show its magnitude. The total of the squares of $(y_i - \bar{y})$ is the total amount of experimental variation, denoted by S_T. Since this represents the changes caused by different experimental conditions, S_T is also called the variation between experiments.

$$S_T = (-5.1)^2 + 11.1^2 + (-6.7)^2 + 0.7^2 = 194.6 \qquad (f=3) \quad (4.6)$$

The total degrees of freedom is 4-1=3 since there are 4 experimental conditions.

The sum of squares caused by a is called variation S_A.

$$S_A = [3.0^2 + (-3.0)^2] \times 2 = 36 \qquad (f=1) \quad (4.7)$$

Similarly, S_B, the sum of squares for b, and S_C, the sum of squares for c, are calculated.

$$S_B = 139.24 \qquad (f=1) \quad (4.8)$$

$$S_C = 19.36 \qquad (f=1) \quad (4.9)$$

Next, the total of S_A, S_B and S_C is calculated.

$$S_A + S_B + S_C = 194.6 \qquad (4.10)$$

It is equal to the between experiment variation S_T=194.6.

$$S_T = S_A + S_B + S_C \qquad (4.11)$$

This is called the decomposition of variation.

Using another expression, the sum of the squares for the total of a, b and c is equal to the total of the sums of the squares of each deviation.

$$\Sigma(a_i + b_i + c_i)^2 = \Sigma a_i^{\ 2} + \Sigma b_i^{\ 2} + \Sigma c_i^{\ 2} \qquad (4.12)$$

By expanding the left side of Equation (4.12), we get the terms of product sum. These terms

$$\Sigma a_i b_i \quad \Sigma a_i c_i \quad \Sigma b_i c_i$$

are equal to zero, which can be seen from Table 4.3.

Table 4.3. Decomposition of Data

No.	$y_i - \bar{y}$	=	a	+	b	+	c
1	-5.1	=	3.0	+	(-5.9)	+	(-2.2)
2	11.1	=	3.0	+	5.9	+	2.2
3	-6.7	=	-3.0	+	(-5.9)	+	2.2
4	0.7	=	-3.0	+	5.9	+	(-2.2)

When these product sums are equal to zero, A, B and C are orthogonal. Orthogonality is the requirement for the decomposition of variation. In order to secure orthogonality, either a multi-way (or full factorial) layout,which is the experiment of conducting all combinations of all factor levels, or an orthogonal array must be used.

In this section, the total of the squares of deviations from the mean is considered as the total variation. In the previous chapter however, total of the squares of data was defined as the total variation. To reconcile these two definitions, consider the decomposition of data as:

$$y_i = \bar{y} + a + b + c \tag{4.13}$$

instead of:

$$y_i - \bar{y} = a + b + c \tag{4.14}$$

Since the total of the squares of $\bar{y}$ is called the general mean, S_m,

$$\text{Total of squares} = S_m + S_A + S_B + S_C \tag{4.15}$$

For the comparison of control factors or the inner factors, it is evaluated as the work performed by a factor. Therefore, there is no need to be concerned about the general mean in many cases, and it is usually removed in the analysis. For this reason, we may consider the fact that the total degrees of freedom is calculated by subtracting the number of data points by 1 as subtracting the degrees of freedom of the general mean. Of course it is all right to use Equation (4.15) to consider the total of the squares as the total variation.

4.2.2 Procedures for the Analysis of Variance

In this section, step-by-step procedures to devise an ANOVA table by calculating variations with the S/N Ratio data are presented:

(1) Preparation of the Supplementary Table.

Calculate level totals for each factor which is assigned to a column of an orthogonal array to obtain Table 4.4.

Table 4.4. Calculation of Level Totals

Factor · Level	Calculation of Totals	Total
A_1	Total of the data including A_1	$y_1 + y_2 = 33.0$
A_2	Total of the data including A_2	$y_3 + y_4 = 21.0$
B_1	Total of the data including B_1	$y_1 + y_3 = 15.2$
B_2	Total of the data including B_2	$y_2 + y_4 = 38.8$
C_1	Total of the data including C_1	$y_1 + y_4 = 22.6$
C_2	Total of the data including C_2	$y_2 + y_3 = 31.4$
Total	Grand Total	$y_1 + \cdots + y_4 = 54.0$

(2) Calculation of Total Variation

Instead of directly calculating the total of $(y-\bar{y})^2$, the following equation is used:

$$S_T = y_1{}^2 + y_2{}^2 + y_3{}^2 + y_4{}^2 - \frac{(y_1 + y_2 + y_3 + y_4)^2}{4} \qquad (f=3) \quad (4.16)$$

The second term on the right side of Equation (4.16) is the general mean, S_m.

$$S_m = \frac{(8.4 + 24.6 + 6.8 + 14.2)^2}{4} = 729 \qquad (f=1) \quad (4.17)$$

$$\begin{aligned} S_T &= 8.4^2 + 24.6^2 + 6.8^2 + 14.2^2 - S_m \\ &= 923.6 - 729 = 194.6 \end{aligned} \qquad (f=3) \quad (4.18)$$

This value agrees with the total of $(y-\bar{y})^2$, the sum of the squared deviations. If the variation after subtracting the mean effect is treated as the total variation, then it is all right to subtract a known constant (called a working mean) from the data or to multiply each observation a constant to simplify the calculations. In such a case, the value calculated from the equation of S_m will be different. In other words, the subtraction of a constant results in loss to the meaning of the general mean; it is only a term for calculation. The value calculated from Equation (4.17) using the data after subtraction of the working mean, is referred to as the correction term, denoted by CF. In the numerical example, there is no merit for subtracting a working mean; however, in general, calculations are easier to calculate when the number of digits becomes smaller after subtraction. Letting the values after subtracting (coding the data) a working mean be y_1', y_2', y_3' and y_4',

$$CF = \frac{(y_1' + y_2' + y_3' + y_4')^2}{4} \qquad (f=1) \quad (4.19)$$

$$S_T = y_1'^2 + y_2'^2 + y_3'^2 + y_4'^2 - CF \qquad (f=3) \quad (4.20)$$

The results from Equations (4.16) and (4.20) should be equal. In many cases, S_m is denoted by CF, even if a working mean is not subtracted.

Assuming the working mean to be 10,

$y_1' = 8.4 - 10 = -1.6$

$y_2' = 24.6 - 10 = 14.6$

$y_3' = 6.8 - 10 = -3.2$

$y_4' = 14.2 - 10 = 4.2$

therefore,

$$CF = \frac{(y_1' + y_2' + y_3' + y_4')^2}{4} = \frac{14^2}{4} = 49 \qquad (f=1) \quad (4.21)$$

$$\begin{aligned} S_T &= y_1'^2 + y_2'^2 + y_3'^2 + y_4'^2 - CF = (-1.6)^2 + \cdots + 4.2^2 - 49 \\ &= 194.6 \end{aligned} \qquad (f=3) \quad (4.22)$$

When a working mean is to be subtracted, the supplementary table must be prepared from coded data.

(3) Variation of Factors

Let the level totals obtained from the supplementary table be A_1 and A_2. The variation of A is:

$$\begin{aligned} S_A &= \frac{A_1^{\,2} + A_2^{\,2}}{\text{No. of replications of } A_1, A_2} - CF(S_m) = \frac{33.0^2 + 21.0^2}{2} - 729.0 \\ &= 36.0 \end{aligned} \qquad (f=1) \quad (4.23)$$

This equation was explained in Chapter 2. S_A in the previous section can be rewritten as:

$$\begin{aligned} S_A &= (a_1^{\,2} + a_2^{\,2}) \times (\text{No. of replications of } A_1, A_2) \\ &= [(\bar{A}_1 - \bar{y})^2 + (\bar{A}_2 - \bar{y})^2] \times (\text{No. of replicationsof } A_1, A_2) \\ &= \left[\left(\frac{A_1}{\text{No. of replications of } A_1} - \frac{\text{Total}}{\text{Total no. of data points}}\right)^2 \right. \\ &\quad \left. + \left(\frac{A_2}{\text{No. of replications of } A_2} - \frac{\text{Total}}{\text{Total no. of data points}}\right)^2\right] \\ &\quad \times (\text{No. of replications of } A_1, A_2) \\ &= \frac{A_1^{\,2} + A_2^{\,2}}{\text{No. of replications of } A_1, A_2} - CF(S_m) \end{aligned} \qquad (4.24)$$

When there are 2 levels, the above equation becomes

$$S_A = \frac{(A_1 - A_2)^2}{\text{Total no. of data points}} = \frac{(33.0 - 21.0)^2}{4} = 36.00 \quad (f=1) \quad (4.25)$$

Similarly,

$$S_B = \frac{(B_1 - B_2)^2}{\text{Total no. of data points}} = \frac{(15.2 - 38.8)^2}{4} = 139.24 \quad (f=1) \quad (4.26)$$

$$S_C = \frac{(C_1 - C_2)^2}{\text{Total no. of data points}} = \frac{(22.6 - 31.4)^2}{4} = 19.36 \quad (f=1) \quad (4.27)$$

(4) Preparation of ANOVA Table

The purpose of decomposition of variation is to distinguish the factors which affect variation from the ones which do not. Comparison is usually made using the variance, V, which is the variation per unit degree of freedom. In the example, the degrees of freedom for each factor is equal to 1; then variation is equal to variance.

$$V_A = S_A = 36.00 \quad (4.28)$$

$$V_B = S_B = 139.24 \quad (4.29)$$

$$V_C = S_C = 19.36 \quad (4.30)$$

Note that the B is the largest and C is the smallest. Can we say that B affects variation? When the conditions of A, B and C are fixed, and we observe that the variation is about the same when the levels of B changes, then the effect of B is not considered to significant. The reference point used to judge the significance of effects is called the between experiment error. When some columns are not assigned with factors, the variance from these columns is between experiment error. When all of the columns of an orthogonal array are assigned with factors, the error term cannot be estimated. When an orthogonal array of sufficient size is used, the smaller variances are combined and used as an estimation of between experiment error. This is called "pooling." But pooling is not an effective strategy for the case when small orthogonal arrays are used such as L_4. In many cases, the number of degrees of freedom from unassigned columns is small, or the number of columns which are pooled in small arrays is small, and therefore the estimation of between experiment error may not be very accurate.

The error variance after pooling is denoted by $V_{(e)}$ which is calculated by:

$$V_{(e)} = \frac{\text{Total variation of pooled factors}}{\text{Total degrees of freedom of pooled factors}} \quad (4.31)$$

The pooled error variance is put either outside the ANOVA table or next to the unpooled error using the symbol (e) and its degrees of freedom, variation, variance, pure variation and degrees of contribution are shown.

It is not appropriate in the case of L_4, but in the example, A and C are pooled for explanation.

$$V_{(e)} = \frac{S_A + S_C}{f_A + f_C} = \frac{36.00 + 19.36}{1 + 1} = 27.68 \tag{4.32}$$

In the previous section, the total variation was decomposed into the variation of A, B and C. Since A and C are considered as the error,

$$S_T = S_B + S_{(e)} \tag{4.33}$$

S_B is the magnitude of variation of the S/N Ratio caused by the level change of B. The magnitude will not be the same if experiments are conducted under the same condition repeatedly, but that it probably distributes around a certain value. Such a central value is called expectation. The expectation of the variation of B, or S_B, is given by:

$$E[S_B] = S_B' + (\text{degrees of freedom of } B) \times V_{(e)} \tag{4.34}$$

where S_B' is called pure variation (true effect). S_B includes S_B' and the error variance multiplied by the degrees of freedom of the error.

$$\begin{aligned} S_B' &= S_B - (\text{degrees of freedom of } B) \times V_{(e)} \\ &= 139.24 - 1 \times 27.68 = 111.56 \end{aligned} \tag{4.35}$$

The error variance, which was subtracted for the calculation of pure variation, is then added to the error to get the pure error variation, $S_{(e)}'$.

$$\begin{aligned} S_{(e)}' &= S_{(e)} + (\text{degrees of freedom of } B) \times V_{(e)} \\ &= 55.36 + 1 \times 27.68 = 83.04 \end{aligned} \tag{4.36}$$

The decomposition of pure variation can then be made.

$$S_T = S_B' + S_{(e)}' = 194.60 \tag{4.37}$$

The portion of pure variation among the total variation is called degrees of contribution and is denoted by ρ. Degrees of contribution of B and the error are calculated by:

$$\rho_B = \frac{S_B'}{S_T} \times 100 = 57.3 \ (\%) \tag{4.38}$$

$$\rho_e = \frac{S_e'}{S_T} \times 100 = 42.7 \ (\%) \tag{4.39}$$

The degrees of contribution is the percentage of causes of variation in terms of S/N Ratios (stability) ranging from 6.8 to 24.6 decibels.

The above results are summarized in the ANOVA table.

Table 4.5. ANOVA Table

Factor	f (Degrees of Freedom)	S (Sum of Squares)	V (Variance)	S' (Pure Variation)	$\rho(\%)$ (Degrees of Contribution)
A	1	36.00	36.00°		
B	1	139.24	139.24	111.56	57.3
C	1	19.36	19.36°		
(e)	(2)	(55.36)	(27.68)	(83.04)	(42.7)
T	3	194.60		194.60	100.0

With an ANOVA table, it is easy to observe and compare variances and degrees of contribution. It is not appropriate to determine the degree of contribution of variance terms which are smaller than the error variance, since a variance which is smaller than the error is about the same as the error. It cannot be used to improve the S/N Ratio. If the degrees of contribution of small factors were calculated, it would become negative, which does not make sense. Such effects are considered as the error and pooled with the error.

4.2.3 Optimum Condition And Estimation

The optimum condition is the condition with the highest S/N Ratio. The inner array is analyzed to determine the optimum conditions. Since the S/N Ratio is always larger the better, $A_1B_2C_2$ is the best combination from supplementary table, Table 4.4. Since this combination is the same as No.2 of L_4, it may not be necessary to estimate. However, the optimum condition may not always exist in the orthogonal array. Also, the level of A_1 was used twice in the array. B_2 and C_2 were both replicated twice. Such information might be utilized to obtain more accurate estimates.

Letting the estimate of $A_1B_2C_2$ be μ,

$$\mu = \text{general mean} + (\text{effect contributed by } A_1) + (\text{effect contributed by } B_2) + (\text{effect contributed by } C_2) \tag{4.40}$$

However, this results in an over-estimate (see Appendix 10), and it is recommended that only larger effects be used. Analysis of variance is used to judge whether an effect is large or not. In this example, only B is judged to be significant. The estimate is made accordingly,

μ = (general mean) + (effect contributed by B_2)

Letting the general mean be $\overline{T}$, the effect contributed by $\overline{B_2}$ is:

$\overline{B}_2 - \overline{T}$

Therefore, μ is calculated as:

$$\mu = \text{general mean} + (\text{effect contributed by } B_2)$$
$$= \overline{T} + (\overline{B}_2 - \overline{T})$$
$$= \overline{B}_2$$

$$\overline{B}_2 = \frac{38.8}{2} = 19.9 \text{ (db)}$$

Thus, the $A_1B_2C_2$ is estimated as 19.9 decibels.

In this example, only the effect of B was considered, but in general, estimation is made by adding significant effects. This is called additivity of factorial effects. The reason of using decibel as the unit of S/N Ratio is to expect additivity. The S/N Ratio before converting to decibel is the reciprocal of variability per unit, which is in the order of square. Table 4.6 shows the anti-log values (reciprocal of σ^2) of S/N Ratios.

Table 4.6. Orthogonal Array $L_4(2^3)$

No. \ Factor · Column	A 1	B 2	C 3	η (anti-log)
1	1	1	1	$y_1 = 6.92$
2	1	2	2	$y_2 = 288.40$
3	2	1	2	$y_3 = 4.79$
4	2	2	1	$y_4 = 26.31$

From Table 4.6, it is obvious that the average of y_1 and y_2, or 147.66, is not an appropriate estimate of A_1. Equation(4.40) cannot be used for these anti-log values. Instead, it is known that the logarithm is a better scale to use. That is why decibel units are used.

The S/N Ratio was estimated as 19.9 decibels. Letting σ^2 be the average of squares of the amount of wear,

$$-10 \log \sigma^2 = 19.9$$

Therefore,

$$\sigma^2 = 0.010\,23 \text{ (mm}^2\text{)}$$

The average amount of wear is its square root, or 0.101 (mm).

The estimation error of S/N Ratio being 19.9 decibels is often made by calculating confidence limits. For such calculation, see Chapter 5 and Appendix.

EXERCISES

Question (4.1). In order to reduce CO content in exhaust gas, CO content was measured when driven on roads. The following factors were assigned to an L_8 orthogonal array and R_1, R_2 and R_3 under a certain mode. The unit are in grams.

No. \ Factor · Column	A	B	C	D	E	F	G	Raw Data		
	1	2	3	4	5	6	7	R_1	R_2	R_3
1	1	1	1	1	1	1	1	1.04	1.20	1.54
2	1	1	1	2	2	2	2	1.42	1.76	2.10
3	1	2	2	1	1	2	2	1.01	1.23	1.52
4	1	2	2	2	2	1	1	1.50	1.87	2.25
5	2	1	2	1	2	1	2	1.28	1.34	2.05
6	2	1	2	2	1	2	1	1.14	1.26	1.88
7	2	2	1	1	2	2	1	1.33	1.42	2.10
8	2	2	1	2	1	1	2	1.33	1.52	2.13

(a) Calculate S/N Ratios in decibel units using smaller-the-better characteristic.

(b) Construct an ANOVA Table using S/N Ratios as the response.

(c) Find the optimum condition and estimate the S/N Ratio at the condition using the factors having larger effect.

(d) Determine the variability of CO (σ^2) at the optimum condition. The specification of CO content is less than 15g and the loss is ¥8000 when the content exceeds the specification. Calculate the loss at the optimum condition. The Loss Function is given by:

$$L = \frac{8{,}000}{15^2}\sigma^2$$

Question (4.2). The specification of pull off force of a car component part is 50 pounds minimum. Control factors

	1	2	3
A (temperature)	500	600	700
B (primary dwell time, sec)	10	15	20
C (force, psi)	40	50	60
D (secondary dwell time, sec)	5	10	15

were assigned to an L_9 orthogonal array and 10 units of the products were made respectively under 9 conditions. Pull off force was measured in pounds. (ITT, U.S.)

No. \ Factor · Column	A 1	B 2	C 3	D 4	Results				η(db)
1	1	1	1	1	46	50	...	42	33.3
2	1	2	2	2	52	51	...	54	34.4
3	1	3	3	3	52	52	...	56	34.9
4	2	1	2	3	52	59	...	52	34.7
5	2	2	3	1	50	54	...	55	34.2
6	2	3	1	2	60	55	...	50	34.7
7	3	1	3	2	50	56	...	50	34.6
8	3	2	1	3	60	58	...	63	35.2
9	3	3	2	1	56	56	...	49	34.8

The S/N Ratio using larger-the-better characteristic was calculated from 10 data points. For example, No. 1 is calculated by:

$$\eta = -10 \log \sigma^2, \quad \sigma^2 = \frac{1}{10}\left(\frac{1}{46^2} + \frac{1}{50^2} + \cdots + \frac{1}{42^2}\right)$$

(a) Make ANOVA table of S/N Ratios.

Note: 3-level orthogonal arrays have not been shown yet, but the procedures are the same as the case of L_4. Use the general equation to calculate variation. To calculate A for example, letting the level totals be A_1, A_2 and A_3,

$$S_A = \frac{A_1{}^2 + A_2{}^2 + A_3{}^2}{\text{No. of replications of } A_i} - CF$$

(b) Compare the process averages of current and the optimum conditions. The 2nd levels are current conditions.

(c) If the loss due to a defective is ¥1000, what is the quality difference between the optimum and current conditions? Use the following Loss Function:

$$L = A\Delta^2\sigma^2$$

CHAPTER 5

EXPERIMENTAL DESIGN USING MIXED-TYPE ORTHOGONAL ARRAYS FOR SMALLER-THE-BETTER AND LARGER-THE-BETTER CHARACTERISTICS

In this chapter, the features of mixed type orthogonal arrays which are used in the experimental design with smaller-the-better and larger-the-better characteristics are explained. The improvement of S/N Ratio means discovering the conditions for stability; it is a method of improving quality.

Only smaller-the-better and larger-the-better characteristics are discussed in this chapter, but the improvement of S/N Ratio has the same meaning for nominal-the-best characteristic.

5.1 INTERACTIONS AND MIXED TYPE ORTHOGONAL ARRAYS

In the previous chapter, an example using the orthogonal array, L_4, was presented, in which factors A, B and C were assigned to columns 1, 2 and 3 respectively. The variation of C was calculated by:

$$S_C = \frac{(C_1 - C_2)^2}{4} = \frac{[(y_1 + y_4) - (y_2 + y_3)]^2}{4} \tag{5.1}$$

where y_1, y_2, y_3 and y_4 denote the outcomes from experiments No.1, 2, 3, and 4 respectively. In the calculation, (y_1+y_4) and (y_2+y_3) represent the level totals of 1 and 2 of columns 3 where factor C was assigned.

In the above equation, (y_1+y_4)-(y_2+y_3) can be rewritten as:

$$(y_1 + y_4) - (y_2 + y_3) = (y_1 - y_2) - (y_3 - y_4) \tag{5.2}$$

Since y_1 is the condition, A_1B_1, and y_2, y_3 and y_4 are conditions A_1B_2, A_2B_1 and A_2B_2 respectively, then:

$$(y_1 - y_2) - (y_3 - y_4) = (A_1B_1 - A_1B_2) - (A_2B_1 - A_2B_2) \tag{5.3}$$

It means:

(difference of B at 1st level of A) – (difference of B at 2nd level of A) (5.4)

showing the difference of the effects of B at different levels of A. This is called the interaction between A and B. In other words, the effect calculated for C also shows the effect of the interaction of A and B. Therefore, the main effect C and interaction $A \times B$ are said to be confounded; because the effects are indistinguishable.

In some orthogonal arrays, the interactions between two columns are confounded with the effects of other columns. Generally, 2^n series (L_4, L_8, L_{16}, etc.) and 3^n series (L_9, L_{27}. etc.) share this property. However, some orthogonal arrays do not behave this way. The orthogonal arrays recommended in quality engineering, L_{12}, L_{18} and L_{36}, which are mixed type orthogonal arrays, belong to this class of orthogonal arrays. The interaction between two columns will not be confounded with any other specific column, but will be allocated in almost equal proportion to all remaining columns. That means one column includes not only the effect assigned to this column but also includes some portion of interactions from other columns. Therefore, the effect associated with a column where a large effect is present will show a comparatively large effect than a column where a small effect is assigned, and thus, it is possible to identify truly large effects. For such a reason, these orthogonal arrays are recommended whenever possible. For a description of these mixed type arrays, see the Appendix.

As described earlier, the presence of an interaction signifies that the effect changes with the level of another factor. For example, the statement "B_1 is better at A_1, but B_2 is better at A_2" is indicative of the interaction, $A \times B$. It is not easy to select the best level. For example, A_2B_2 was selected when the interaction is significant. The best condition might depend upon different days or different equipment or different people who conduct the experiment. Thus, we can rely on a consistent effect (a large main effect), but it is not reliable when an effect is dependent upon other conditions.

Do we then care about interactions in quality engineering? The answer is that we rarely consider interactions between control factors, and the ones between noise factors, but we do consider all interactions between inner control factors and outer noise factors. But instead of analyzing in the form of interaction, characteristics called S/N Ratio is used. The difference in the S/N Ratio between different levels of a control factor is because there is an interaction between a control factor and a noise factor (and the signal factor).

For experiments when it is appropriate to calculate S/N Ratio, when control factors and noise factors are assigned in one orthogonal array, it probably does not work in many cases. When very large interactions do exist, a success can not be guaranteed even if a mixed type orthogonal array is used. That is why the inner factors and the outer factors must be distinguished.

It is important for the levels to be carefully set when the inner factors may be inter-related. In a heat treatment experiment, for example, *A* represents heat treatment temperature, and *B* is heat treatment time. It is obvious that time must be set shorter at a high temperature, and longer at a low temperature. But if the levels of *A* and *B* are set independently, it is obvious that some combinations may be inappropriate. In such a case, the levels of *A* and *B* may be set as shown in Table 5.1. (The reader should refer to Chapter 5 of Reference (1), for more information.)

Table 5.1. Level of *B* (time) at different Level of *A*

	B_1	B_2	B_3
A_1 (500°)	5	6	7
A_2 (700°)	4	5	6
A_3 (900°)	3	4	5

Even if the above precautions are taken to avoid inappropriate combinations, we may not totally avoid the interaction between control factors. What we try to do is to avoid not selecting the optimum condition even when there exist interactions to a certain extent by using mixed type orthogonal arrays.

At the stage of designing an experiment, selection of factors,classification of factors (including the selection of orthogonal arrays),setting of levels and selection of S/N Ratios are important to obtain appropriate conclusions.

5.2 DESIGN OF EXPERIMENT FOR SMALLER-THE-BETTER AND LARGER-THE-BETTER CHARACTERISTICS

5.2.1 Layout of Inner Array

First, the inner factors, which mostly consist of control factors, and their levels are listed. As described in Chapter 4, up to 11 2-level factors can be assigned to an L_{12}.

If the majority of the factors are 3-level factors, and there are:

(1) up to seven 3-level factors, or

(2) one 6-level factor and up to six 3-level factors ,

then it is better to use mixed type L_{18} array. If there are a few 2-level factors mixed with 3-levels factors, then a combination design as explained in Chapter 6, may be used by treating two 2-level factors as a single 3-level factor, then L_{18} can be used even if quite a few 2-level factors are tested.

When there are many 3-level factors and we cannot fit an L_{18}, then the L_{36} may be used to assign up to thirteen 3-level factors. When the selection of factors and levels is simple, a layout is achieved simply by corresponding factors to the columns of an orthogonal array. However, not all experiments can be assigned to an L_{12}, L_{18} and L_{36}. For example, there may exist a practical need to test a 4-level factor, in which case, the use of these arrays is not possible. Or, perhaps only a small number of experimental runs is allowed, etc. In such cases, the layout techniques which are introduced in Chapter 6 should be used. Sometimes L_8 may be used when there are up to seven 2-level factors, or L_{16} may be used when there are say twelve to fifteen 2-level factors without modifying arrays.

5.2.2 Layout of Outer Factors for Efficient Experimentation

For smaller-the-better or larger-the-better characteristics, it is desirable that the output of products does not vary despite the variation in the conditions of product application, environment or manufacturing. Therefore, data must be collected at various conditions. If there is an error factor of 4-level, four sets of data must be collected for each of the inner conditions. It is the same for the case to investigate the difference at different locations of a product unit.

When there are many noise factors, it is ideal that those noise factors be assigned to an orthogonal array. If there are five noise factors assigned to an L_{18} array for example, the data for eighteen noise conditions must be collected for each of the inner condition. In robust designs, to improve the S/N Ratio, we frequently use large inner arrays. With a large outer array , the whole experiment would become very large. In order to maintain a manageable size, the technique of compound noise factors should be considered.

Compounding of noise factors involves the evaluation of two noise conditions—the worst condition and the standard condition under various environmental conditions.

N_1 = worst condition

N_2 = standard condition

Of course, the worst condition might correspond to a few conditions, and sometimes the standard condition is not tested. In the case of inner noise, all component parts are set close to their limits of specification.

After noise factors are compounded, there will be only two or three outer conditions. The S/N Ratio thus obtained would be much worse than actual situation. It is quite all right since our objective is to discover a high robust condition.

5.3 LAYOUT AND ANALYSIS OF L_{12}

5.3.1 Layout and Results (Example)

In order to reduce the wear of a certain product through design, eleven factors with 2-levels were assigned to L_{12} array. Those factors from *A* to *K* were assigned to columns 1 through 11 respectively. For the noise factors, two conditions associated with the most wearing and the moderate wearing were evaluated in the life test. The results are presented in Table 5.2. The first levels in the inner array correspond to current design conditions.

Table 5.2. Layout of L_{12} and Results

Factor · Column	*A*	*B*	*C*	*D*	*E*	*F*	*G*	*H*	*I*	*J*	*K*	Amount of Wear (μm)				η (db)
No.	1	2	3	4	5	6	7	8	9	10	11	N_1		N_2		
1	1	1	1	1	1	1	1	1	1	1	1	39	19	9	10	-27.12
2	1	1	1	1	1	2	2	2	2	2	2	21	20	3	16	-24.42
3	1	1	2	2	2	1	1	1	2	2	2	57	74	33	17	-34.02
4	1	2	1	2	2	1	2	2	1	1	2	42	27	13	24	-29.08
5	1	2	2	1	2	2	1	2	1	2	1	37	35	9	29	-29.44
6	1	2	2	2	1	2	2	1	2	1	1	68	85	38	64	-36.38
7	2	1	2	2	1	1	2	2	1	2	1	21	10	5	2	-21.54
8	2	1	2	1	2	2	2	1	1	1	2	38	23	17	4	-27.55
9	2	1	1	2	2	2	1	2	2	1	1	30	34	30	24	-29.46
10	2	2	2	1	1	1	1	2	2	1	2	72	46	24	40	-33.75
11	2	2	1	2	1	2	1	1	1	2	2	10	6	1	2	-15.47
12	2	2	1	1	2	1	2	1	2	2	1	30	12	8	0	-24.42

In an L_{12}, there are eleven columns. When there are less than eleven factors, some columns are not assigned. Such unassigned columns are ignored at the stage of experiment. Since these columns can be used to estimate the error, they are called error columns. In this example, there are no error columns.

Since the amount of wear is smaller-the-better and there are no negative values, a smaller-the-better characteristic is used. Four results under N_1 and N_2 denoted by n_1, n_2, n_3 and n_4 are used to calculate S/N Ratios as shown in Table 5.2. For example, the S/N Ratio of No.1 is calculated as:

$$\eta = -10 \log \left[\frac{1}{4} (39^2 + 19^2 + 9^2 + 10^2) \right]$$
$$= -27.12 \text{ (db)} \tag{5.5}$$

5.3.2 Analysis of Variance for the Comparison of S/N Ratios

First, a supplementary table is prepared by calculating the level totals of column 1 through column 11 in L_{12}. The level total of A which is assigned to column 1 is the total of No.1 through No. 6 for the first level of A and the total of No. 7 through No. 12 for the second level of A. Similar level totals are calculated for B through K to obtain Table 5.3.

Table 5.3. Level Totals

	A	*B*	*C*	*D*	*E*	*F*	*G*
1	-180.46	-164.11	-149.97	-166.70	-158.68	-169.96	-169.26
2	-152.19	-168.54	-182.68	-165.95	-173.97	-162.72	-163.39

	H	*I*	*J*	*K*
1	-164.96	-150.20	-183.34	-168.36
2	-167.69	-182.45	-149.31	-164.29

Since all level totals are the total of 6 results, comparison of S/N Ratios can be made from this table. The combination yielding the highest value of the S/N Ratio is $A_2B_1C_1D_2E_1F_2G_2H_1I_1J_2K_2$. For estimating the amount of wear, if the following calculation were made by using the additivity as explained in the previous chapter, it would result in an over-estimation.

$$\mu = \overline{T} + (\overline{A_2} - \overline{T}) + (\overline{B_1} - \overline{T}) + (\overline{C_1} - \overline{T}) + (\overline{D_2} - \overline{T})$$
$$+ (\overline{E_1} - \overline{T}) + (\overline{F_2} - \overline{T}) + (\overline{G_2} - \overline{T}) + (\overline{H_1} - \overline{T})$$
$$+ (\overline{I_1} - \overline{T}) + (\overline{J_2} - \overline{T}) + (\overline{K_2} - \overline{T}) \tag{5.6}$$

In order to get an appropriate estimate, some methods may be used:

(1) Estimate using about half of the larger factors.

(2) Estimate using a discount coefficient.

(3) Estimate using the factors with a variance ratio of more than 2 (which are significant in the ANOVA table).

(4) Estimate using the significant factors from the F test in ANOVA.

Method (1) does not require analysis of variance. That may make the beginner feel insecure. Methods (2), (3) and (4) require the use of an ANOVA table. In this book however, ANOVA tables are always constructed no matter which method will be used. The methods of F-tests and discount coefficients, are discussed in Appendix 6 and 10 respectively.

First, a decomposition of the variation is performed to prepare an ANOVA table. Then, total variation which is the total amount of variation of S/N Ratios is calculated by:

$$S_T = (-27.12)^2 + (-24.42)^2 + \cdots + (-24.42)^2 - \frac{[(-27.12) + \cdots + (-24.42)]^2}{12} = 369.3059 \qquad (f=11) \quad (5.7)$$

There is no strict rule about the number of digits of precision to calculate for the variation and variance terms, but as a rule of thumb, raw data are squared and used for analysis. In this example, S/N Ratios are calculated up to the second decimal point. Therefore, after they are squared, up to the fourth decimal point is calculated.

Next, the variation terms for each column is calculated. Letting the level totals of A be A_1 and A_2 respectively,

$$S_A = \frac{A_1{}^2 + A_2{}^2}{6} - \frac{(\text{Total})^2}{12} = \frac{(A_1 - A_2)^2}{12} = \frac{(-180.46 + 152.19)^2}{12} = 66.5994 \qquad (f=1) \quad (5.8)$$

Similarly,

$$\begin{cases} S_B = 1.6354 \\ S_C = 89.1622 \\ S_D = 0.0470 \\ S_E = 19.4820 \\ S_F = 4.3319 \\ S_G = 2.8713 \\ S_H = 0.6213 \\ S_I = 86.6716 \\ S_J = 96.5034 \\ S_K = 1.3804 \end{cases} \qquad (5.9)$$

The total variation, S_T, is equal to:

$$S_A + S_B + \cdots + S_K = 369.306 \qquad (5.10)$$

The ANOVA table is shown in Table 5.4.

Table 5.4. ANOVA Table

Factor	f	S	V	ρ(%)
A	1	66.599 4	66.599 4	17.5
B	1	1.635 4	1.635 4°	
C	1	89.162 2	89.162 2	23.7
D	1	0.047 0	0.047 0°	
E	1	19.482 0	19.482 0	4.8
F	1	4.331 9	4.331 9°	
G	1	2.871 3	2.871 3°	
H	1	0.621 3	0.621 3°	
I	1	86.671 6	86.671 6	23.0
J	1	96.503 4	96.503 4	25.6
K	1	1.380 4	1.380 4°	
(e)	(6)	(10.887 3)	(1.814 55)	(5.4)
T	11	369.305 9		100.0

A, C, E, I and J make up about half of the larger variances terms. The factors which seem to have large effects are called significant factors. These significant factors can be used for estimation purposes. The degrees of contribution to the overall sum of squares may also be calculated for evaluation. For this calculation, the factors other than A, C, E, I and J are pooled to get (e). Mark "o"is put to the variance of those pooled factors.

$$V_{(e)} = \frac{S_B + S_D + S_F + S_G + S_H + S_K}{6} = \frac{10.887\,3}{6} = 1.814\,55 \tag{5.11}$$

The degrees of contribution of A is calculated using the pooled error variance.

$$\rho_A = \frac{S_A - f_A \times V_{(e)}}{S_T} \times 100 = \frac{66.5994 - 1 \times 1.81455}{369.305\,9} \times 100$$
$$= 17.5\ (\%) \tag{5.12}$$

Similar calculations are made for the other significant terms. The degrees of contribution of the pooled error, (e), is calculated as:

$$\rho_{(e)} = \frac{S_{(e)} + (f_A + f_C + f_E + f_I + f_J) \times V_{(e)}}{S_T} \times 100$$
$$= \frac{10.887\,3 + 5 \times 1.814\,55}{369.3059} \times 100 = 5.4 \tag{5.13}$$

5.3.3 Estimation of the Optimum Condition and Evaluation of Quality Improvement

In the above example, the optimum condition is $A_2B_1C_1D_2E_1F_2G_2$ $H_1I_1J_2K_2$. In order to avoid over estimation, the larger five factors may be used to estimate the process average.

$$\begin{aligned}\mu &= \overline{T} + (\overline{A}_2 - \overline{T}) + (\overline{C}_1 - \overline{T}) + (\overline{E}_1 - \overline{T}) + (\overline{I}_1 - \overline{T}) + (\overline{J}_2 - \overline{T}) \\ &= \overline{A}_2 + \overline{C}_1 + \overline{E}_1 + \overline{I}_1 + \overline{J}_2 - 4 \times \overline{T} \\ &= \frac{-152.19 + (-149.97) + (-158.68) + (-150.20) + (-149.31)}{6} \\ &\quad - 4 \times \frac{(-332.65)}{12} = -15.84 \text{ (db)}\end{aligned} \tag{5.14}$$

Although the estimation of the confidence limits is not very important, it may be calculated by:

$$\pm\sqrt{F_6{}^1(0.05) \times V_{(e)} \times \frac{1}{n_e}} = \pm 1.90 \tag{5.15}$$

$$F_6{}^1(0.05) = 5.99$$

$$V_{(e)} = 1.814\ 55$$

$$n_e = \frac{\text{total number of experiments}}{\text{total of degrees of freedom used in the estimation of } \mu}$$

$$= \frac{18}{\underset{T\ A\ C\ E\ I\ J}{1+1+1+1+1+1}} = 3 \tag{5.16}$$

n_e is called the effective number of replications to indicate how many averages are used to calculate μ.

If the S/N Ratio obtained from a confirmatory experiment under the optimum condition falls within the range of:

$$\pm\sqrt{F_6{}^1(0.05) \times V_{(e)} \times \left(\frac{1}{n_e} + 1\right)} = \pm 3.80 \tag{5.17}$$

we may conclude that the reproducibility is confirmed. For the discussion on confirmation run, see Appendix 11.

For more detailed discussion on the confidence limits and the effective number of replications , see Appendix 11.

The first level of all factors indicates current conditions. The same factors are used for the estimation of this process average.

$$\mu_0 = \overline{A}_1 + \overline{C}_1 + \overline{E}_1 + \overline{I}_1 + \overline{J}_1 - 4\overline{T}$$
$$= \frac{-180.46 + (-149.97) + (-158.68) + (-150.20) + (-183.34)}{6}$$
$$-4 \times \frac{(-332.65)}{12} = -26.22 \text{ (db)} \tag{5.18}$$

By using the optimum condition, there will be a gain of 10.38 decibels compared with current condition. Its anti-log value is 10.9, meaning that the square of the average amount of wear is 1/10.9.

When the amount of wear at any location exceeds 200 μm, there is trouble in the marketplace resulting in an average loss of ¥80,000. The current amount of wear in square terms is estimated as $(60\mu m)^2$. To estimate the quality improvement in using the optimum condition, we start with the Loss Function:

$$L = \frac{A}{\Delta^2}\sigma^2 = \frac{80{,}000}{200^2}\sigma^2 = 2\sigma^2 \tag{5.19}$$

Since the average square of the amount of wear is $(60\mu m)^2$, put $\sigma^2 = 60^2$ into Equation (5.19),

$$L_0 = 2 \times 60^2 = 7{,}200 \ \text{(¥/unit)} \tag{5.20}$$

From the experiment, it was estimated that the average square of the amount of wear would be reduced to 1/10.9 of current condition if the optimum condition were applied. It is known that the gain from S/N Ratios has a high reproducibility when the results are compared in the market. The Loss Function under the optimum condition in the market would be:

$$L = L_0 \times \frac{1}{10.9} = 7{,}200 \times \frac{1}{10.9} = 661 \ \text{(¥/unit)} \tag{5.21}$$

It means that the quality improvement would be L_0-L=6,539 (¥). Even for the case when the cost increases when the optimum conditions are used, the condition should be used as long as the cost increase is less than ¥6,539.

5.4 Layout of L_{18} and Its Analysis

5.4.1 Calculation of the S/N Ratio for Larger-the-better Characteristics

In order to improve the binding strength of a heat press binding process, a 2-level factor, A, and 3-level factors, B, C, D, E, F and G, were studied. The current condition is $A_1B_2C_2D_2E_2F_2G_2$. Since there is one

2-level column and seven 3-level columns in an L_{18} array, *A* may be assigned to column 1 and factors *B, C, D, E, F,* and *G* assigned to other columns. Table 5.5 shows the assignment with column 6 used as an error term.

Table 5.5. Layout and Results

No. \ Factor · Column	*A* 1	*B* 2	*C* 3	*D* 4	*E* 5	*e* 6	*F* 7	*G* 8	K_1	K_2	η(db)
1	1	1	1	1	1	1	1	1	0.46	0.51	-6.3
2	1	1	2	2	2	2	2	2	0.61	0.69	-3.8
3	1	1	3	3	3	3	3	3	0.55	0.30	-8.6
4	1	2	1	1	2	2	3	3	0.65	0.71	-3.4
5	1	2	2	2	3	3	1	1	0.71	0.79	-2.5
6	1	2	3	3	1	1	2	2	0.83	0.81	-1.7
7	1	3	1	2	1	3	2	3	0.95	1.05	0.0
8	1	3	2	3	2	1	3	1	0.78	0.95	-1.4
9	1	3	3	1	3	2	1	2	0.95	0.81	-1.2
10	2	1	1	3	3	2	2	1	0.58	0.55	-5.0
11	2	1	2	1	1	3	3	2	0.63	0.62	-4.1
12	2	1	3	2	2	1	1	3	0.66	0.70	-3.4
13	2	2	1	2	3	1	3	2	0.76	0.78	-2.3
14	2	2	2	3	1	2	1	3	0.74	0.91	-1.8
15	2	2	3	1	2	3	2	1	0.72	0.59	-3.8
16	2	3	1	3	2	3	1	2	0.92	0.99	-0.4
17	2	3	2	1	3	1	2	3	0.69	1.00	-1.9
18	2	3	3	2	1	2	3	1	0.91	0.77	-1.6

Binding strength (kgf) was measured at two conditions, K_1 and K_2, after binding under all 18 conditions. Since binding strength is a larger-the-better characteristic, the S/N Ratio for a larger-the-better characteristic is used. For example, No.1 is calculated as:

$$\eta = -10 \log \frac{1}{2}\left(\frac{1}{0.46^2} + \frac{1}{0.51^2}\right) = -6.3 \text{ (db)} \tag{5.22}$$

Others are calculated in similar fashion.

5.4.2 Calculation of Variations

There are 36 data points, and 18 S/N Ratios are calculated. Table 5.56 shows the supplementary table.

$$CF = \frac{(-53.2)^2}{18} = 157.24 \qquad (f=1) \tag{5.23}$$

$$S_T = (-6.5)^2 + (-3.8)^2 + \cdots + (-1.6)^2 - CF = 77.62 \qquad (f{=}17) \quad (5.24)$$

Table 5.6. Level Totals

Level \ Factor	*A*	*B*	*C*	*D*	*E*	*F*	*G*	Col. 6
1	-28.9	-31.2	-17.4	-20.7	-10.5	-15.6	-20.6	-17.0
2	-24.3	-15.5	-15.5	-13.6	-16.2	-16.2	-13.5	-16.8
3		- 6.5	-20.3	-18.9	-21.5	-21.4	-19.1	-19.4
Total	-53.2	-53.2	-53.2	-53.2	-53.2	-53.2	-53.2	-53.2

$$S_A = \frac{[(-28.9) - (-24.3)]^2}{18} = 1.18 \qquad (f{=}1) \quad (5.25)$$

B is a 3-level factor; therefore the equations in Chapters 2 and 4 will be used:

$$\begin{aligned} S_B &= \frac{B_1{}^2 + B_2{}^2 + B_3{}^2}{\text{No. of replications of } B_1, B_2 \text{ and } B_3} - CF \\ &= \frac{(-31.2)^2 + (-15.5)^2 + (-6.5)^2}{6} - 157.24 \\ &= 52.08 \end{aligned} \qquad (f{=}2) \quad (5.26)$$

Similarly,

$$\begin{cases} S_C = 1.95 \\ S_D = 4.54 \\ S_E = 3.59 \\ S_F = 3.39 \\ S_G = 4.67 \\ S_6 = 0.69 \end{cases} \qquad (5.27)$$

The total of the variation terms for columns 1 through 8 is calculated.

$$S_A + S_B + S_C + \cdots + S_6 = 72.09 \qquad (f{=}15) \quad (5.28)$$

This does not agree with the total variation, S_T (77.62). This difference is not due to a calculation mistake. Note that there are 15 degrees of freedom shown in Equation (5.28), but that the total degrees of freedom for an L_{18} is 17. The reason for this disagreement is explained in the following section:

5.4.3 Features of L_{18} and Interactions

One of the features of an L_{18} is that the interaction of the factors assigned to columns 1 and 2 can be obtained regardless of the other columns. The difference between the total and explained variation in the above example is just the interaction between columns 1 and 2. Since A and B are assigned to columns 1 and 2, respectively, the difference between S_T and Equation (5.28) is equal to the variation of interaction $A\times B$.

To explain this, let the effect of the first level of A and B be a_1 and b_1 respectively. There are 3 experimental runs at A_1B_1 and the average of these results is denoted by y_{11}. In y_{11}, each level of the other factors $C, D, \cdots, G$ occurs once; therefore, the effects of those factors are canceled out. Thus, the difference between y_{11} and the grand average shows the effect of A_1B_1.

$$y_{11} - \overline{T} = a_1 + b_1 \tag{5.29}$$

This is calculated as:

$$y_{11} = \frac{(-6.3) + (-3.8) + (-8.6)}{3} = -6.233 \tag{5.30}$$

$$\overline{T} = \frac{-53.2}{18} = -2.956 \tag{5.31}$$

$$a_1 = \frac{\text{Total of } A_1}{9} - \overline{T} = -0.255 \tag{5.32}$$

$$b_1 = \frac{\text{Total of } B_1}{6} - \overline{T} = -2.244 \tag{5.33}$$

Therefore,

$$y_{11} - \overline{T} = -3.277 \tag{5.34}$$

$$a_1 + b_1 = -2.499 \tag{5.35}$$

The two do not agree. If such calculations are made for the 6 combinations of A and B, it will be noticed that $(y_{ij}-\overline{T})$ and (a_i+b_j) are different from one and another. Such differences are probably caused by different combinations of A and B. The total of these differences squared and multiplied by 3 (the number of replications) shows the magnitude of the interaction, denoted by A×B.

$$S_{A\times B} = (\text{Total of squares of } [y_{ij} - \overline{T} - (a_i + b_j)]) \times (\text{No. of replications of } A_iB_j) \tag{5.36}$$

Actually, the variation of interaction, $S_{A\times B}$, is calculated from the totals of A,B combinations.

Table 5.7. Totals of A and B Combinations

A_1B_1	A_1B_2	A_1B_3	A_2B_1	A_2B_2	A_2B_3
-18.7	-7.6	-2.6	-12.5	-7.9	-3.9

$$S_{A\times B} = \frac{A_1B_1{}^2 + A_1B_2{}^2 + \cdots + A^2B_3{}^2}{\text{No. of replications of } A_iB_j} - CF - S_A - S_B$$

$$= \frac{(-18.7)^2 + (-7.6)^2 + \cdots + (-3.9)^2}{3} - CF - S_A - S_B$$

$$= 5.53 \qquad (f\text{=}2) \quad (5.37)$$

The degrees of freedom for $A\times B$ is:

$$\begin{aligned}&(\text{degrees of freedom of } A \times \text{degrees of freedom of } B)\\ &= 1 \times 2 = 2\end{aligned} \qquad (5.38)$$

From the above calculations,

$$\begin{aligned} S_A + S_B + S_C + \cdots + S_6 + S_{A\times B} &= 72.09 + 5.53 \\ &= 77.62 \end{aligned} \qquad (f\text{=}17) \quad (5.39)$$

It now agrees with the total variation, S_T.

In many cases, interactions are considered as error.

$$S_e = S_6 + S_{A\times B} = 0.69 + 5.53 = 6.22 \qquad (f\text{=}4) \quad (5.40)$$

We may calculate the error from the beginning as:

$$S_e = S_T - S_A - S_B - S_C - S_D - S_E - S_F - S_G = 6.22 \qquad (f\text{=}4) \quad (5.41)$$

However, the calculation details are shown to facilitate the reader's understanding of interaction terms.

5.4.4 ANOVA and Estimation

Table 5.8 show the ANOVA table for this example.

From the results, only factor B has a large effect.

$$\begin{cases} \overline{B}_1 = \dfrac{-31.2}{6} = -5.20 \\ \overline{B}_2 = \dfrac{-15.5}{6} = -2.50 \\ \overline{B}_3 = \dfrac{-6.5}{6} = -1.08 \end{cases} \qquad (5.42)$$

Table 5.8. ANOVA Table

Factor	f	S	V	$\rho(\%)$
A	1	1.18	1.18°	
B	2	52.08	26.04	62.7
C	2	1.95	0.98°	
D	2	4.54	2.27°	
E	2	3.59	1.80°	
F	2	3.39	1.70°	
G	2	4.67	2.34°	
e	4	6.22	1.552	
(e)	(15)	(25.53)	(1.702)	(37.3)
T	17	77.62		

The optimum condition is found from the levels resulting in the largest S/N Ratios, but only the larger effect, B, is used for estimation purposes.

$$\hat{\mu} = \overline{T} + (\overline{B}_3 - \overline{T}) = \overline{B}_3 = -1.08 \text{ (db)} \tag{5.43}$$

For the estimation of current condition, only B is used for the same reason.

$$\hat{\mu}_0 = \overline{B}_1 = -5.20 \tag{5.44}$$

The gain is:

$$\hat{\mu} - \hat{\mu}_0 = 4.12 \text{ (db)} \tag{5.45}$$

The anti-log value of the gain, 4.12 decibels, is 2.58. The binding strength in terms of the reciprocal of the square is improved to 1/2.58.

5.4.5 Evaluation of Quality Improvement

When the loss due to a larger-the-better characteristic becomes smaller than Δ is ¥A, the Loss Function as described in Chapter 1 is:

$$L = A\Delta^2\sigma^2 \tag{5.46}$$

where σ^2 is given by the average of the square of the reciprocal observations.

A lower specification limit of binding strength is 0.8 kgf and the loss associated with an out-of-specification product is ¥10,000. By changing the condition of B from B_1 to B_3, the gain would be:

$$\begin{aligned} L(B_1) - L(B_3) &= 10{,}000 \times 0.8^2 \times [\sigma^2(B_1) - \sigma^2(B_3)] \\ &= 10{,}000 \times 0.8^2 \times (3.31 - 1.28) \\ &= 12{,}992 \text{ (¥/unit)} \end{aligned} \tag{5.47}$$

where $\sigma^2(B_1)$ was calculated by:

$$-10 \log \sigma^2 (B_1) = \text{db of } B_1 = -5.20 \quad (5.48)$$
$$\sigma^2(B_1) = 3.31 \quad (5.49)$$

Similarly,

$$\sigma^2(B_3) = 1.28 \quad (5.50)$$

The binding force is estimated by:

$$\begin{cases} y(B_1) = 0.55 \ \text{(kgf)} \\ y(B_2) = 0.88 \ \text{(kgf)} \end{cases} \quad (5.51)$$

5.5 THE CASE WHEN THERE IS AN INDICATIVE FACTOR AS AN OUTER FACTOR

5.5.1 Layout of Factors and Results

In order to improve audible noise of oil pumps for cars, a 2-level factor A and 3-level factors B, C, D, E, F and G were assigned to an L_{18} array. Noise conditions were measured for each of the 18 design conditions by setting the number of revolutions at K_1=1000, K_2=1500 and K_3=2000 (rpm). Sound pressure was used as the output, and the S/N Ratio of the smaller-the-better characteristic was calculated. It may be natural to the consider noise level as an error factor. But it is obvious that the noise level is different at different number of revolutions. Also in practice, specification is different at different speeds. For such reasons,the number of revolutions is considered as an indicative factor and a S/N Ratio is calculated from the two results at each level of K. The results are shown in Table 5.9.

5.5.2 ANOVA Table

In the case, an indicative factor is used outside of the orthogonal array. The effect of the indicative factor, and interactions between the indicative factor and inner factors are generally presented in the ANOVA table. Those factorial effects and degrees of freedom are shown in Table 5.11.

Table 5.9. Layout and Results

Factor · Column / No.	*A*	*B*	*C*	*e*	*D*	*E*	*F*	*G*	η(db) -30			Total
	1	2	3	4	5	6	7	8	K_1	K_2	K_3	
1	1	1	1	1	1	1	1	1	-0.1	-5.0	-11.0	-16.1
2	1	1	2	2	2	2	2	2	6.5	1.1	- 5.0	2.6
3	1	1	3	3	3	3	3	3	7.7	2.0	- 4.5	5.2
4	1	2	1	1	2	2	3	3	7.7	2.0	- 3.1	6.6
5	1	2	2	2	3	3	1	1	1.4	-7.0	-14.5	-20.0
6	1	2	3	3	1	1	2	2	3.2	-3.1	- 6.0	- 6.0
7	1	3	1	2	1	3	2	3	2.4	-3.4	- 8.1	- 9.1
8	1	3	2	3	2	1	3	1	4.0	-1.1	- 5.6	- 2.7
9	1	3	3	1	3	2	1	2	2.0	-1.6	- 5.0	- 4.6
10	2	1	1	3	3	2	2	1	7.7	1.4	- 7.0	2.1
11	2	1	2	1	1	3	3	2	5.9	-0.1	- 5.0	0.8
12	2	1	3	2	2	1	1	3	5.4	1.1	- 6.0	0.5
13	2	2	1	2	3	1	3	2	7.1	2.4	- 3.1	6.4
14	2	2	2	3	1	2	1	3	3.2	-1.1	- 6.0	- 3.9
15	2	2	3	1	2	3	2	1	1.1	-6.0	-12.5	-17.4
16	2	3	1	3	2	3	1	2	5.9	0.5	- 6.0	0.4
17	2	3	2	1	3	1	2	3	5.9	1.4	- 4.0	3.3
18	2	3	3	2	1	2	3	1	4.9	-0.1	- 5.0	- 0.2

Table 5.10. Factorial Effects and Degrees of Freedom

Factor	*f*
A	1
B	2
C	2
D	2
E	2
F	2
G	2
e_1	4
K	2
$A \times K$	2
$B \times K$	4
$C \times K$	4
$D \times K$	4
$E \times K$	4
$F \times K$	4
$G \times K$	4
e_2	8
T	53

Table 5.11 is a supplementary table.

First, the correction factor is calculated. Notice that the total number of data points is equal to 54.

$$CF = \frac{(-52.2)^2}{54} = 50.46 \qquad (f{=}1) \quad (5.52)$$

$$\begin{aligned} S_T &= (-0.1)^2 + (-5.0)^2 + \cdots + (-5.0)^2 - CF \\ &= 1{,}517.18 \end{aligned} \qquad (f{=}53) \quad (5.53)$$

$$S_A = \frac{(-44.2 + 8.0)^2}{54} = 24.27 \qquad (f{=}1) \quad (5.54)$$

$$\begin{aligned} S_B &= \frac{(-4.9)^2 + (-34.4)^2 + (-12.9)^2}{18} - CF \\ &= 25.86 \end{aligned} \qquad (f{=}2) \quad (5.55)$$

The denominator, 18, is the number of replications of B_1,B_2 and B_3. Similarly,

$$\begin{cases} S_C = 5.11 & (f{=}2) \\ S_D = 24.51 & (f{=}2) \\ S_E = 51.54 & (f{=}2) \\ S_F = 103.87 & (f{=}2) \\ S_G = 114.33 & (f{=}2) \end{cases} \qquad (5.56)$$

$$\begin{aligned} S_{e1} &= S_{\text{col.4}} + S_{A\times B} \\ &= (\text{variation between 18 experiments}) - S_A - S_B - \cdots - S_G \\ &= 21.10 \end{aligned} \qquad (f{=}4) \quad (5.57)$$

$$\begin{aligned} S_K &= \frac{K_1{}^2 + K_2{}^2 + K_3{}^2}{\text{No. of replications}} - CF = \frac{81.9^2 + (-16.7)^2 + (-117.4)^2}{18} - CF \\ &= 1{,}103.39 \end{aligned} \qquad (f{=}2) \quad (5.58)$$

$$\begin{aligned} S_{A\times K} &= \frac{A_1K_1{}^2 + A_1K_2{}^2 + \cdots + A_2K_3{}^2}{\text{No. of replications of } A_iK_j} - CF - S_A - S_K \\ &= \frac{34.8^2 + (-16.2)^2 + \cdots + (-54.6)^2}{3} - CF - S_A - S_K \\ &= 1.57 \end{aligned} \qquad (f{=}2) \quad (5.59)$$

$$\begin{cases} S_{B\times K} = 9.20 & (f=4) \\ S_{C\times K} = 1.70 & (f=4) \\ S_{D\times K} = 5.31 & (f=4) \\ S_{E\times K} = 9.60 & (f=4) \\ S_{F\times K} = 2.01 & (f=4) \\ S_{G\times K} = 9.25 & (f=4) \end{cases} \tag{5.60}$$

Table 5.11. Supplementary Table

	K_1	K_2	K_3	Total
A_1	34.8	-16.2	-62.8	-44.2
A_2	47.1	- 0.5	-54.6	- 8.0
B_1	33.1	- 0.5	-38.5	- 4.9
B_2	23.7	-12.9	-45.2	-34.4
B_3	25.1	- 4.3	-33.7	-12.9
C_1	30.7	- 2.1	-38.3	- 9.7
C_2	26.9	- 6.8	-40.1	-2.0
C_3	24.3	- 7.8	-39.0	-22.5
D_1	19.5	-12.9	-41.1	-34.5
D_2	30.6	- 2.4	-38.2	-10.0
D_3	31.8	- 1.4	-38.1	- 7.7
E_1	25.5	- 4.4	-35.7	-14.6
E_2	32.0	1.7	-31.1	2.6
E_3	24.4	-14.0	-50.6	-40.2
F_1	17.8	-13.1	-48.5	-43.8
F_2	26.8	- 8.7	-42.6	-24.5
F_3	37.3	5.1	-26.3	16.1
G_1	19.0	-17.8	-55.6	-54.4
G_2	30.6	- 0.9	-30.1	- 0.4
G_3	32.3	2.0	-31.7	2.6
(Col. 4)$_1$	22.5	- 9.3	-40.6	-27.4
(Col. 4)$_2$	27.7	- 5.9	-41.7	-19.9
(Col. 4)$_3$	31.7	- 1.5	-35.1	- 4.9
Total	81.9	-16.7	-117.4	-52.2

S_{e2} is calculated by subtracting the variations calculated above from S_T

$$S_{e2} = 4.56 \tag{5.61}$$

Table 5.12 shows the ANOVA table.

In this ANOVA table, there are two kinds of error. Error e_1 is the experimental error as discussed before. e_2 is the interaction between e_1

and K (revolution). As a rule, the effects of the inner factors is judged by comparing those effects with e_1. The effects of the indicative factor and the interactions between the indicative factor and the inner factors are compared with e_2. However, we need to know whether it is necessary to distinguish e_1 and e_2. To do this, pool about half of the inner factors (including e_1) with V_{e1} to get $V_{(e1)}'$, then pool the ones from the indicative factor and interactions between the indicative factor and the inner factors whose magnitude is about the same as V_{e2} to get $V_{(e2)}$. If $V_{(e1)}$ is several times larger than $V_{(e2)}$, the two error terms are distinguishable. But if they are about the same [e.g., $V_{(e1)}$ is less than twice the magnitude of $V_{(e2)}$], they are pooled together. In this example, $V_{(e1)}$ is significantly larger and is not pooled. $B \times K$, $E \times K$ and $G \times K$ are also significant. An F-test maybe used to test their significance. Calculation of the degrees of contribution when there are two kinds of error is explained in Chapter 8 and in Appendix 13.

Table 5.12. ANOVA Table

Factor	f	S	V
A	1	24.27	24.27
B	2	25.86	12.93°
C	2	5.11	2.56°
D	2	24.51	12.26°
E	2	51.54	25.77
F	2	103.87	51.93
G	2	114.33	57.17
e_1	4	21.10	5.275
K	2	1,103.39	551.69
$A \times K$	2	1.57	0.78°
$B \times K$	4	9.20	2.30
$C \times K$	4	1.70	0.43°
$D \times K$	4	5.31	1.33°
$E \times K$	4	9.60	2.40
$F \times K$	4	2.01	0.50°
$G \times K$	4	9.25	2.31
e_2	8	4.56	0.57
(e_1)	(10)	(76.58)	(7.658)
(e_2)	(22)	(15.15)	(0.689)
T	53	1,517.18	

5.5.3 Optimum Condition and Estimation

From the ANOVA table, it is known that interactions $B \times K$, $E \times K$ and $G \times K$ are significant, but they are not large enough to reverse the effects of the main effects B, E and G, which may be noticed from Table 5.11. Therefore, the optimum condition can be determined to be $A_2B_1C_1D_3$ $E_2F_3G_3$. In this case, revolution, K, is irrelevant to the determination of the optimum condition.

Next, the noise level at different levels of K is estimated. For this calculation, the larger effects, A, E, F, G, K, $B{\times}K$, $E{\times}K$ and $G{\times}K$ are used. The estimation is made in two ways: ignoring interactions and using interactions.

[Case When Interactions are Ignored]

Calculation is made without forgetting that a working mean has been subtracted.

$$\begin{aligned}\hat{\mu}(K_1) &= 30 + \overline{T} + (\overline{A_2} - \overline{T}) + (\overline{E_2} - \overline{T}) + (\overline{F_3} - \overline{T}) + (\overline{G_3} - \overline{T}) \\ &\quad + (\overline{K_1} - \overline{T}) = \overline{A_2} + \overline{E_2} + \overline{F_3} + \overline{G_3} + \overline{K_1} - 4\overline{T} \\ &= 30 + \frac{-8.0}{27} + \frac{2.6}{18} + \frac{16.1}{18} + \frac{2.6}{18} + \frac{81.9}{18} - 4 \times \frac{-52.2}{54} \\ &= 30 + 9.3 = 39.3 \text{ (db)} \end{aligned} \tag{5.62}$$

$$\begin{aligned}\hat{\mu}(K_2) &= 30 + \overline{A_2} + \overline{E_2} + \overline{F_3} + \overline{G_3} + \overline{K_2} - 4\overline{T} \\ &= 30 + \frac{-8.0}{27} + \frac{2.6}{18} + \frac{16.1}{18} + \frac{2.6}{18} + \frac{-16.7}{18} - 4 \times \frac{-52.2}{54} \\ &= 30 + 3.8 = 33.8 \text{ (db)} \end{aligned} \tag{5.63}$$

Similarly,

$$\hat{\mu}(K_3) = 30 + (-1.8) = 28.2 \text{ (db)} \tag{5.64}$$

[Case When Interactions are Considered]

In this case, interactions are added to Equation (5.62)

$$\begin{aligned}\mu(K_1) &= 30 + \overline{T} + (\overline{A_2} - \overline{T}) + (\overline{E_2} - \overline{T}) + (\overline{F_3} - \overline{T}) + (\overline{G_3} - \overline{T}) \\ &\quad + (\overline{K_1} - \overline{T}) + (\overline{B_1K_1} - \overline{B_1} - \overline{K_1} + \overline{T}) \\ &\quad + (\overline{E_2K_1} - \overline{E_2} - \overline{K_1} + \overline{T}) + (\overline{G_3K_1} - \overline{G_3} - \overline{K_1} + \overline{T}) \\ &= 30 + \overline{A_2} + \overline{F_3} + \overline{B_1K_1} - \overline{B_1} + \overline{E_2K_1} + \overline{G_3K_1} - 2\overline{K_1} - \overline{T} \\ &= 30 + \frac{-8.0}{27} + \frac{16.1}{18} + \frac{33.1}{6} - \frac{-4.9}{18} + \frac{32.0}{6} + \frac{32.3}{6} \\ &\quad -2 \times \frac{81.9}{27} - \frac{-52.2}{54} \\ &= 30 + 12.0 = 42.0 \text{ (db)} \end{aligned} \tag{5.65}$$

$$\begin{aligned}\mu(K_2) &= 30 + \overline{A_2} + \overline{F_3} + \overline{B_1K_2} - \overline{B_1} + \overline{E_2K_2} + \overline{G_3K_2} - 2\overline{K_2} - \overline{T} \\ &= 30 + \frac{-8.0}{27} + \frac{16.1}{18} + \frac{-0.5}{6} - \frac{-4.9}{18} + \frac{1.7}{6} + \frac{2.0}{6} \end{aligned}$$

$$-2 \times \frac{16.7}{27} - \frac{-52.2}{54}$$
$$= 30 + 3.6 = 33.6\,(\text{db}) \tag{5.66}$$

$$\mu(K_3) = 30 + \overline{A}_2 + \overline{F}_3 + \overline{B_1K_3} - \overline{B}_1 + \overline{E_2K_3} + \overline{G_3K_3} - 2\overline{K}_3 - T$$
$$= 30 + \frac{-8.0}{27} + \frac{16.1}{18} + \frac{-38.5}{6} - \frac{-4.9}{18} + \frac{-31.1}{6} + \frac{-31.7}{6}$$
$$-2 \times \frac{-117.4}{27} - \frac{-52.5}{54}$$
$$= 30 + (-6.3) = 23.7\,(\text{db}) \tag{5.67}$$

At K_1 (1000 rpm), the result is better when interactions are considered than the case when they are ignored. But at K_3 (2000 rpm), the result is reversed. To determine which is the better estimate, a confirmatory experiment must be conducted. The confidence limits for the confirmatory experiment are not shown here. The limits become wider when interactions are ignored, since those ignored interactions will be pooled with e_2.

The current condition is the first level; their estimation is made and shown in Table 5.13. In the table, results are also converted into the units of audible noise level, in decibel units.

Table 5.13. Comparison of Optimum and Current Condition

		Without Considering Interactions			Considering Interactions		
		K_1	K_2	K_3	K_1	K_2	K_3
S/N Ratio,	Optimum	39.3	33.8	28.2	42.0	33.6	23.7
(db)	Current	30.5	25.0	19.4	34.0	24.6	14.2
Noise,	Optimum	54.7	60.2	65.8	52.0	60.4	70.3
(db)	Current	63.5	69.0	74.6	60.0	69.4	79.8

Note: db of noise = $20 \log [\text{Pa}/(2 \times 10^{-5})$
S/N Ratio = $-10 \log \text{Pa}^2$

EXERCISES

Question (5.1). In order to improve eccentricity of a drilling process, control factors $A, B, \ldots, H$ were assigned to L_{18} with 4 combinations of noise factors X and Y. The eccentricity was measured twice for each combination.

(a) Calculate the S/N Ratios of No.1 and No.2 (to the first decimal point).

(b) Do an analysis of variance using the S/N Ratio as a performance characteristic.

(c) Determine the optimum condition, and estimate its S/N Ratio.

(d) To determine the improvement in eccentricity, compare the optimum with current condition $A_2B_2C_2D_2E_2F_2G_2H_2$.

Factor	Column	Level
A: Cutting tool I	1	2
B: Cutting tool II	2	3
C: Cutting tool III	3	3
D: Feeding speed	4	3
E: RPM	5	3
F: Cutting amount I	6	3
G: Cutting amount II	7	3
H: Cutting oil amount	8	3
X: Base type	?	2
Y: Machine	?	2

L_{18}									Results 10mm								db
	A	*B*	*C*	*D*	*E*	*F*	*G*	*H*	X_1Y_1		X_1Y_2		X_2Y_1		X_2Y_1		S/N Ratio
1	1	1	1	1	1	1	1	1	8	7	16	15	12	18	11	11	
2	1	1	2	2	2	2	2	2	10	11	16	10	9	8	16	13	
3	1	1	3	3	3	3	3	3	11	6	11	13	10	4	13	8	-20.0
4	1	2	1	1	2	2	3	3	3	12	7	11	5	8	13	6	-18.9
5	1	2	2	2	3	3	1	1	5	2	13	6	3	6	5	3	-15.9
6	1	2	3	3	1	1	2	2	8	15	10	15	9	12	16	7	-21.6
7	1	3	1	2	1	3	2	3	8	3	3	2	7	5	5	5	-14.2
8	1	3	2	3	2	1	3	1	9	15	14	8	12	8	12	7	-20.8
9	1	3	3	1	3	2	1	2	8	3	6	7	9	6	9	3	-16.6
10	2	1	1	3	3	2	2	1	12	15	14	15	16	13	19	12	-23.3
11	2	1	2	1	1	3	3	2	14	20	15	8	13	13	11	6	-22.4
12	2	1	3	2	2	1	1	3	16	12	16	15	18	16	17	13	-23.8
13	2	2	1	2	3	1	3	2	16	14	17	19	16	16	17	17	-24.4
14	2	2	2	3	1	2	1	3	12	10	17	10	11	14	8	11	-21.5
15	2	2	3	1	2	3	2	1	15	7	4	12	9	4	6	12	-19.5
16	2	3	1	3	2	3	1	2	7	6	8	6	5	12	7	7	-17.5
17	2	3	2	1	3	1	2	3	6	13	14	14	15	9	13	10	-21.7
18	2	3	3	2	1	2	3	1	17	16	14	11	16	18	17	14	-23.8

Question (5.2). In order to find the heat treatment condition which minimizes the variation of dimension and hardness, the following factors were assigned to an L_{18} array. From two lots of material, one unit was used to measure hardness at 16 locations, and dimensions at 12 locations.

From 32 results of hardness, the S/N Ratio and sensitivity were calculated with the use of the following equations:

The S/N Ratios of hardness and level totals are shown in the tables.

$$\eta = 10 \log \frac{\frac{1}{32}(S_m - V_e)}{V_e}$$

Factor		Level		
A: Quenching cooling speed		w/fan	w/o fan	—
B: Maximum heating temperature	(°C)	1,050	1,000	950
C: Heating time	(min.)	40	30	20
D: Annealing temperature	(°C)	650	630	610
E: Annealing time	(min.)	90	75	60
F: Annealing cooling speed		Fan 1	w/o fan	Fan 2
G: Heating speed	(°C/min.)	12	8	4
H: Secondary preheat time	(min.)	60	45	30

$$S = 10 \log \frac{1}{32}(S_m - V_e)$$

Factor · Column / No.	*A* 1	*B* 2	*C* 3	*D* 4	*E* 5	*F* 6	*G* 7	*H* 8	η(db)
1	1	1	1	1	1	1	1	1	30.6
2	1	1	2	2	2	2	2	2	33.8
3	1	1	3	3	3	3	3	3	29.1
4	1	2	1	1	2	2	3	3	25.8
5	1	2	2	2	3	3	1	1	33.5
6	1	2	3	3	1	1	2	2	31.5
7	1	3	1	2	1	3	2	3	27.3
8	1	3	2	3	2	1	3	1	26.6
9	1	3	3	1	3	2	1	2	26.5
10	2	1	1	3	3	2	2	1	23.0
11	2	1	2	1	1	3	3	2	17.9
12	2	1	3	2	2	1	1	3	19.7
13	2	2	1	2	3	1	3	2	20.4
14	2	2	2	3	1	2	1	3	20.4
15	2	2	3	1	2	3	2	1	19.4
16	2	3	1	3	2	3	1	2	6.2
17	2	3	2	1	3	1	2	3	7.1
18	2	3	3	2	1	2	3	1	7.1
	Total								405.9

Supplementary Table

	1	2	3
A	264.7	141.2	—
B			
C	133.3	139.3	133.3
D	127.3	141.8	136.8
E	134.8	131.5	139.6
F	135.9	136.6	133.4
G	136.9	142.1	126.9
H			

(a) Do an analysis of variance and determine the optimum condition.

(b) What would be the gain in performance at the optimum condition compared with the current condition of $A_2B_2C_2D_2E_2F_2G_2$?

(c) The specification of hardness is 48±3.0 and the loss of failing to meet the specification is ¥800. Show the gain by using the optimum condition.

Hint: Put m=48 to equation $\eta=10\log(m^2/\sigma^2)$, determine σ^2 and put it to the Loss Function equation, $L=(A/\Delta^2)\sigma^2$.

CHAPTER 6

SIMPLE LAYOUT TECHNIQUES FOR MIXED TYPE ORTHOGONAL ARRAYS

Sometimes the levels of factors in an experiment may not be able to be fitted exactly to a mixed type orthogonal array. Some layout techniques which can be applied to mixed type arrays are described in this chapter.

6.1 THE CASE WITH DUMMY TREATMENT

6.1.1 Dummy Treatment

In L_{18} array, there are one 2-level and seven 3-level columns. In the example given in the previous chapter, all factors and levels were fitted exactly to the array. But if there are two 2-level and six 3-level factors, can we still use an L_{18} array? The answer is yes if a dummy treatment technique is used.

With a dummy treatment, a 2-level factor can be assigned to a 3-level column—it is formally treated as if it were a 3-level factor. With a dummy treatment, one of the levels is repeated. For example, consider an experiment with A and B as 2-level factors and C through H as 3-level factors. In this case, A is assigned to column 1, B is assigned to column 2, while repeating the more important level of B, and factors C through H are assigned to the rest of the columns as before. If B_1 is more important, the three levels are:

$B_1 = B_1 \quad B_2 = B_2 \quad B_3 = B_1$

When a level is dummy treated, we put a prime (′) notation to the dummy treated level, such as 1′. The variation of the dummy treated factor is calculated by:

$$S_B = \frac{(B_1 + B_1')^2}{12} + \frac{B_2{}^2}{6} - \frac{(B_1 + B_1' + B_2)^2}{18} \qquad (f=1) \quad (6.1)$$

Since B_1 and B_1' are the same level, their results must be equal. If not, it is caused by error, and from the difference, part of the error is calculated. It is called the error between dummy levels and is calculated by:

$$S_e = \frac{(B_1 - B_1')^2}{12} \qquad (f=1) \quad (6.2)$$

6.1.2 Application of a Dummy Treatment in L_{18}

In order to reduce NO_x content from diesel engines, 2-level factors A, G, 3-level factors B, C, D, E, F, and H were assigned to an L_{18} array. NO_x content was measured in ppm at three driving conditions. A smaller-the-better characteristic was used for NO_x content. The layout and S/N Ratios are shown in Table 6.1.

Table 6.1. Layout and Results

No. \ Factor · Column	A 1	B 2	C 3	D 4	E 5	F 6	G 7	H 8	(+40) η(db)
1	1	1	1	1	1	1	1	1	-1.0
2	1	1	2	2	2	2	2	2	0.3
3	1	1	3	3	3	3	1′	3	1.0
4	1	2	1	1	2	2	1′	3	1.4
5	1	2	2	2	3	3	1	1	0.3
6	1	2	3	3	1	1	2	2	-1.1
7	1	3	1	2	1	3	2	3	1.1
8	1	3	2	3	2	1	1′	1	-0.8
9	1	3	3	1	3	2	1	2	0.6
10	2	1	1	3	3	2	2	1	0.5
11	2	1	2	1	1	3	1′	2	1.6
12	2	1	3	2	2	1	1	3	0.8
13	2	2	1	2	3	1	1′	2	0.4
14	2	2	2	3	1	2	1	3	1.9
15	2	2	3	1	2	3	2	1	1.5
16	2	3	1	3	2	3	1	2	2.1
17	2	3	2	1	3	1	2	3	0.6
18	2	3	3	2	1	2	1′	1	1.7

A dummy treatment was applied to factor G, with

$G_1 = G_1 \quad G_2 = G_2 \quad G_3 = G_1'$

The current condition is $A_1B_2C_2D_2E_2F_2G_2H_2$.

First, a supplementary table is prepared.

Table 6.2. Level Totals

Factor \ Level	1	2	3	Total
A	1.8	11.1		12.9
B	3.2	4.4	5.3	12.9
C	4.5	3.9	4.5	12.9
D	4.7	4.6	3.6	12.9
E	4.2	5.3	3.4	12.9
F	-1.1	6.4	7.6	12.9
G	4.7	2.9	(G1′)5.3	12.9
H	2.2	3.9	6.8	12.9

Since the results have been subtracted by a working mean of 40 decibels, the total squared and divided by 18 is not called the general mean, but rather, the correction factor, *CF*.

$$CF = \frac{12.9^2}{18} = 9.245 \qquad (f{=}1) \quad (6.3)$$

$$S_T = \text{ total of the squares} - CF = 15.445 \qquad (f{=}17) \quad (6.4)$$

$$S_A = \frac{(1.8 - 11.1)^2}{18} = 4.805 \qquad (f{=}1) \quad (6.5)$$

$$S_B = \frac{3.2^2 + 4.4^2 + 5.3^2}{6} - CF = 0.37 \qquad (f{=}2) \quad (6.6)$$

$$\left\{\begin{array}{ll} S_C = 0.040 & (f{=}2) \\ S_D = 0.123 & (f{=}2) \\ S_E = 0.303 & (f{=}2) \\ S_F = 7.410 & \end{array}\right. \quad (6.7)$$

$$S_G = \frac{(4.7 + 5.3)^2}{12} + \frac{2.9^2}{6} - CF = 0.49 \qquad (f{=}1) \quad (6.8)$$

$$S_H = 1.803 \qquad (f{=}2) \quad (6.9)$$

$$S_e = S_T - S_A - \cdots - S_H = 0.101 \qquad (f{=}3) \quad (6.10)$$

6.1.3 Decomposition Into Detailed Degrees of Freedom for the Selection of the Optimum Condition

After arranging the results to make the ANOVA table, the optimum condition will be determined and compared with the current condition.

From the supplementary table, the optimum condition is $A_2B_3C_1D_1$ $E_2F_3G_1H_3$. It looks like that F_3 is not quite different from current condition, F_2; the effect of F is mostly contributed by the difference between F_1 and F_2F_3. If so, the current level F_2 may be selected. For clarification, the variation of F may be decomposed into the following,

$$S_F = S_F' + S_{F2'(in)} \tag{6.11}$$

where S_F' is the variation of F_1' and F_2'. F_1' is the same as F_1. F_2' is F_2 and F_3 combined. $S_{F2'(in)}$ is the variation of F_2', which consists of F_2 and F_3.

The calculation of S_F' is the same as the calculation for the dummy treatment.

$$S_F' = \frac{F_1{}^2}{6} + \frac{(F_2 + F_3)^2}{12} - \frac{(F_1 + F_2 + F_3)^2}{18}$$

$$= \frac{(-1.1)^2}{6} + \frac{(6.4 + 7.6)^2}{12} - \frac{12.9^2}{18} = 7.290 \qquad (f{=}1) \tag{6.12}$$

$S_{F2'(in)}$ is calculated by:

$$S_{F2}'\,(\text{in}) = \frac{(F_2 - F_3)^2}{12} = \frac{(6.4 - 7.6)^2}{12} = 0.120 \qquad (f{=}1) \tag{6.13}$$

The total of Equations (6.12) and (6.13) is:

$$S_F' + S_{F2}'\,(\text{in}) = 7.41 \tag{6.14}$$

This agrees with the S_F which is calculated from Equation (6.7). In other words, S_F is decomposed into $S_{F'}$ and $S_{F'(in)}$.

Such decomposition may not always be necessary, but it is better to do so if we want to estimate the improvement compared with current situation.

Table 6.3 shows the ANOVA table. $F_2'{}_{(in)}$ may be pooled with the error. Since there is no difference between F_2 and F_3, the estimate is:

$$\overline{F_2F_3} = \frac{F_2 + F_3}{12} = \frac{6.4 + 7.6}{12} = 7.0\,(\text{db}) \tag{6.15}$$

The optimum condition is $A_2B_3C_1D_1E_2F_2(F_3)G_1H_3$. For the estimation, only larger effects A, F, G and H are used. The optimum levels for F and G are the same as the current levels. Therefore, the gain is:

$$\frac{11.1 - 1.8}{9} + \frac{6.8 - 3.9}{6} = 1.03 + 0.48$$

$$= 1.51\ (\text{db}) \tag{6.16}$$

Table 6.3. ANOVA Table

Factor	f	S	V	$\rho(\%)$
A	1	4.805	4.805	30.5
B	2	0.370	0.185°	
C	2	0.040	0.020°	
D	2	0.123	0.061°	
E	2	0.303	0.152°	
F { F'	1	7.290	7.290	46.6
F { F_2' (in)	1	0.120	0.120	
G	1	0.490	0.490	2.6
H	2	1.803	0.901	10.5
e	3	0.101	0.034	
(e)	(12)	(1.057)	(0.088)	(9.8)
T	17	15.445		100.0

6.2 COMBINATION DESIGN

Dummy treatments as described in the previous section is a method to assign a factor whose number of levels is less than the number of levels of a column in an orthogonal array. Combination design is the method to use when there are a few factors with which dummy treatment techniques can be applied. For example, if we want to dummy treat 2-level factors, A and B, two 3-level columns are required. But, only one 3-level column is necessary if the following three combinations are used:

$(AB)_1 = A_1B_1$

$(AB)_2 = A_2B_1$

$(AB)_3 = A_1B_2$

(AB) is called a combination factor, and the method using a combination factor is called combination design. By using combination design, two 2-level factors can be assigned to a 3-level column. In the extreme case, fifteen 2-level factors can be assigned to L_{18} array. However, combined factors are not orthogonal to each other; therefore, the total variation cannot be decomposed into every factorial effect. It must be understood that combination design is not used unless it is unavoidable.

For example, factors A, B and C are 2-level factors and D, E, F, G and H are 3-level factors. If B and C are dummy treated, all factors cannot be assigned to L_{18}. But by applying combination design, it is possible. In such a case, combination design is appropriate. Which two factors should be combined, i.e., AB or AC or BC? The answer is if engineering knowledge is available, combine a factor which is known to have a smaller effect. This will be explained in the following numerical example.

2-level factors A and B are combined as follows, and each combination has r repetitions. When (AB) is assigned to a 3-level column in an L_{18} array, r is equal to 6.

$(AB)_1 = A_1B_1$

$(AB)_2 = A_2B_1$

$(AB)_3 = A_1B_2$

The variation of the combined factor, (AB) is:

$$S_{(AB)} = \frac{(AB)_1{}^2 + (AB)_2{}^2 + (AB)_3{}^2}{r} - CF \qquad (f=2) \quad (6.17)$$

This is caused by the difference between the three combinations. In the analysis of variance, it is important to decompose the variation into the effect of the individual factors. In combination design however, it is impossible to decompose to:

$$S_{(AB)} = S_A + S_B \qquad (6.18)$$

because A and B are not orthogonal.

The first level of A is in combinations $(AB)_1$ and $(AB)_3$ and the second level is in combination $(AB)_2$. The former includes B_1 and B_2, but the latter includes only B_1. Therefore,

$$\frac{(AB)_1 + (AB)_3}{2r} - \frac{(AB)_2}{r} = \frac{A_1B_1 + A_1B_2}{2r} - \frac{A_2B_1}{r} \qquad (6.19)$$

is not a fair comparison of B. The effect of A may be compared at the fixed level of B: $(AB)_1=A_1B_1$ and $(AB)_2=A_2B_1$. The variation of A is therefore:

$$S_A = \frac{(A_1B_1)^2 + (A_2B_1)^2}{r} - \frac{(A_1B_1 + A_2B_1)^2}{2r} \qquad (6.20)$$

or

$$S_A = \frac{(A_1B_1 - A_2B_1)^2}{2r} \qquad (6.21)$$

This is the effect of A when B is fixed at B_1; the effect might not be held when B is at B_2. However, it is considered to be the effect of A.

In the same way, the effect of B is calculated from $(AB)_1=A_1B_1$ and $(AB)_3=A_1B_2$ where A is fixed.

$$S_B = \frac{(A_1B_1)^2 + (A_1B_2)^2}{r} - \frac{(A_1B_1 + A_1B_2)^2}{2r} \qquad (6.22)$$

or

$$S_B = \frac{(A_1B_1 - A_1B_2)^2}{2r} \tag{6.23}$$

6.3 NUMERICAL EXAMPLE FOR COMBINATION DESIGN

6.3.1 Layout and Results

In order to improve bonding strength, the following control factors:

A: 2 levels *B*: 3 levels *C*: 3 levels *D*: 3 levels

E: 3 levels *F*: 3 levels *G*: 2 levels *H*: 2 levels

I: 3 levels

were assigned to an L_{18} array. Bonding strength was measured at the two worst noise conditions, N_1 and N_2. The assignment was made as follows:

Column	1	2	3	4	5	6	7	8
Factor	*A*	*B*	*C*	*D*	*E*	*F*	(*GH*)	*I*

$(GH)_1 = G_1H_1 \quad (GH)_2 = G_2H_1 \quad (GH)_3 = G_1H_2$

The S/N Ratio for a larger-the-better characteristic was used. Table 6.4 shows the layout and the S/N Ratios. The current condition is $A_1B_2C_2D_2E_2F_2G_2H_2I_2$.

6.3.2 Analysis of Variance

Prepare a supplementary table as shown in Table 6.5.

$$CF = \frac{357.4^2}{18} = 7{,}096.38 \qquad (f=1) \tag{6.24}$$

$$S_T = 7{,}245.88 - CF = 149.50 \qquad (f=17) \tag{6.25}$$

$$S_A = \frac{(176.1 - 181.3)^2}{18} = 1.50 \qquad (f=1) \tag{6.26}$$

$$S_B = \frac{102.6^2 + 126.1^2 + 128.7^2}{6} - CF = 68.90 \qquad (f=2) \tag{6.27}$$

Similarly,

$$\begin{cases} S_C = 50.07 & (f=2) \\ S_D = 5.89 & (f=2) \\ S_E = 7.68 & (f=2) \\ S_F = 8.33 & (f=2) \end{cases} \tag{6.28}$$

Table 6.4. Layout and Results

No. \ Factor · Column	A 1	B 2	C 3	D 4	E 5	F 6	(GH) 7	I 8	η(db)
1	1	1	1	1	1	1	1	1	12.3
2	1	1	2	2	2	2	2	2	18.9
3	1	1	3	3	3	3	3	3	19.1
4	1	2	1	1	2	2	3	3	19.5
5	1	2	2	2	3	3	1	1	21.0
6	1	2	3	3	1	1	2	2	23.2
7	1	3	1	2	1	3	2	3	18.2
8	1	3	2	3	2	1	3	1	20.7
9	1	3	3	1	3	2	1	2	23.2
10	2	1	1	3	3	2	2	1	18.1
11	2	1	2	1	1	3	3	2	14.7
12	2	1	3	2	2	1	1	3	19.5
13	2	2	1	2	3	1	3	2	18.6
14	2	2	2	3	1	2	1	3	21.2
15	2	2	3	1	2	3	2	1	22.6
16	2	3	1	3	2	3	1	2	20.4
17	2	3	2	1	3	1	2	3	22.2
18	2	3	3	2	1	2	3	1	24.0

Table 6.5. Level Totals

	Total		Total		Total		Total
A_1	176.1	B_1	102.6	C_1	107.1	D_1	114.5
A_2	181.3	B_2	126.1	C_2	118.7	D_2	120.2
		B_3	128.7	C_3	131.8	D_3	122.7
E_1	113.6	F_1	117.6	G_1H_1	117.6	I_1	118.7
E_2	121.6	F_2	124.9	G_2H_1	123.2	I_2	119.0
E_3	122.2	F_3	116.0	G_1H_2	116.6	I_3	119.7
						Total	357.4

Next, S_G and S_H are calculated. The variation between G_1H_1 and G_2H_1, where H_1 is common, is considered as the effect of G,

$$S_G = \frac{(G_1H_1 - G_2H_1)^2}{12} = \frac{(117.6 - 123.2)^2}{12} = 2.61 \qquad (f=1) \quad (6.29)$$

Similarly, S_H is calculated from G_1H_1 and G_1H_2 where G is fixed.

$$S_H = \frac{(G_1H_1 - G_1H_2)^2}{12} = \frac{(117.6 - 116.6)^2}{12} = 0.08 \qquad (f=1) \quad (6.30)$$

The variation of the three levels of the combined factor is:

$$S_{(GH)}=\frac{117.6^2+123.2^2+116.6^2}{6}-CF=4.21 \qquad (f{=}2) \quad (6.31)$$

The results show that S_G+S_H is not equal to $S_{(GH)}$, since G and H are not orthogonal. Next, S_I and S_e are calculated.

$$S_I=\frac{118.7^2+119.0^2+199.7^2}{6}-CF=0.08 \qquad (f{=}2) \quad (6.32)$$

$$\begin{aligned} S_e &= S_T-S_A-S_B-S_C-S_D-S_E-S_F-S_{(GH)}-S_I \\ &= 2.84 \end{aligned} \qquad (f{=}2) \quad (6.33)$$

Table 6.6. ANOVA Table

Factor	f	S	V
A	1	1.50	1.50°
B	2	68.90	34.45
C	2	50.07	25.04
D	2	5.89	2.94°
E	2	7.68	3.84
F	2	8.33	4.16
(GH)	2	4.21	2.11°
(G)	(1)	(2.61)	(2.61)
(H)	(1)	(0.08)	(0.08)
I	2	0.08	0.04°
e	2	2.84	1.42
(e)	(9)	(14.52)	(1.61)
T	17	149.50	

It is seen that the variance of combined factor (GH) is small, and it seems that both G and H are small. Observing the effects of G and H which are calculated separately, it is seen that H is very small. Therefore, by neglecting the effect of H, the effect of G can be recalculated. Since effect H is insignificant, the three levels of the combined factor, $(GH)_1$, $(GH)_2$ and $(GH)_3$ are the conditions of G_1, G_2, and G_3, respectively. That is, the same as if factor G is dummy treated. Its variation is calculated as:

$$\begin{aligned} S_G &= \frac{(G_1H_1+G_1H_2)^2}{12}+\frac{G_2H_1{}^2}{6}-CF \\ &= \frac{(117.6+116.6)^2}{12}+\frac{123.2^2}{6}-CF \\ &= 4.13 \end{aligned} \qquad (f{=}1) \quad (6.34)$$

The calculation may be made as:

$$S_G = S_{(GH)} - S_H = 4.13 \qquad (f=1) \quad (6.35)$$

These results are summarized in Table 6.7, which does not contain the combination factor.

Table 6.7. ANOVA Table

Factor	f	S	V
A	1	1.50	1.50°
B	2	68.90	34.45
C	2	50.07	25.04
D	2	5.89	2.94°
E	2	7.68	3.84
F	2	8.33	4.16
G	1	4.13	4.13
I	2	0.08	0.04°
e	3	2.92	0.97
(e)	(8)	(10.39)	(1.30)
T	17	149.50	

It is desirable that in combination design, either one of the combined factors remains, or both are insignificant. However, it cannot always be expected to happen this way.

6.3.3 Optimum Condition And Its Estimation

The optimum condition is $A_2B_3C_3D_3E_3F_2G_2H_1I_3$. The S/N Ratio at the optimum level of some factors such as B, D, E and I is close to the current level and their differences might be insignificant. To confirm this, set new levels as follow:

$B_1' = B_1 \quad B_2' = B_2, \quad B_3$

$D_1' = D_1 \quad D_2' = D_2, \quad D_3$

$E_1' = E_1 \quad E_2' = E_2, \quad E_3$

$I_1' = I_1 \quad I_2' = I_2, \quad I_3$

Variations B, D, E and I are then decomposed to:

$$\begin{cases} S_B' = S_{B'}' + S_{B2'(in)} \\ S_D' = S_{D'}' + S_{D2'(in)} \\ S_E' = S_{E'}' + S_{E2'(in)} \\ S_I' = S_{I'}' + S_{I2'(in)} \end{cases} \tag{6.36}$$

For example,

$$S_{B'} = \frac{102.6^2}{6} + \frac{(126.1 + 128.7)^2}{12} - CF = 68.34 \qquad (f=1) \tag{6.37}$$

$$S_{B2'(in)} = \frac{(126.1 - 128.7)^2}{12} = 0.56 \qquad (f=1) \tag{6.38}$$

Similar calculations are made to obtain Table 6.8.

Table 6.8. ANOVA Table with Fine Decomposition

Factor		f	S	V
	A	1	1.50	1.50°
B	B'	1	68.34	68.34
	B_2'(in)	1	0.56	0.56°
	C	2	50.07	25.04
D	D'	1	5.37	5.37
	D_2' (in)	1	0.52	0.52°
E	E'	1	7.65	7.65
	E_2'(in)	1	0.03	0.03°
	F	2	8.33	4.16
	G	1	4.13	4.13
I	I'	1	0.04	0.04°
	I_2' (in)	1	0.04	0.04°
	e	3	2.92	0.97
	(e)	(6)	(5.61)	(0.935)
	T	17	149.50	

From Table 6.8, B_2, D_2 and E_2 may be used as the optimum levels. Since there is no difference between B_2 and B_3, its estimate is:

$$\overline{B_2B_3} = \frac{B_2 + B_3}{12} = 21.2 \tag{6.39}$$

$\overline{D_2D_3}$ and $\overline{E_2E_3}$ are calculated in the same way. The estimate of the optimum condition is:

$$\begin{aligned}\hat{\mu} &= \overline{B_2B_3} + \overline{C}_3 + \overline{E_2E_3} + \overline{F}_2 + \overline{G}_2 - 4\overline{T} \\ &= \frac{126.1 + 128.7}{12} + \frac{131.8}{6} + \frac{121.6 + 122.2}{12} + \frac{124.9}{6} + \frac{123.2}{6} - 4 \times \frac{357.4}{18} \\ &= 25.44 \text{ (db)}\end{aligned} \tag{6.40}$$

This is the S/N Ratio of bonding strength at the optimum condition. Letting the average of square of reciprocal of the strength(y) be σ^2,

$$-10 \log \sigma^2 = 25.44 \tag{6.41}$$

$$\sigma^2 = 0.002\,858 \tag{6.42}$$

$$\therefore \; y = 18.71 \text{ (kgf)} \tag{6.43}$$

This is the estimate from the compounded noise factor. The optimum condition is probably worse than normal condition. In order to estimate the optimum bonding strength at normal conditions, the S/N Ratio of the current condition is estimated from experimental results. Then the gain of S/N Ratio at the optimum condition is converted into anti-log values. This figure describes the reduction of variability (the square of reciprocal of bonding strength) in terms of a fraction. From this figure and current bonding strength (average of square of reciprocal), the variability at the optimum condition can be estimated.

The current condition is $A_1B_2C_2D_2E_2F_2G_2H_2I_2$. Among the significant factors, the factor whose optimum level is different from current one is C; its gain is:

$$\overline{C}_3 - \overline{C}_2 = \frac{131.8 - 118.7}{6} = 2.18 \text{ (db)} \tag{6.44}$$

It is 1.65 in anti-log units; the reduction is 1/1.65.

If current bonding strength from actual data (not from experiment) is 25.8 kgf, its square of reciprocal, σ^2 is:

$$\sigma^2 \text{ (current)} = \frac{1}{25.8^2} = 0.001\,5 \tag{6.45}$$

The optimum is:

$$\sigma^2 \text{ (optimum)} = 0.001\,5 \times \frac{1}{1.65} = 0.000\,9 \tag{6.46}$$

It is equivalent to 33.17 kgf.

If the specification of bonding strength is 15 kgf at the minimum and the loss of out-of-specification product is ¥1200, the quality at the optimum condition is:

$$L = A\Delta^2\sigma^2 = 1{,}200 \times 15^2 \times 0.000\,9 = 243\ (¥) \tag{6.47}$$

The loss at current condition is:

$$L\ (\text{current}) = 1{,}200 \times 15^2 \times 0.001\,5 = 405\ \ (¥) \tag{6.48}$$

6.4 MULTI-LEVEL ARRANGEMENT

The objective of multi-level arrangements is to utilize the interaction column of any two columns. This technique is not very applicable to mixed type orthogonal arrays, wherein interactions are distributed evenly to other columns in the array. With the exception of the interaction of column 1 (2 levels) and 2 (3 levels) of an L_{18} array, their combinations,

(1 1) (1 2) (1 3) (2 1) (2 2) (2 3),

can be considered as a 6-level column as shown in Table 6.9.

Table 6.9. L18 Array with a 6-level Column

No. \ Column	1 2	3	4	5	6	7	8
1	1	1	1	1	1	1	1
2	1	2	2	2	2	2	2
3	1	3	3	3	3	3	3
4	2	1	1	2	2	3	3
5	2	2	2	3	3	1	1
6	2	3	3	1	1	2	2
7	3	1	2	1	3	2	3
8	3	2	3	2	1	3	1
9	3	3	1	3	2	1	2
10	4	1	3	3	2	2	1
11	4	2	1	1	3	3	2
12	4	3	2	2	1	1	3
13	5	1	2	3	1	3	2
14	5	2	3	1	2	1	3
15	5	3	1	2	3	2	1
16	6	1	3	2	3	1	2
17	6	2	1	3	1	2	3
18	6	3	2	1	2	3	1

This 6-level column may be used to:

(1) Assign a 6-level factor.

(2) Assign a 4-level or a 5-level factor using dummy treatment technique.

(3) Arrange a 6-level combined factor. For example, combine a 2-level factor A and 5-level factor B such as:

$(AB)_1 = A_1B_1$

$(AB)_2 = A_2B_1$

$(AB)_3 = A_1B_2$

$(AB)_4 = A_1B_3$

$(AB)_5 = A_1B_4$

$(AB)_6 = A_1B_5$

or combine 3-level factor A, 2-level factor B and 3-level factor C such as:

$(ABC)_1 = A_1B_1C_1$

$(ABC)_2 = A_2B_1C_1$

$(ABC)_3 = A_3B_1C_1$

$(ABC)_4 = A_1B_2C_1$

$(ABC)_5 = A_1B_1C_2$

$(ABC)_6 = A_1B_1C_3$

The variation of a 6-level factor is calculated as:

$$S_A = \frac{A_1{}^2 + A_2{}^2 + A_3{}^2 + A_4{}^2 + A_5{}^2 + A_6{}^2}{r} - CF \tag{6.49}$$

where A_1, A_2,... A_6 are level totals. Numerical examples will be omitted here.

EXERCISES

Question (6.1). In order to improve the precision of LSI process, the following factors will be studied:

A: mask dimension	2 levels
B: viscosity	2 levels
C: spin speed	3 levels
D: baking temperature	2 levels
E: baking time	3 levels
F: aperture	3 levels
G: exposure time	3 levels
H: development time	3 levels
I: plasma etching time	3 levels

(a) Assign the main effects to an L_{18}.

(b) When B and D are 3-levels, can it be assigned to L_{18}?

Question (6.2). To improve the kneader for powder preparation, the following factors were assigned to an L_{18}.

A: puddle combination	6 levels
B: B heater temperature	3 levels
C: C heater temperature	3 levels
D: rpm	3 levels
E: feed quantity	3 levels

A was assigned to the 6 level column by combining columns 1 and 2.

Noise factors were compounded and assigned outside the L_{18} array.

N_1 = standard condition

N_2 = worst condition

The softening point was measured as shown in the following table.

(a) Softening point is a larger-the-better characteristic. Determine the S/N Ratio in decibel units and make an ANOVA table.

(b) Find the optimum condition and estimate its S/N Ratio.

(c) The specification of softening point is 129° minimum. When it is below 129°, the loss per kg is ¥1000. What is the quality improvement at the optimum condition comparing current condition, $A_2B_2C_2D_2E_2$?

No. \ Factor · Column	A (1∩2)	B 3	C 4	D 5	E 6	e 7	e 8	Softening Point C° N_1	N_2
1	1	1	1	1	1	1	1	125.8	126.4
2	1	2	2	2	2	2	2	128.2	127.4
3	1	3	3	3	3	3	3	131.4	130.0
4	2	1	1	2	2	3	3	127.6	125.6
5	2	2	2	3	3	1	1	128.8	129.0
6	2	3	3	1	1	2	2	130.2	129.4
7	3	1	2	1	3	2	3	129.4	129.2
8	3	2	3	2	1	3	1	128.4	128.4
9	3	3	1	3	2	1	2	125.8	125.6
10	4	1	3	3	2	2	1	130.0	129.4
11	4	2	1	1	3	3	2	129.0	128.2
12	4	3	2	2	1	1	3	127.6	126.8
13	5	1	2	3	1	3	2	127.2	125.4
14	5	2	3	1	2	1	3	132.0	130.6
15	5	3	1	2	3	2	1	129.6	127.6
16	6	1	3	2	3	1	2	131.8	131.2
17	6	2	1	3	1	2	3	125.6	126.0
18	6	3	2	1	2	3	1	130.4	128.8

CHAPTER 7

QUALITY IMPROVEMENT FOR NOMINAL-THE-BEST CHARACTERISTICS

Quality improvement by parameter design can be applied not only to product design but to various problems in manufacturing processes. The S/N Ratio with signal factor as explained in Chapter 3 is effective for the improvement of processing or measurement studies.

7.1 EXPERIMENT DESIGN FOR NOMINAL-THE-BEST CHARACTERISTICS

7.1.1 Layout of the Inner Array

The layout for the inner array is exactly as described in the cases for smaller-the-better and larger-the-better characteristics. After listing the factors and their levels, they should be assigned to a mixed array whenever possible.

7.1.2 Compounding of Noise Factors for Efficient Experiments

The way noise factors are treated with nominal-the-best characteristics depends on the objectives of the experiment. It is important to consider a sufficient number of noise factors, and to use continuous variables to set the following levels based on environmental conditions:

$$K_1 = m - \sqrt{\frac{3}{2}}\,\sigma \tag{7.1}$$

$$K_2 = m \tag{7.2}$$

$$K_3 = m + \sqrt{\frac{3}{2}}\,\sigma \tag{7.3}$$

where

m is the average of noise factor, K

σ is the standard deviation of noise factor, K.

By setting such levels, it is possible to simulate variability similar to the case when the noise factor varies at random with the mean m and standard deviation (σ) around m. (See Appendix 15.)

However, if an experiment is to be conducted for improvement, there are many inner conditions most of the case. Therefore, it is desirable that the number of outer conditions be kept as small as possible in order to keep the required number of experiments to a manageable size. For this purpose, only major noise factors should be considered, and these factors should be compounded. The compounding of noise factors for nominal-the-best characteristics is achieved by setting the three levels of the compound noise factors whose qualitative trends are known:

N_1: the combination of noise factor conditions which makes the result small (called the "worst condition of negative side")

N_2: standard condition

N_3: the combination of noise factor conditions which makes the result large (called the "worst condition of positive side")

In many cases, only two conditions N_1 and N_3 are set.

Suppose noise factors K, L, P and Q have the following qualitative effects to the output, y:

K: When K increases, y increases and never decreases.

L: When L increases, y decreases and never increases.

P: When P increases, y decreases and never increases.

Q: When Q increases, y increases and never decreases.

The compounded three levels will be:

$N_1 = K_1L_3P_3Q_1$

$N_2 = K_2L_2P_2Q_2$

$N_3 = K_3L_1P_1Q_3$

where the lower levels of K, L, P and Q have smaller values.

The S/N Ratios obtained from such compounded noise will be worse than the actual condition, but there is no problem in the determination of the better control factor levels. Also this is a good approach because a control factor level which is highly significant under compounded noise factors is truly a good one.

7.1.3 Layout of Outer Conditions With a Signal Factor

The objective in the assignment of outer conditions is to provide a layout for the calculation of data for various types of S/N Ratios. It is appropriate to have about 5 levels for the signal factor. However, when an L_{18} is used as the outer array, it is better to have either 3 or 6 levels.

Try to use compounded noise factors as much as possible. If there are noise factors whose qualitative trend is unknown, it is recommended that a simple preliminary experiment be conducted to study their behavior. If a large scale experiment is affordable, then factors which cannot be compounded maybe assigned together with the compounded factor in an orthogonal array.

When it is easy to conduct experiments, noise factors may not be compounded in many cases. Thus, there are many different options available. The following are some typical patterns:

[The Case Without Compounding]

1. Collect the data for all factorial combinations of the noise factors. This is the case when the number of combinations of the outer factors is about 10 or less, but this is not frequently used. It may also be used when data collection is easy, even there are a large number of combinations to test for.
2. Assign noise factors to a small orthogonal array such as an L_9. When there is a signal factor, one set of data from the orthogonal array will be collected for each signal factor level.
3. Alternatively, the signal factor and noise factors may be assigned to one orthogonal array. L_{18} is generally used in this case.

Whichever outer conditions layout is used, a S/N Ratio for each inner run is calculated. The sample results are shown in Table 7.1, after calculation of the signal factor level totals.

Table 7.1. The Case Without Compounding

Repetitions	M_1	M_2	$\cdots$	M_k	
1	y_{11}	y_{21}	$\cdots$	y_{k1}	
2	y_{12}	y_{22}	$\cdots$	y_{k2}	
.	.	.	.	.	
.	.	.	.	.	
.	.	.	.	.	
r_0	y_{r0}	y_{r0}	$\cdots$	y_{r0}	
Total	y_1	y_2	$\cdots$	y_k	T

[The Case With Compounding]

With compound noise strategies, the following options are available.

1. Collect the data for all combinations between the signal factor and the compounded noise factor. Sometimes, r_0 repetitions are made for each combination.
2. Assign the signal factor and compounded noise factor with those noise factors which cannot be compounded together into a single orthogonal array.

With the use of compounded noise factors, the results are summarized in Table 7.2, and a S/N Ratio is calculated similar to the case of compounding as shown in Chapter 3.

Table 7.2. Results with Compounding

Repetitions	M_1	M_2	···	M_k	
N_1	y_{11}	y_{21}	···	y_{k1}	N_1
N_2	y_{12}	y_{22}	···	y_{k2}	N_2
N_3	y_{13}	y_{23}	···	y_{k3}	N_3
	y_1	y_2	···	y_k	T

y_{ij} is the total of r_0 data points. In many cases, r_0=1.

7.2 Experimental Design for the Evaluation of Processing

7.2.1 Factors and Levels

In this section, an experiment* to optimize the processability of a cutting process is illustrated, with control factors:

A: cutting speed	A_1=70	A_2=90	A_3=110	(m/mm)
B: feeding speed	B_1=0.05	B_2=0.075	B_3=0.1	(mm/rev)
C: corner radius	C_1=0.4	C_2=0.6	C_3=0.8	(mm)
D: horizontal scooping angle	D_1=10	D_2=10	D_3=14	(deg)
E: tool shape	E_1=slanting		E_2=single blade	
F: cutting oil	F_1=yes	F_2=no		

To compare processability, processed material type, K, which is an indicative factor, was assigned.

K: processed material K_1=S45C K_2=SUS304

These factors were assigned to an L_{18} inner orthogonal array. K was assigned to column 1, A through F were assigned to columns 2 through 7, respectively. E and F were dummy treated with:

* *Shizuoka Kogyo Shikenzyo*

$E_1 = E_1 \quad E_2 = E_1' \quad E_3 = E_2$

$F_1 = F_1 \quad F_2 = F_1' \quad F_3 = F_2$

No factor was assigned to column 8; it is reserved for the error in this study. The amount cut was used as the signal factor with,

M: cutting amount $M_1 = 0.1 \quad M_2 = 0.3 \quad M_3 = 0.5$ (mm)

6 measured locations were considered as the error factor, R.

From each of the 18 cutting conditions, the cutting was made with cutting amounts of M_1=0.1, M_2=0.3 and M_3=0.5 (mm), and 6 locations were measured, to obtain the table presented in Table 7.3.

Table 7.3. Form of Data Collection

Factor · Column / No.	K 1	A 2	B 3	C 4	D 5	E^* 6	F^* 7	e 8	M_1 $R_1\ R_2 \cdots R_6$	M_2 $R_1\ R_2 \cdots R_6$	···	M_4 $R_1\ R_2 \cdots R_6$
1	1	1	1	1	1	1	1	1				
2	1	1	2	2	2	2'	2	2				
3	1	1	3	3	3	2	2	3				
.					.						···	
.					.							
.					.							
18	2	3	3	2	1	2'	2	1				

* $E_1 = E_1, E_2 = E_1', E_3 = E_2$

$F_1 = F_1, F_2 = F_1', F_3 = F_2$

Table 7.4. Cutting Amount of Experiment No. 1

Unit: mm

	M_1		M_2		M_3	
	0.098	0.098	0.294	0.296	0.495	0.495
	0.097	0.098	0.288	0.287	0.493	0.495
	0.093	0.092	0.288	0.297	0.489	0.488
Total	0.576		1.750		2.955	

7.2.2 S/N Ratios from the Outer Conditions

Table 7.4 displays the results of experimental run No.1 in the L_{18} array.

The aimed cutting amount, M, and the actual cut amount, y, are expressed by the following proportional equation to calculate the S/N Ratios:

$$y = \beta M \tag{7.4}$$

First, total variation, S_T, is calculated as:

$$S_T = 0.098^2 + 0.098^2 + \cdots + 0.488^2 = 2.021\,241 \qquad (f=18) \quad (7.5)$$

The S/N Ratio is calculated as:

$$S_\beta = \frac{(M_1 y_1 + M_2 y_2 + M_3 y_3)^2}{r_0(M_1{}^2 + M_2{}^2 + M_3{}^2)}$$

$$= \frac{(0.1 \times 0.576 + 0.3 \times 1.750 + 0.5 \times 2.955)^2}{6 \times (0.1^2 + 0.3^2 + 0.5^2)}$$

$$= 2.020\,958\,1 \qquad (f=1) \quad (7.6)$$

$$S_e = S_T - S_\beta = 0.000\,282\,9 \qquad (f=17) \quad (7.7)$$

$$V_e = \frac{S_e}{17} = 0.000\,016\,64 \qquad (7.8)$$

$$\eta = 10 \log \frac{\frac{1}{r}(S_\beta - V_e)}{V_e} = 10 \log \frac{\frac{1}{2.1}(2.020\,958\,1 - 0.000\,016\,64)}{0.000\,016\,64}$$

$$= 10 \log 578\,34 = 47.6\,(\text{db}) \qquad (7.9)$$

where

$$r = 6 \times (0.1^2 + 0.3^2 + 0.5^2) = 2.1$$

The S/N Ratios of No.2 and others are calculated similarly to obtain Table 7.5.

Table 7.5. S/N Ratios

No. \ Factor · Column	K 1	A 2	B 3	C 4	D 5	E^* 6	F^* 7	e 8	η(db) -50
1	1	1	1	1	1	1	1	1	- 3.3
2	1	1	2	2	2	2	2	2	11.5
3	1	1	3	3	3	3	3	3	- 0.1
4	1	2	1	1	2	2	3	3	3.4
5	1	2	2	2	3	3	1	1	8.2
6	1	2	3	3	1	1	2	2	6.8
7	1	3	1	2	1	3	2	3	1.8
8	1	3	2	3	2	1	3	1	- 2.6
9	1	3	3	1	3	2	1	2	4.9
10	2	1	1	3	3	2	2	1	- 6.1
11	2	1	2	1	1	3	3	2	- 5.6
12	2	1	3	2	2	1	1	3	5.8
13	2	2	1	2	3	1	3	2	- 1.2
14	2	2	2	3	1	2	1	3	2.9
15	2	2	3	1	2	3	2	1	2.6
16	2	3	1	3	2	3	1	2	1.0
17	2	3	2	1	3	1	2	3	8.5
18	2	3	3	2	1	2	3	1	3.8

7.2.3 Calculation of Variation Terms

The total number of data points in this example is 324, but the analysis is made using 18 S/N Ratios. Table 7.6 is a supplementary table.

Table 7.6. Level Totals

Level \ Factor	K	A	B	C	D	E^*	F^*	Col. 8
1	30.6	2.2	- 4.4	10.5	6.4	14.0	19.5	2.6
2	11.7	22.7	22.9	29.9	21.7	20.4	25.1	17.4
3		17.4	23.8	1.9	14.2	7.9	- 2.3	22.3
Total	42.3	42.3	42.3	42.3	42.3	42.3	42.3	42.3

* $E_1 = E_1, E_2 = E_1', E_3 = E_2$

$F_1 = F_1, F_2 = F_1', F_3 = F_2$

$$CF = \frac{42.3^2}{18} = 99.41 \qquad (f=1) \quad (7.10)$$

$$S_T = (-3.3)^2 + 11.5^2 + \cdots + 3.8^2 - CF = 409.30 \qquad (f=17) \quad (7.11)$$

$$S_K = \frac{(30.6 - 11.7)^2}{18} = 19.84 \qquad (f=1) \quad (7.12)$$

$$S_A = \frac{2.2^2 + 22.7^2 + 17.4^2}{6} - 99.41 = 37.74 \qquad (f=2) \quad (7.13)$$

Similarly,

$$\begin{cases} S_B = 85.63 \\ S_C = 68.57 \\ S_D = 19.51 \end{cases} \qquad (7.14)$$

S_E is then determined.

$$S_E = \frac{(E_1 + E_1')^2}{12} + \frac{E_2{}^2}{6} - CF = \frac{(14.0 + 20.4)^2}{12} + \frac{7.9^2}{6} - CF$$
$$= 9.61 \qquad (f=1) \quad (7.15)$$

Similarly,

$$S_F = 67.23 \qquad (f=1) \quad (7.16)$$

The error variation is:

$$S_e = S_T - S_K - S_A - S_B - \cdots - S_F = 101.17 \qquad (f=6) \quad (7.17)$$

This term includes the variations due to column 8 effects, interaction $A \times B$, and the differences between dummy levels (E_1 and E_1) and (F_1 and F_1').

The ANOVA Table is shown in Table 7.7.

Table 7.7. ANOVA Table

Factor	f	S	V	ρ(%)
K	1	19.84	19.84°	
A	2	37.74	18.87°	
B	2	85.63	42.815	13.3
C	2	68.57	34.285	9.1
D	2	19.51	9.755°	
E	1	9.61	9.61°	
F	1	67.23	67.23	12.6
e	6	101.17	16.861	
(e)	(12)	(187.87)	(15.656)	(65.0)
T	17	409.30		100.0

The optimum condition is $A_2B_3C_2D_2E_1F_1$, without consideration of the indicative factor, K. For E, E_1 is better since $E_1{=}E_1$, $E_2{=}E_1'$ and $E_3{=}E_2$.

F_1 is better for the same reason. The larger factors considered are B, C and F, and the estimated condition for the optimum is:

$$\hat{\mu} = \overline{B}_3 + \overline{C}_2 + \overline{F}_1 - 2\overline{T}$$
$$= 50 + \frac{22.9 + 23.8}{12} + \frac{29.9}{6} + \frac{19.5 + 21.5}{12} - 2 \times \frac{42.3}{18}$$
$$= 57.6 \text{ (db)} \qquad (7.18)$$

Since there is not a significantly large difference between B_2 and B_3, B_2B_3 is used in Equation (7.18). Since the indicative factor, K, is not significant, there is no necessity to determine the S/N Ratios of the two different processed materials separately.

The confidence limits are calculated by:

$$\pm\sqrt{F_{12}{}^{1}(0.05) \times V_e \times \frac{1}{n_e}} = \pm\sqrt{4.75 \times 15.656 \times \frac{5.5}{18}} = \pm 4.77$$

where

$$n_e = \frac{\text{Total degrees of freedom (including the general mean)}}{\underset{m}{1} + \underset{B}{2} + \underset{C}{2} + \underset{F}{0.5}} = \frac{18}{5.5}$$

The degrees of freedom for the calculation of confidence limits when dummy treatment is used (called the effective degrees of freedom) is:

$$\frac{\text{No. of levels before dummy treatment}}{\text{No. of dummy level replicated}} - 1$$

F_1 is repeated twice; therefore, 3/2 - 1 = 0.5. For the dummy treatment calculation, see Appendix 16.

7.2.4 Proportional Coefficient

The optimum condition in the previous section is determined from S/N Ratio only, but actually, we need to know the proportional relationship between the aimed cutting amount (signal) and the actual amount cut under the optimum condition. Such estimates are generally made as follows:

First, the numerator of the equation for the S/N Ratio in terms of 10 times its logarithm is calculated as the sensitivity, S. For No.1,

$$S = 10 \log \frac{1}{r}(S_\beta - V_e) = 10 \log \frac{1}{2.1}(2.020\,958\,1 - 0.000\,016\,64)$$
$$= -0.167 \text{ (db)} \qquad (7.19)$$

The sensitivity for all of the 18 runs is calculated, and the factors which affect sensitivity are determined. These factors may be found from the ANOVA table.

The ANOVA table is omitted here, but factors A and C are significant with respect to sensitivity.

$$\begin{cases} \overline{A}_1 = -0.149 \text{ (db)} \\ \overline{A}_2 = -0.061 \text{ (db)} \\ \overline{A}_3 = -0.035 \text{ (db)} \\ \overline{C}_1 = -0.158 \text{ (db)} \\ \overline{C}_2 = -0.017 \text{ (db)} \\ \overline{C}_3 = -0.070 \text{ (db)} \\ \overline{T} \ = -0.082 \text{ (db)} \end{cases} \tag{7.20}$$

Using the optimum condition from S/N Ratio analysis, $A_2B_3C_2D_2E_1F_1$, sensitivity is estimated from the factors which have a significant effect on β.

$$\begin{aligned} S &= \overline{A}_2 + \overline{C}_2 - \overline{T} = (-0.061) + (-0.017) - (-0.082) \\ &= 0.004 \text{ (db)} \end{aligned} \tag{7.21}$$

Its anti-log value is 1.001. Since it is ideally equal to 1, condition $A_2B_3C_2D_2E_1F_1$ may be considered as the optimum condition. It is a distinctive feature in nominal-the-best characteristic that the optimum condition is determined not only from the S/N Ratio, but also considering sensitivity.

When sensitivity deviates from the target, change the level of the factor which affects sensitivity but does not affect the S/N Ratio, such as factor A should be changed. The change of such a factor can adjust sensitivity without affecting the S/N Ratio significantly. In some cases, a factor which was not assigned in the experiment can be used to adjust sensitivity. In that case, the optimum condition can be determined only from the S/N Ratio. For sensitivity adjustment, refer to discussion in Chapter 9.

EXERCISES

Question (7.1). To improve the deformation of the following heating cylinder part in an injection molding machine due to heat treatment, control factors A through H were assigned to columns 1 through 8 respectively in an L_{18} array.

Two spots of different dimensions, I_1 and I_2, were measured before and after heat treatment.

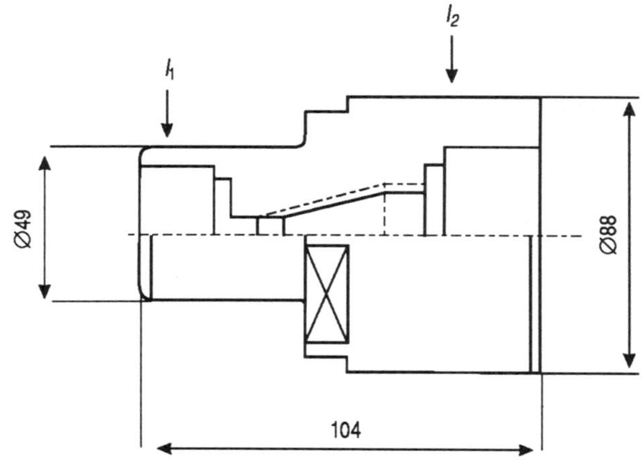

Control Factors for Heat Treatment (L_{18})

Factor \ Level	1	2	3
A: quenching cooling speed	with fan	w/o fan	—
B: maximum heating temperature (°C)	1,050	1,000	950
C: maximum heat holding time (min)	40	30	20
D: annealing temperature (°C)	650	630	610
E: annealing time (min)	90	75	60
F: annealing cooling speed (°C/mim)	with fan	w/o fan	w/o fan
G: heat rise temperature (°C/min)	12	8	4
H: secondary preheating time (min)	60	45	30

In each cell, four repetitions was made using four lots of material.

(a) The results of No.1 is shown as follows. Consider before/after heat treatment and material lots as noise factors and calculate S/N Ratios of I_1 and I_2 in decibel units.

Spot	I_1		I_2	
Before/After \ Material Lot	Before Heat Treatment	After Heat Treatment	Before Heat Treatment	After Heat Treatment
R_1	49.21	49.24	88.24	88.31
R_2	49.21	49.24	88.22	88.29
R_3	49.22	49.25	88.21	88.29
R_4	49.20	49.24	88.24	88.31

(b) The S/N Ratios except No.1 are shown below. Perform an analysis of variance using I as the outer indicative factor and estimate the error variance of dimension under the optimum condition. The supplementary table shows level totals except the S/N Ratio of No. 1.

Layout and Results

No. \ Factor · Column	A 1	B 2	C 3	D 4	E 5	F 6	G 7	H 8	I_1	I_2
1	1	1	1	1	1	1	1	1		
2	1	1	2	2	2	2	2	2	66.06	66.00
3	1	1	3	3	3	3	3	3	67.00	65.74
4	1	2	1	1	2	2	3	3	71.69	66.18
5	1	2	2	2	3	3	1	1	68.70	66.23
6	1	2	3	3	1	1	2	2	66.75	66.67
7	1	3	1	2	1	3	2	3	65.98	68.67
8	1	3	2	3	2	1	3	1	67.69	69.89
9	1	3	3	1	3	2	1	2	70.99	67.42
10	2	1	1	3	3	2	2	1	64.55	67.20
11	2	1	2	1	1	3	3	2	64.75	67.98
12	2	1	3	2	2	1	1	3	64.37	68.80
13	2	2	1	2	3	1	3	2	61.98	74.61
14	2	2	2	3	1	2	1	3	65.48	75.36
15	2	2	3	1	2	3	2	1	65.53	72.11
16	2	3	1	3	2	3	1	2	71.50	65.53
17	2	3	2	1	3	1	2	3	70.29	71.50
18	2	3	3	2	1	2	3	1	68.49	70.29

Supplementary Table (except No.1)

	I_1	I_2		I_1	I_2
A_1	544.86	607.00	E_1	331.45	413.96
A_2	596.94	634.67	E_2	406.84	411.03
			E_3	403.51	416.68
B_1	326.73	409.31	F_1	331.08	422.35
B_2	400.13	417.57	F_2	407.26	407.75
B_3	414.94	414.79	F_3	403.46	411.57
C_1	335.70	416.96	G_1	341.04	421.09
C_2	402.97	414.03	G_2	399.16	408.10
C_3	403.13	410.68	G_3	401.60	412.48
D_1	343.25	404.05	H_1	334.96	405.09
D_2	395.58	423.76	H_2	402.03	417.27
D_3	402.97	413.86	H_3	404.81	419.31

Question (7.2). Following control factors are listed to improve the cutting process:

	1.	2.	3.	4.
A: lathe				*N*
B: tool	*K* company	*L* company	*M* company	company
C: scooping angle	P20	M20	Thermet	
(deg.)	0	4	8	
D: escape angle (deg.)	2	4	5	
E: cutting blade	0	15	30	
angle (deg.)				
F: corner radius (mm)	0.4	0.7	1.0	
G: cutting speed	80	110	140	
(m/min.)	chuck	both	single	
H: setting method		centers	chuck	

(a) Assign these factors to an L_{18} array.

(b) The signal and noise factors will be assigned to outside using an L_{18} array. Show the degrees of freedom of factors for the analysis of variance of the S/N Ratios.

CHAPTER 8

DECOMPOSITION OF THE CAUSES OF VARIATION AND TOLERANCE DESIGN

After the optimum condition is determined using the methods described in the previous chapters, experiments can be conducted to decompose the causes of error. That is, to further improve quality by controlling the causes of error. In this chapter, analysis of variance for the decomposition of the causes of error is explained.

8.1 PARAMETER DESIGN AND TOLERANCE DESIGN

So far, discussion has focused on the methods of experimentation for the improvement of stability. When there are various causes of variation, it is important to design a system which is not affected by those causes. As described in Chapter 1 and later chapters, control factors are assigned to the inner orthogonal array, whereas noise factors are assigned to the outer orthogonal array. Experiments with all combinations of the inner and outer conditions (called direct product design) will be conducted. If the design results in too many experiments when there are many inner factors and noise factors, a method called compounding noise factors is recommended to improve the efficiency of the experiment. If noise factors are compounded, the variation developed in the experiment often becomes larger than actual situation. Also the effect of individual noise factor cannot be observed.

If satisfactory improvement cannot be achieved in parameter design, further improvement is made by directly controlling the causes. In this case, it is necessary to analyze the relationship between noise factors and the output characteristics. For this purpose, an experiment may be conducted under the optimum condition which was determined at parameter design stage to know the effects of noise factors. This information is used in tolerance design to study the effects of the noise factors and how much to control them. The methods can also be applied to product design and manufacturing to study the relationship between processing methods and causes of variation.

8.2 Experimental Design for Decomposition of Effects of Noise Factors

When a signal factor is included in experimentation, it is recommended to select 3 to 6 levels for the signal factor. As described in Chapter 7, the levels of noise factors which are continuous variables and to be determined from environmental conditions is set as follows:

$$K_1 = m - \sqrt{\frac{3}{2}}\,\sigma \tag{8.1}$$

$$K_2 = m \tag{8.2}$$

$$K_3 = m + \sqrt{\frac{3}{2}}\,\sigma \tag{8.3}$$

where m is the average of error factor, K.

σ is the standard deviation of error factor, K

There are three ways to assign factors. Case (3) is generally used:

1. Collect the data for all combinations. This is the case when all combinations is in the neighbor of 10. However, this method is not frequently used. It may be used if data collection is easy.
2. Assign noise factors to a small orthogonal array such as an L_9. When there is a signal factor, collect data from an orthogonal array for each level of the signal factor.
3. Assign the signal factor together with noise factors in an orthogonal array.

L_{18} is generally used as a mixed type orthogonal array. When L_{18} is used, it is appropriate to set 3 to 6 levels for the signal factor. Dummy treatment of combination design may be used. But the numbers of replications for the signal factor are different and we must be careful to calculate the proportional or linear effect. The case when the numbers of replications is different will be discussed in Chapter 12.

8.3 DECOMPOSITION OF THE EFFECTS OF NOISE FOR THE DRIVEABILITY OF CARS

8.3.1 Layout and Results

In order to study the effects of noise on the driveability of a car which was made under a certain specification, noise factor levels of K, L,N, P and Q were set as follows:

K: turning direction (2 levels)	K_1 = right K_2 = left
L: road condition (3 levels)	L_1 = asphalt surface
	L_2 = dry asphalt surface
	L_3 = dry cement surface
N: load condition (3 levels)	N_1 = front N_2 = standard
	N_3 = rear
P: tire type (3 levels)	P_1 =
	P_2 =
	P_3 = radial tire
Q: front tire pressure (3 levels)	$Q_1 = 6\ \text{kgf/cm}^2$ for right and left
	Q_2 = 3 for right, 6 for left
	Q_3 = 3 for right and left

In the case of a nominal-the-best characteristics, where adjustment is made by the rotating angle of a steering wheel to right and left direction when driving a car, a signal factor is used in the experiment. Three levels of the signal factor were chosen, considering a low speed of 10 (km/h), which is the speed of driving into a garage, parking or making a hairpin turn:

Signal factor (M) M_1 = 560 (degree)
M_2 = 700
M_3 = 840

For studying the effect of speeds, indicative factor levels:

$C_1 = 5 \quad C_2 = 10 \quad C_3 = 15$ (km/h)

were set.

The signal factor and noise factors were assigned to an L_{18} orthogonal array and turning radius at different speed was measured. The results are shown in Table 8.1.

Table 8.1. Layout and Results

No. \ Factor · Column	K 1	L 2	M 3	e 4	N 5	P 6	Q 7	e 8	Turning Radius (unit:m) C_1	C_2	C_3	Total
1	1	1	1	1	1	1	1	1	14.4	14.8	15.1	44.3
2	1	1	2	2	2	2	2	2	11.3	11.5	11.4	34.2
3	1	1	3	3	3	3	3	3	9.2	9.2	9.0	27.4
4	1	2	1	1	2	2	3	3	13.9	14.5	14.6	43.0
5	1	2	2	2	3	3	1	1	10.9	10.9	10.7	32.5
6	1	2	3	3	1	1	2	2	9.9	10.2	10.9	31.0
7	1	3	1	2	1	3	2	3	15.1	15.6	16.5	47.2
8	1	3	2	3	2	1	3	1	11.8	11.7	12.4	35.9
9	1	3	3	1	3	2	1	2	8.4	9.0	9.0	26.4
10	2	1	1	3	3	2	2	1	12.8	13.2	13.0	39.0
11	2	1	2	1	1	3	3	2	11.3	12.7	12.7	36.7
12	2	1	3	2	2	1	1	3	8.5	8.8	9.0	26.3
13	2	2	1	2	3	1	3	2	12.9	13.3	13.2	39.4
14	2	2	2	3	1	2	1	3	10.5	11.4	11.8	33.7
15	2	2	3	1	2	3	2	1	9.0	9.1	9.6	27.7
16	2	3	1	3	2	3	1	2	13.3	13.3	13.6	40.2
17	2	3	2	1	3	1	2	3	9.9	10.0	10.2	30.1
18	2	3	3	2	1	2	3	1	9.5	9.9	10.5	29.9
								Total	202.6	209.1	213.2	624.8

Table 8.2. Level Totals

	C_1	C_2	C_3	Total		C_1	C_2	C_3	Total
K_1	104.9	107.4	109.6	321.9	N_1	70.7	74.6	77.5	222.8
K_2	97.7	101.7	103.6	303.0	N_2	67.8	68.9	70.6	207.3
—	—	—	—	—	N_3	64.1	65.6	65.1	194.8
L_1	67.5	70.2	70.2	207.9	P_1	67.4	68.8	70.8	207.0
L_2	67.1	69.4	70.8	207.3	P_2	66.4	69.5	70.3	206.2
L_3	68.0	69.5	72.2	209.7	P_3	68.8	70.8	72.1	211.7
M_1	82.4	84.7	86.0	253.1	Q_1	66.0	68.2	69.2	203.4
M_2	65.7	68.2	69.2	203.1	Q_2	68.0	69.6	71.6	209.2
M_3	54.5	56.2	58.0	168.7	Q_3	68.6	71.3	72.4	212.3
$(4)_1$	55.9	70.1	71.2	208.2	$(8)_1$	68.4	69.5	71.3	209.3
$(4)_2$	68.2	70.0	71.3	209.5	$(8)_2$	67.1	70.0	70.8	207.9
$(4)_3$	67.5	69.0	70.7	207.2	$(8)_3$	67.1	69.5	71.1	207.7
					Total	202.6	209.1	213.2	624.9

8.3.2 Analysis of Variance

Basically there is no difference in the analysis of variance methods used to analyze the significance of control factors, or the decomposition of error factors. The difference is what to be judged from the results. In the analysis of error causes, the objective is not to find the optimum condition, but to determine the direction of the research for the future and to conduct tolerance design.

Since we are not interested in the general mean, it is calculated as the correction factor.

$$CF = S_m = \frac{624.9^2}{18 \times 3} = 7{,}231.48 \qquad (f=1) \quad (8.4)$$

$$S_T = 14.4^2 + 14.8^2 + \cdots + 10.5^2 - CF = 239.07 \qquad (f=53) \quad (8.5)$$

$$S_K = \frac{(K_1 - K_2)^2}{54} = \frac{(321.9 - 303.0)^2}{54} = 6.62 \qquad (f=1) \quad (8.6)$$

$$S_L = \frac{L_1{}^2 + L_2{}^2 + L_3{}^2}{18} - CF = 0.17 \qquad (f=2) \quad (8.7)$$

For the signal factor, the linear effect, S_{Ml}, is calculated.

$$S_{Ml} = \frac{(-M_1 + M_3)^2}{18 \times 2} = \frac{(-253.1 + 168.7)^2}{36} = 197.87 \qquad (f=1) \quad (8.8)$$

$$S_N = \frac{N_1{}^2 + N_2{}^2 + N_3{}^2}{18} - CF = 21.86 \qquad (f=2) \quad (8.9)$$

$$S_P = 0.98 \qquad (f=2) \quad (8.10)$$

$$S_Q = 2.27 \qquad (f=2) \quad (8.11)$$

Variations from No.1 through No.18, S_{T1}, is calculated by:

$$S_{T1} = \frac{1}{3}(44.3^2 + 34.2^2 + \cdots + 29.9^2) - CF = 233.10 \qquad (f=17) \quad (8.12)$$

$$\therefore\ S_{e1} = S_{T1} - S_K - S_L - S_{Ml} - S_N - S_P - S_Q = 3.33 \qquad (f=7) \quad (8.13)$$

In S_{e1}, variations of columns 4 and 8, interaction $K \times L$ and portion of M caused by the deviation from linearity are included.

$$S_C = \frac{C_1{}^2 + C_2{}^2 + C_3{}^2}{18} - CF = 3.17 \qquad (f=2) \quad (8.14)$$

The following interactions can be calculated, although they may be pooled together as the error.

$$S_{K\times C}=\frac{K_1C_1{}^2+\cdots+K_2C_3{}^2}{9}-S_K-S_C=0.07 \tag{8.15}$$

$$S_{L\times C}=0.31 \tag{8.16}$$

$$\begin{aligned}S_{Ml\times C}&=\frac{L(C_1)^2+L(C_2)^2+L(C_3)^2}{6\times 2}-S_{Ml}\\&=\frac{(-27.9)^2+(-28.5)^2+(-28.0)^2}{12}-197.87\\&=0.02\end{aligned} \tag{8.17}$$

$$S_{N\times C}=1.56 \tag{8.18}$$

$$S_{P\times C}=0.13 \tag{8.19}$$

$$S_{Q\times C}=0.08 \tag{8.20}$$

$$S_{e2}=S_T-S_{T1}-S_C-S_{K\times C}-\cdots-S_{Q\times C}=0.63 \tag{8.21}$$

The ANOVA table is shown in Table 8.3.

Table 8.3. ANOVA Table

Factor	f	S	V
K	1	6.62	6.62
L	2	0.17	0.085°
M_l	1	197.87	197.87
N	2	21.86	10.93
P	2	0.98	0.49°
Q	2	2.27	1.135°
e_1	7	3.33	0.476
C	2	3.17	1.59
K×C	2	0.07	0.035°
L×C	4	0.31	0.077°
M_l×C	2	0.02	0.01°
N×C	4	1.56	0.39
P×C	4	0.13	0.04°
Q×C	4	0.08	0.02°
e_2	14	0.73	0.052
T	53	239.07	

V_{e1} is 9 times the magnitude of V_{e2}; it is shown in the table without being pooled, and there are two error types. Pool L, P and Q with V_{e1} and pool KxC, LxC, M_l×C, P×C and Q×C with V_{e2} to get a rearranged ANOVA table as shown in Table 8.4.

Table 8.4. Rearranged ANOVA Table

Factor	f	S	V	ρ(%)	Effects Compared With 100 as the Signal Factor
K	1	6.62	6.62	2.6	3.2
M_l	1	197.87	197.87	82.5	100.0
N	2	21.86	10.93	8.7	10.5
e_1	13	6.75	0.519	3.4	4.1
C	2	3.17	1.59	1.3	1.6
$N{\times}C$	4	1.56	0.39	0.6	0.7
e_2	30	1.24	0.04	0.9	1.1
T	53	239.07		100.0	

When there two error types, the degrees of contribution is calculated as follows:

$$\rho_K = \frac{S_K - f_K \times V_{e1}}{S_T} \times 100 = \frac{6.62 - 0.519}{239.07} \times 100 = 26 \tag{8.22}$$

$$\rho_{Ml} = \frac{S_{Ml} - f_{Ml} \times V_{e1}}{S_T} \times 100 = 82.5 \tag{8.23}$$

$$\rho_N = \frac{S_N - f_N \times V_{e1}}{S_T} \times 100 = 8.7 \tag{8.24}$$

$$\rho_{e1} = \frac{S_{e1} + (f_K + f_{Ml} + f_N)V_{e1} - (f_K + f_{Ml} + f_N + f_{e1})V_{e2}}{S_T} \times 100$$
$$= 3.4 \tag{8.25}$$

$$\rho_C = \frac{S_C - f_C V_{e2}}{S_T} \times 100 = 1.3 \tag{8.26}$$

$$\rho_{N\times C} = \frac{S_{N\times C} - f_{N\times C} V_{e2}}{S_T} \times 100 = 0.6 \tag{8.27}$$

$$\rho_{e2} = \frac{S_{e2} + (f_K + f_{Ml} + f_N + f_{e1} + f_{e2})V_{e2}}{S_T} \times 100 = 0.9 \tag{8.28}$$

For reference, the ratios calculated by letting the signal factor be 100 are shown under the right column. It is seen that the effect of load position is the largest. If $N{\times}C$ is considered to be significant, the impact of load position significantly affects turning radius as speed changes. However, it is impossible to fix the load at a certain position at all times, therefore, no direct countermeasures can be taken.

Understanding the impact of noise factors related to operating conditions or environmental conditions provides information useful for future designs. But for improvement, it is recommended that many controllable factors be investigated by conducting parameter design which has been explained in previous chapters, rather than investigating countermeasures for noise.

8.4 Tolerance Design for Component Parts

8.4.1 Levels and Layout for Tolerance Design

When the variation of component parts affects the output to a certain level, it is possible to improve quality by upgrading or screening component parts. This will increase cost as a result. Thus, it is necessary to trade off cost with quality. Tolerance design is a method to deal with this, and the decomposition of causes of variation. For example, in the following, we show the design parameters for the circuit elements of a temperature adjusting device:

$A = 3.9\ (\text{k}\,\Omega)$

$B = 7.5\ (\text{k}\Omega\ )$

$C = 1.0\ (\text{k}\,\Omega)$

$D = 3.3\ (\text{k}\,\Omega)$

$E = 360\ (\text{k}\Omega\ )$

$F = 5.0\ (\text{V})$

$G = 10.0\ (\text{V})$

One of the quality characteristics in a temperature adjuster is thermister-resistance; its specification is 1.54±0.10 (kΩ). The loss when the quality becomes out of specification is ¥10,000. Currently, secondary grade elements are used with standard deviation of about 3%. In order to quantify which component has a larger effect, experimental levels are set as:

$$\begin{aligned} &\text{First level:} && \text{Nominal value} - \sqrt{\frac{3}{2}}\,\sigma \\ &\text{Second level:} && \text{Nominal} \\ &\text{Third level:} && \text{Nominal} + \sqrt{\frac{3}{2}}\,\sigma \end{aligned} \tag{8.29}$$

for an L_{18}, it is rather cumbersome to assemble 18 circuits by changing the levels of components slightly. Instead CAD (computer-aided design) may be used for simulation. The thermister-resistance is given by the following equation:

$$y = \frac{\dfrac{CD}{C+D} B(FE+GA)}{F(AB+AE+BE) - B \times (FE+GA)} \tag{8.30}$$

For simplicity, $\sqrt{3/2}$ was approximated to be 5% to get the levels shown in Table 8.5.

Table 8.5. Three levels of Noise Factors

Factor \ Level	1	2	3
A	3.9×0.95	3.9	3.9×1.05
B	7.5×0.95	7.5	7.5×1.05
C	1.0×0.95	1.0	1.0×1.05
D	3.3×0.95	3.3	3.3×1.05
E	360.0×0.95	360.0	360.0×1.05
F	5.0×0.95	5.0	5.0×1.05
G	10.0×0.95	10.0	10.0×1.05

Factors A through G were assigned to columns 2 through 8 of an L_{18} array and thermister-resistance, y, was calculated from Equation (8.30).

8.4.2 Analysis of Variance

With a target of 1.54 KΩ, the deviations from the target are shown in Table 8.6. Table 8.7 is a supplementary table.

$$S_T = (-0.077)^2 + \cdots + (-0.025)^2 = 0.196\,180 \tag{8.31}$$

$$S_m = \frac{0.035^2}{18} = 0.000\,069 \tag{8.32}$$

Table 8.6. (Calculated value)-(target)

No. \ Factor · Column	A 1	B 2	C 3	D 4	E 5	F 6	G 7	8	Results -1.54
1	1	1	1	1	1	1	1	1	-0.077
2	1	1	2	2	2	2	2	2	0.079
3	1	1	3	3	3	3	3	3	0.243
4	1	2	1	1	2	2	3	3	-0.135
5	1	2	2	2	3	3	1	1	0.014
6	1	2	3	3	1	1	2	2	0.123
7	1	3	1	2	1	3	2	3	-0.162
8	1	3	2	3	2	1	3	1	-0.023
9	1	3	3	1	3	2	1	2	-0.035
10	2	1	1	3	3	2	2	1	0.069
11	2	1	2	1	1	3	3	2	-0.009
12	2	1	3	2	2	1	1	3	0.178
13	2	2	1	2	3	1	3	2	-0.064
14	2	2	2	3	1	2	1	3	0.049
15	2	2	3	1	2	3	2	1	0.008
16	2	3	1	3	2	3	1	2	-0.092
17	2	3	2	1	3	1	2	3	-0.106
18	2	3	3	2	1	2	3	1	-0.025

Table 8.7. Level Totals

	A	B	C	D	E	F	G
1	0.483	-0.461	-0.354	-0.101	0.031	0.037	-0.034
2	-0.005	0.004	0.020	0.015	0.002	0.011	0.002
3	-0.443	0.492	0.369	0.121	0.002	-0.013	0.067
Total	0.035	0.035	0.035	0.035	0.035	0.035	0.035

It is recommended that the effects of A through G be decomposed into their linear and residual effect (quadratic effect) so that the effect of the output resistance change after the variability of component parts is reduced can be studied. As the three levels are equally-spaced, an orthogonal polynomial response equation, as explained in Chapter 3, can be used to calculate the linear effects. The linear and quadratic portions of A are calculated from Appendix 2 as:

$$S_{Al} = \frac{(W_1A_1 + W_2A_2 + W_3A_3)^2}{r_0\lambda^2 S} = \frac{(-A_1 + A_3)^2}{6 \times 2}$$

$$= \frac{(0.483 + 0.443)^2}{12} = 0.071\,456 \tag{8.33}$$

$$S_{Aq} = \frac{(W_1A_1 + W_2A_2 + W_3A_3)^2}{r_0\lambda^2 S} = \frac{(A_1 - 2A_2 + A_3)^2}{6 \times 6}$$

$$= \frac{[0.483 - 2 \times (-0.005) + (-0.443)]^2}{36} = 0.000\,069 \tag{8.34}$$

The total of S_{Al} and S_{Aq} is:

$$S_A = \frac{A_1^{\ 2} + A_2^{\ 2} + A_3^{\ 2}}{6} - S_m = 0.071\,525 \tag{8.35}$$

S_e was calculated by:

$$S_e = S_T - S_m - S_{Al} - S_{Aq} - \cdots - S_{Gq} = 0.000\,091 \qquad (f{=}3) \tag{8.36}$$

Summarizing the above results, Table 8.8 is obtained. Table 8.9 is a rearranged ANOVA table.

Table 8.8. ANOVA Table for Tolerance Design

Factor		f	S	V
m		1	0.000 069°	
A	l	1	0.071 456	
	q	1	0.000 069°	
B	l	1	0.075 684	
	q	1	0.000 015°	
C	l	1	0.043 561	
	q	1	0.000 017°	
D	l	1	0.004 107	
	q	1	0.000 003°	
E	l	1	0.000 070°	
	q	1	0.000 027°	
F	l	1	0.000 208	
	q	1	0.000 000°	
G	l	1	0.000 850	
	q	1	0.000 023°	
e		3	0.000 091	0.000 030
T		18	0.196 180	0.010 899

Table 8.9. Rearranged ANOVA Table

Factor	f	S	V	ρ(%)
A_l	1	0.071 456	←	36.4
B_l	1	0.075 684	←	38.6
C_l	1	0.043 561	←	22.2
D_l	1	0.004 107	←	2.1
F_l	1	0.000 208	←	0.1
G_l	1	0.000 850	←	0.4
e	12	0.000 314	0.000 026	0.2
T	18	0.196 180	0.010 899	100.0

8.4.3 Selection of Optimum Component Levels

Since the levels of A through G are set at $m-\sqrt{3/2}\,\sigma, m, m+\sqrt{3/2}\,\sigma$,

$$V_T=\frac{S_T}{18}=\frac{0.196\,180}{18}=0.010\,899 \qquad (8.37)$$

V_T is considered to be the variability when all elements are second grade. See Appendix 12 for more details.

The quality loss when out of specification is ¥10,000. Current loss is calculated by:

$$L_{\text{current}}=\frac{10{,}000}{0.1^2}\times V_T=1{,}000{,}000\times 0.010\,899$$
$$=10{,}899(\text{¥/unit}) \qquad (8.38)$$

The cost of primary grade elements is ¥400 higher, but the standard deviation is 1/3 of the secondary grade components. The degree of contribution of A_l is 36.4%. If A is upgraded to first grade, its variance would be reduced to:

$$\left(\frac{1}{3}\right)^2\times 0.364\times V_T=0.040\,4\times 0.010\,899=0.000441 \qquad (8.39)$$

In other words, the amount of reduction by A would be:

$$[1-\left(\frac{1}{3}\right)^2]\times 0.364\times V_T=0.324\times V_T=0.003\,531 \qquad (8.40)$$

In monetary units, the amount of reduction is:

$$L_A=\frac{10{,}000}{0.1^2}\times 0.003\,531=3{,}531\ (\text{¥/unit}) \qquad (8.41)$$

Therefore, it would be much more profitable to upgrade by spending ¥400. Such calculations are summarized in Table 8.10.

Table 8.10. Judgment of Upgrading

Component	ρ	$q=[1-\left(\frac{1}{3}\right)^2]\times\frac{\rho}{100}$	$L=\frac{10{,}000}{0.1^2}\times qV_T$	Upgrading
A	36.4	0.324	3,531	O
B	38.6	0.343	3,738	O
C	22.2	0.197	2,147	O
D	2.1	0.019	207	X
F	0.1	0.000 89	10	X
G	0.4	0.003 6	39	X

From the table, it is better to change *A*, *B* and *C* to first grade and leave *D*, *F* and *G* as current second grade. *E* was pooled because its effect is insignificant, so it remains as second grade.

The loss after changing *A*, *B* and *C* to first grade is:

$$L_0=\frac{10{,}000}{0.1^2}\times V_T\times[(0.364+0.386+0.222)\times\left(\frac{1}{3}\right)^2$$
$$+0.021+0.001+0.004+0.002]$$
$$=1{,}000{,}000\times0.010\,899\times0.136=1{,}482\ (\yen/\text{unit}) \tag{8.42}$$

By adding the cost increase,

$$L=L_0+1{,}200=2{,}682\ (\yen/\text{unit}) \tag{8.43}$$

Comparing with current loss, this is a quality improvement of

$$L_{\text{current}}-L=10{,}899-2{,}682=8{,}217\ (\yen/\text{unit}) \tag{8.44}$$

EXERCISES

Question (8.1). In a deposition process, temperature affects the surface potential of a certain product. Surface potential, y, was measured 10 times at different depositing temperature, x. A working mean 800 (°C) has been subtracted from depositing temperature, x, and the target value 550 (V) has been subtracted from surface potential, y.

x (°C)	18	20	25	22	26	20	19	24	20	22
y (V)	-50	-32	60	24	70	6	-60	55	-20	30

(a) Construct the analysis of variance table after decomposing the total of squares of deviations from the target into:
(general mean) + (linear effect of x) + (error)
(for the variation of the linear term of x, refer to the linear equation in Chapter 1.)

(b) The specification of surface potential is 550±80 (V). The loss due to be out of specification is ¥8000 per product. Calculate the current loss.

(c) The annual cost of controlling the depositing temperature to 1/3 of current variance is ¥8,000,000. What would be profit assuming the annual production is 50,000 units?

Question (8.2). Parameter design was conducted to reduce the swing of a rotating axle, and an optimum condition of $A_1B_2C_2D_3E_3F_2$ was determined.

In order to decompose the effects of noise factors, the following noise factors:

K: radial bearing pressure	K_1=-1	K_2=0	K_3=1 (kgf/cm^2)
L: thrust bearing pressure	L_1=-1	L_2=0	L_3=1 (kgf/cm^2)
O: load (large)	O_1=0	O_2=10	O_3=20 (kgf)
P: load (small)	P_1=0	P_2=2	P_3=4 (kgf)

were assigned to an L_9 and swing was measured twice each.

Layout and Results

Factor · Column / No.	K 1	L 2	O 3	P 4	Swing (μm)		Total
1	1	1	1	1	0.08	0.15	0.23
2	1	2	2	2	0.22	0.19	0.41
3	1	3	3	3	0.19	0.21	0.40
4	2	1	2	3	0.18	0.14	0.32
5	2	2	3	1	0.23	0.19	0.42
6	2	3	1	2	0.09	0.05	0.14
7	3	1	3	2	0.22	0.25	0.47
8	3	2	1	3	0.06	0.10	0.16
9	3	3	2	1	0.13	0.08	0.21

(a) Calculate S_T from 18 data points and construct an ANOVA table by decomposing it into the following portions:

ANOVA Table

Factor	f
m	1
$K\{l$	1
$\;\;\;q$	1
$L\{l$	1
$\;\;\;q$	1
$O\{l$	1
$\;\;\;q$	1
$P\{l$	1
$\;\;\;q$	1
e	9
T	18

(b) Make comments from the results of this ANOVA table.

Question (8.3). After conducting the parameter design of a constant current circuit with factors *A* through *K*, tolerance design was conducted by setting each noise factor levels as follows:

1st level $m - \sqrt{3/2}\,\sigma$

2nd level m

3rd level $m + \sqrt{3/2}\,\sigma$

where m is the nominal value and is the standard deviation of current grade (2nd grade). The factors were assigned to an L_{36} orthogonal array and the total of the squares of deviations from the target (3.6mA) into the following components:

The specification is 3.6±0.1mA. When a product becomes out of specification, adjustment is made at a cost of ¥80.

(a) Calculate the loss of current product.

(b) Currently, 2nd grade elements are used for *A*, *B* and *C*. The 1st grade element costs ¥5 higher but variability can be reduced to 1/3 in terms of standard deviation.
If elements *D* through *K* are screened and reduce the standard deviation to 1/2, the cost increases by ¥50 each.
Determine the best grade for *A*, *B* and *C*, also decide if screening *D* through *K* is profitable. Estimate how much the quality could be improved after the optimum selection.

Factor		f	$S(\times 10^{-6})$
m		1	12
A	l	1	55,196
	q	1	4
B	l	1	85,530
	q	1	32
C	l	1	88,979
	q	1	11
D	l	1	12,901
	q	1	0
E	l	1	12,330
	q	1	19
F	l	1	37
	q	1	2
G	l	1	214,921
	q	1	0
H	l	1	13,239
	q	1	11
I	l	1	2,710
	q	1	0
J	l	1	415
	q	1	29
K	l	1	16,299
	q	1	1,394
e		13	474
T		36	504,332

CHAPTER 9

PARAMETER DESIGN AND TOLERANCE DESIGN USING SIMULATION MODELING

In this chapter, we show if there exists a functional relationship between factor levels and the output, it is possible to conduct parameter design through simulation, thus saving time and money for experimentation.

9.1 INNER NOISE AND STABILITY DESIGN

In the previous chapters, examples in the study of outer noise factors to improve stability were given. But there is another type of noise factor—inner noise—which affects stability such as the case when the characteristic of a component part deviates from target value or the characteristic value drifts during usage. It is important to consider inner noise factors in design.

For example, consider an electric circuit which converts 100 V alternating current into 115 V direct current. Component parts in the circuit, A and B, affect the output voltage as shown in Figure 9.1. It is seen that combination A=20 and B=40 is one of the possible solutions to get an output voltage of 115 V. However, the nominal value of A is set at 20, it may not stay at 20 if a low grade component part is used, or if the component quickly deteriorates. As a result, the output voltage will not be maintained at 115 V. If A is set at 40, the output voltage is hardly affected when the component deviates from 40. When A is set at 40, the output voltage is 125 V instead of 115 V. To adjust the voltage to 115 volts, the value of B may be changed to 55 for example. No matter where B is set,the effect to the output per unit change of B is the same. In other words, A is the type of factor which affects stability, while B affects the mean only. B is therefore referred to as the adjustment factor.

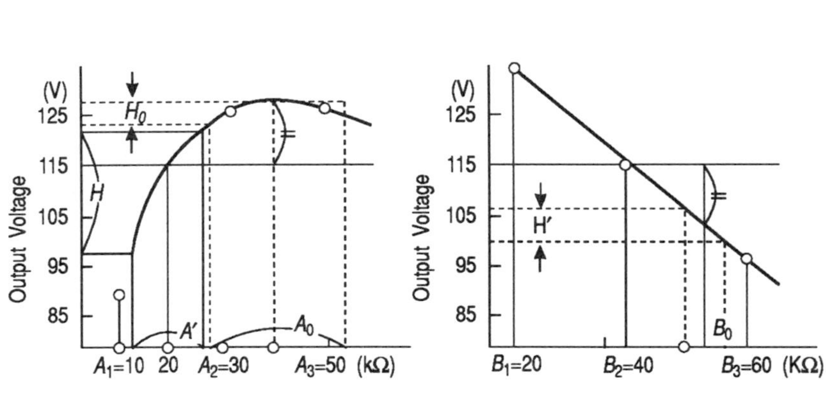

Figure 9.1. Parameter Levels and Output Voltage

Having combination A=20 and B=40, the tolerance for both A and B needs to be tight. But with combination A=40 and B=55, the tolerance of A can be loose. It is obvious that the latter combination is a better design.

For nominal-the-best characteristics, we need to find out which factors affect stability and which factors are for adjustment. For this purpose, the S/N Ratio and sensitivity are analyzed. Factors which affect S/N Ratio affect stability, and factors which do not affect the S/N Ratio but affect sensitivity are used as adjustment factors. The same approach applies for the case when inner noise factors are considered as the primary source of noise. To determine S/N Ratios, outer experiments are necessary. But it is cumbersome to systematically prepare combinations of component parts with slight variation. For this reason, repetition is made to substitute inner noise.

If theoretical equations of models are available, it is possible to plan a rational design for simulation. Theoretical models include CAD, simulation models or finite element method, etc., which have a wide variety of application.

What parameter design is aiming for is stability. It is desirable to set wide control factor level intervals. But there is limitation for parameter design experiments. In simulation however, it is possible to set level intervals so wide that exceed our common sense, and quite often,very useful conclusions have been obtained because of that.

9.2 DIE CASTING MOLD DESIGN

9.2.1 Introduction

Die casting mold design is a very important part of casting engineering. It requires an understanding of the relationship between the die and casting machine. The design and manufacturing of dies require years of engineering experience.

Die casting mold design includes:

1. Mold size design
2. Cooling mechanism design
3. Extrusion device design
4. Mold dividing method design
5. Design of casting plunger, runner, gate, molten metal storage and air vent

Part of the casting method in (5) is discussed in this section.

A die casting machine is a system consisting of a machine and mold to produce products by operating an oil-pressed plunger in an oil-pressed cylinder, pressing and filling molten aluminum alloy into metal cavities within the solidifying time of the alloy. (See Figure 9.2)

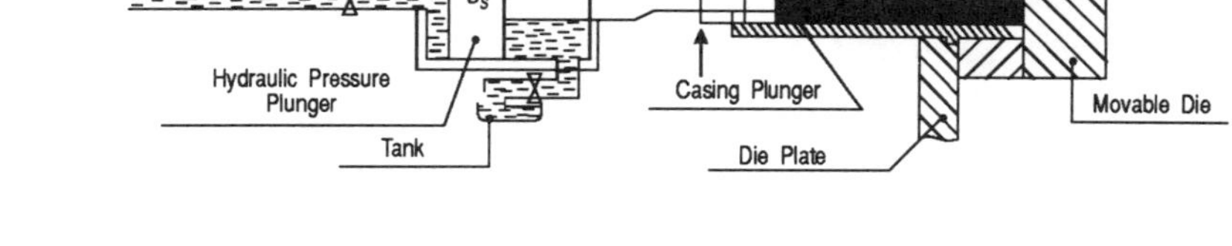

Figure 9.2. Die Casting Machine

The time to fill an aluminum alloy into cavities must be less than the solidifying time, which is functionally related to casting temperature, mold temperature and product thickness, etc. The product discussed in this case has the following conditions:

1. It is considered that the appropriate filling time is 0.04±0.01 (sec). *
2. When the time is shorter than 0.03 (sec), burning occurs and mold repairing cost increases.
3. When the time is slower than 0.05 (sec), molten metal crease defect occurs.

9.2.2 Equations to Determine Filling Time

The filling time for a system with a die casting machine and a mold is calculated by the following equations:

$$X_1 = \frac{P_a S_s}{V^2} \tag{9.1}$$

$$X_2 = \frac{\gamma S_c{}^3}{2gk^2a^2} \tag{9.2}$$

$$X_3 = PS_c \tag{9.3}$$

$$\frac{Q}{tg} \leq S_c \sqrt{\frac{X_1 V^2 - X_3}{X_1 + X_2}} \tag{9.4}$$

where
X_1 = casting loss
P_a = oil plunger cross section area
V = plunger speed during dry cycle
X_2 = gate loss
r = molten metal weight per unit volume
S_c = casting plunger cross section area
g = gravity acceleration
k = gate flow coefficient
a = gate cross section area
X_3 = cavity loss
P = cavity inner pressure
Q = product volume (total volume of product and molten metal)
t_g = solidifying time
t = filling time

* The solidifying time of aluminum, tg, is determined either from experimental data or calculated by the following equation:

$$tg = \frac{0.132\,21\, e^2}{(\theta_a - \theta_m)^2}[48{,}305 + 0.057\,6(\theta_a - 580)^2]$$

where θ_a : Aluminum alloy casting temperature
θ_m: Die temperature
e : Product thickness

However, it is known in this product.

$$t \le tg \tag{9.5}$$

$$t = \frac{Q}{S_c\sqrt{\dfrac{X_1V^2 - X_3}{X_1 + X_2}}} \tag{9.6}$$

Table 9.1 illustrates the selection of factors and levels used in the equations.

Table 9.1. Illustration of Factors and Levels

Factor	Symbol	Meaning of Symbol	Explanation
A	P_a	Inner pressure of accumulator (kgf/cm^2)	P_a and S_s are determined by die casting machine. Changing the average of parameter means selecting the machine. $A = P_aS_s$
	S_s	Oil plunger cross section area (cm^2)	
B	V	Oil press plunger speed during dry cycle (cm/s)	Controllable by adjusting oil circuit valve.
	r	Molten metal unit volume weight (kg/cm^3)	2.564×10^{-3}
C	S_e	Casting plunger cross section area (cm^2)	Some standard chips for casting plungers prepared and selectively used. It is the selection of chips.
	g	Acceleration of gravity	980
D	k	Gate flow coefficient	Depends on the shapes of feeder, land, complexity of inside cavity and wear. 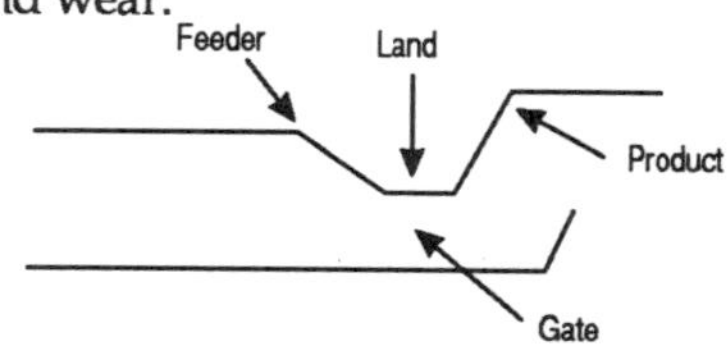
E	a	Gate cross section area (cm^2)	Width depends on product, and thickness restricted by the method of gate cutting after casting, which affect cost.
F	P	Cavity inner pressure (kgf/cm2)	Depends on product shape, complexity, overflow and air vent.
	Q	Product volume (cm^3)	Total volume of product and overflow. 101.4cm^3 in this case.
		Filling time (sec)	Time sufficient for solidifying. 0.04 ±0.01 in this case.

9.2.3 Parameter Design

In parameter design, our objective is to determine the levels of factors *A* through *F* which minimize the effects of inner noise. Factors A through *F* are therefore control factors. In parameter design, the intervals between levels of the inner factors should be as wide as possible, as shown in Table 9.2.

Table 9.2. Factors and Levels

Factor	Level		
	1	2	3
A	4,000 (No. 1 machine)	5,600 (No. 2 machine)	7,200 (No. 3 machine)
B	80	150	220
C	ø40	ø50	ø60
D	0.4	0.6	0.8
E	0.25	0.5	0.75
F	30	45	60

The levels of noise factors are set as follows. If there are outer noise factors in the model, they are also set and included in the study. In this example, only inner noise terms are studied. Based on the control factor levels, noise levels were set as shown in Table 9.3, with explanation in Table 9.4.

Table 9.3. Noise Factors and Levels

Factor	Level (%)		
	1	2	3
A′	-10	0	10
B′	-10	0	10
C′	- 0.4	0	0.4
D′	-10	0	10
E′	-10	0	10
F′	-10	0	10

Control factors *A* through *F* are assigned to columns 2 through 7 of the inner array L_{18}, and noise factors *A′* through *F′* reassigned to columns 2 through 7 of the outer array L_{18}. The layout is shown in Table 9.5.

Table 9.4. Noise Factors and Their Explanation

Factor	Explanation
A′	Varies by temperature and volume change of nitrogen gas in accumulator, and by volume change of oil during pressurizing. The gas is supplied from high pressure bomb after blown through.
B′	Change of oil viscosity and machine contact part due to temperature.
C′	Varies by abrasion and water cooling device. When clearance increases from an initial clearance of 0.05mm to 0.2mm (one side), molten metal backs through the clearance.
D′	Varies by abrasion (generally becomes larger).
E′	Varies by abrasion (cost changes by gate \cutting and after treatment).
F′	Varies by separating agent, amount of cooling water, and mold polishing.

Table 9.5. Layout

Outer Array

No.	1	2 *A′*	3 *B′*	4 *C′*	5 *D′*	6 *E′*	7 *F′*	8
1	1	1	1	1	1	1	1	1
2	1	1	2	2	2	2	2	2
3	1	1	3	3	3	3	3	3
4	1	2						
5					1	3	3	
				2	2	1	1	3
		2	1	2	3	1	3	2
	2	2	2	3	1	2	1	3
15	2	2	3	1	2	3	2	1
16	2	3	1	3	2	3	1	2
17	2	3	2	1	3	1	2	3
18	2	3	3	2	1	2	3	1

Inner Array

	e 1	*A* 2	*B* 3	*C* 4	*D* 5	*E* 6	*F* 7	*e* 8
1	1	1	1	1	1	1	1	1
2	1	1	2	2	2	2	2	2
3	1	1	3	3	3	3	3	3
4	1	2	1	1	2	2	3	3
5	1	2	2	2	3	3	1	1
6	1	2	3	3	1	1	2	2
7	1	3	1	2	1	3	2	3
8	1	3	2	3	2	1	3	1
9	1	3	3	1	3	2	1	2
10	2	1	1	3	3	2	2	1
11	2	1	2	1	1	3	3	2
12	2	1	3	2	2	1	1	3
13	2	2	1	2	3	1	3	2
14	2	2	2	3	1	2	1	3
15	2	2	3	1	2	3	2	1
16	2	3	1	3	2	3	1	2
17	2	3	2	1	3	1	2	3
18	2	3	3	2	1	2	3	1

Noise factor levels are determined as the deviations from control factor levels, so that their actual levels change from inner array conditions. For example, the levels of No.1 of the inner array are:

$$A=4000 \quad B=80 \quad C=40 \quad D=0.4 \quad E=0.25 \quad F=30$$

The noise factor levels of this combination are shown in Table 9.6.

The filling times of No.1 of the inner array and 18 conditions of the outer array are calculated. A portion of the data is shown in Table 9.7.

Table 9.6. Noise Factor Levels of No.1 of Inner Array, $A_1B_1C_1D_1E_1F_1$

Level	1	2	3
A'	$4{,}000 \times 0.9$	4,000	$4{,}000 \times 1.1$
B'	80×0.9	80	80×1.1
C'	$ø\,40 \times 0.996$	ø 40	$ø\,40 \times 1.004$
D'	0.4×0.9	0.4	0.4×1.1
E'	0.25×0.9	0.25	0.25×1.1
F'	30×0.9	30	30×1.1

Table 9.7. Calculated Results of Outer Conditions of No.1 of Inner Array

No. \ Factor · Column	1	A' 2	B' 3	C' 4	D' 5	E' 6	F' 7	8	Filling Time
1	1	1	1	1	1	1	1	1	0.148
2	1	1	2	2	2	2	2	2	0.129
3	1	1	3	3	3	3	3	3	0.114
.	.	.	.	.	.	.	.	.	.
.	.	.	.	.	.	.	.	.	.
.	.	.	.	.	.	.	.	.	.
17	2	3	2	1	3	1	2	3	0.125
18	2	3	3	2	1	2	3	1	0.120

A S/N Ratio is calculated from these results.

$$S_T = 0.148^2 + 0.129^2 + 0.114^2 + \cdots + 0.120^2 = 0.294\,695 \tag{9.7}$$

$$S_m = \frac{(0.148 + 0.129 + \cdots + 0.120)^2}{18} = 0.293\,276 \tag{9.8}$$

$$S_e = 0.001\,419 \tag{9.9}$$

$$V_e = 0.000\,083\,5 \tag{9.10}$$

$$\eta = 10 \log \frac{\frac{1}{18}(S_m - V_e)}{V_e} = 22.9\,(\text{db}) \tag{9.11}$$

S/N Ratios for other inner conditions are calculated as shown in Table 9.8. The results shown in the table were calculated by computer, accounting for the minor differences from the results shown in Equation (9.11).

Table 9.8. S/N Ratio and Sensitivity

No. \ Factor	*e* 1	*A* 2	*B* 3	*C* 4	*D* 5	*E* 6	*F* 7	*e* 8	S/N Ratio (db)	Sensitivity *S* (db)
1	1	1	1	1	1	1	1	1	22.641	-17.882
2	1	1	2	2	2	2	2	2	22.154	-25.962
3	1	1	3	3	3	3	3	3	20.395	-30.192
4	1	2	1	1	2	2	3	3	21.403	-19.088
5	1	2	2	2	3	3	1	1	21.951	-28.245
6	1	2	3	3	1	1	2	2	17.619	-20.244
7	1	3	1	2	1	3	2	3	21.861	-22.750
8	1	3	2	3	2	1	3	1	18.953	-24.030
9	1	3	3	1	3	2	1	2	21.980	-27.953
10	2	1	1	3	3	2	2	1	21.890	-24.064
11	2	1	2	1	1	3	3	2	22.071	-23.735
12	2	1	3	2	2	1	1	3	18.769	-23.742
13	2	2	1	2	3	1	3	2	22.325	-21.590
14	2	2	2	3	1	2	1	3	19.604	-25.585
15	2	2	3	1	2	3	2	1	21.957	-27.715
16	2	3	1	3	2	3	1	2	22.040	-25.813
17	2	3	2	1	3	1	2	3	22.380	-24.140
18	2	3	3	2	1	2	3	1	20.679	-27.419

Sensitivity is given by:

$$S = 10 \log \frac{1}{18}(S_m - V_e) \tag{9.12}$$

9.2.4 ANOVA

In tables 9.9 and 9.10, supplementary tables for the S/N Ratio and sensitivity, respectively, are presented. Analysis of variance tables for S/N Ratio, η, and sensitivity, *S*, are shown in Tables 9.11 and 9.12.

Table 9.9. Level Totals for S/N Ratio

	1	2	3
A	127.920	124.859	127.893
B	132.160	127.113	121.399
C	132.432	127.739	120.501
D	124.475	125.276	130.921
E	122.687	127.710	130.275
F	126.985	127.861	125.826

Table 9.10. Level Totals for Sensitivity

	1	2	3
A	-145.577	-142.467	-152.105
B	-131.187	-151.697	-157.265
C	-140.513	-149.708	-149.928
D	-137.615	-146.350	-156.184
E	-131.628	-150.071	-158.450
F	-149.220	-144.875	-146.054

Table 9.11. ANOVA Table for S/N Ratio

Factor	*f*	*S*	*V*
A	2	1.032 2	0.516 1°
B	2	9.662 2	4.831 1
C	2	12.042 4	6.021 2
D	2	4.114 5	2.057 3
E	2	4.965 9	2.483 0
F	2	0.347 4	0.173 7°
e	5	4.845 1	0.969 02
(*e*)	(9)	(6.224 7)	(0.691 63)
T	17	37.009 7	

Table 9.12. ANOVA Table for Sensitivity

Factor	*f*	*S*	*V*
A	2	8.065 5	4.032 7°
B	2	62.873 3	31.436 6
C	2	9.624 2	4.812 1°
D	2	28.767 7	14.383 9
E	2	62.765 2	31.382 6
F	2	1.683 1	0.841 5°
e	5	12.855 5	2.571 1
(*e*)	(11)	(32.228 3)	(2.929 8)
T	17	186.634 5	

The optimum condition for the S/N Ratio is $A_3B_1C_1D_3E_3F_2$. First, filling time is estimated before the estimation of the S/N Ratio. Since the target is 0.04 seconds, it is desirable to have a sensitivity of:

$$10 \log m^2 = 10 \log 0.04^2 = -27.96 \text{ (db)} \tag{9.13}$$

The sensitivity at the optimum condition is calculated by considering the effects of B_1, D_3 and E_3.

$$\hat{\mu}_s = \frac{-131.187 + (-156.184) + (-158.450)}{6} - 2 \times \frac{-440.149}{18}$$
$$= -25.40 \text{ (db)} \qquad (9.14)$$

This is equivalent to a filling time of 0.054 seconds, which is a little longer than the target. It is necessary to adjust to the target by changing the levels of factors which affect sensitivity. In this example, however, none of the factors affect only sensitivity without affecting the S/N Ratio.

Factors *D* and *E* are shown to have comparatively small effects on the S/N Ratio, but their best levels are selected to be D_3E_3, which are the conditions producing the shorter filling time. Therefore, we have to change the levels of *B* or *C*. If *B* is changed to B_2, the sensitivity is calculated as:

$$\hat{\mu}_s = \frac{-151.697 + (-156.184) + (-158.450)}{6} - 2 \times \frac{-440.149}{18}$$
$$= -28.82 \text{ (db)} \qquad (9.15)$$

It is close enough to the target.

By calculating the filling time of $A_3B_2C_1D_3E_3F_2$, the result is -25 decibels which is longer than the target. Therefore, *C* was changed to C_2 and the optimum condition was determined as $A_3B_2C_2D_3E_3F_2$.

9.2.5 Tolerance Design

The objective of parameter design is to select the best combination of factors. In contrast, the objective of tolerance design is to tighten the variability caused by inner and outer disturbance.

In parameter design $A_3B_2C_2D_3E_3F_2$ was selected as the optimum condition. In order to study the variability when those factor levels vary around their mid values, inner noises of *A* through *F* are used as noise factors. Letting *m* be the mid-value and σ be the standard deviation, the three levels are set at the following values:

1st level: $m - \sqrt{3/2}\,\sigma$

2nd level: m (9.16)

3rd level: $m + \ 3/2$

The standard deviation includes the variability such as using low grade component parts.

If the levels in Table 9.3 were set using Equation (9.16), the levels of *A*, *B*, *C*, *D*, *E*, and *F* are calculated and shown in Table 9.13.

Table 9.13. 3 Levels for Tolerance Design

Level \ Factor	1	2	3
A	6,480	7,200	7,920
B	135	150	165
C	ø 49.8	50	50.2
D	0.72	0.80	0.88
E	0.675	0.75	0.825
F	40.5	45.0	49.5

Filling time was calculated from Table 9.13 to get Table 9.14.

Table 9.14. Filling Time at the Optimum Condition

No. \ Factor	1	*A* 2	*B* 3	*C* 4	*D* 5	*E* 6	*F* 7	8	Filling Time	(Difference from 0.04 second) × 10^4
1	1	1	1	1	1	1	1	1	0.043 5	35
2	1	1	2	2	2	2	2	2	0.038 8	-12
3	1	1	3	3	3	3	3	3	0.035 0	-50
4	1	2	1	1	2	2	3	3	0.042 8	28
5	1	2	2	2	3	3	1	1	0.037 6	-24
6	1	2	3	3	1	1	2	2	0.035 8	-42
7	1	3	1	2	1	3	2	3	0.041 9	19
8	1	3	2	3	2	1	3	1	0.038 2	-18
9	1	3	3	1	3	2	1	2	0.034 5	-55
10	2	1	1	3	3	2	2	1	0.042 2	22
11	2	1	2	1	1	3	3	2	0.039 4	- 6
12	2	1	3	2	2	1	1	3	0.035 7	-43
13	2	2	1	2	3	1	3	2	0.042 6	26
14	2	2	2	3	1	2	1	3	0.038 1	-19
15	2	2	3	1	2	3	2	1	0.035 1	-49
16	2	3	1	3	2	3	1	2	0.041 1	11
17	2	3	2	1	3	1	2	3	0.038 2	-18
18	2	3	3	2	1	2	3	1	0.035 3	-47

Analysis of variance was conducted from the data after subtracting the target of 0.04 seconds. It is recommended that the deviations, be decomposed into their linear effect and residue (quadratic effect). Table 9.15 is a supplementary table. In the table, each result was multiplied by 10^4.

Table 9.15. Level Totals

	1	2	3	Total
A	-54	-80	-108	-242
B	141	-97	-286	-242
C	-65	-81	- 96	-242
D	-60	-83	- 99	-242
E	-60	-83	- 99	-242
F	-95	-80	- 67	-242

$$S_T = 35^2 + (-12)^2 + \cdots + (-47)^2 = 19{,}164 \qquad (f{=}18) \quad (9.17)$$

$$S_m = \frac{[35 + (-12) + \cdots + (-47)]^2}{18} = \frac{(-242)^2}{18} = 3{,}254 \qquad (f{=}1) \quad (9.18)$$

$$S_A = \frac{(-54)^2 + (-80)^2 + (-108)^2}{6} - S_m = 243 \qquad (f{=}2) \quad (9.19)$$

The linear effect of A is calculated by using the coefficients of orthogonal polynomial equation described in Chapters 3 and 8.

$$S_{Al} = \frac{(-A_1 + A_3)^2}{2\times 6} = \frac{[-(-54) + (-108)]^2}{12} = 243 \qquad (f{=}1) \quad (9.20)$$

Therefore,

$$S_{Ares} = S_A - S_{Al} = 0 \qquad (f{=}1) \quad (9.21)$$

S_{Ares} may also be calculated by the following equation:

$$S_{Aq} = \frac{(A_1 - 2A_2 + A_3)^2}{6\times 6}$$

$$= \frac{[(-54) - 2\times(-80) + (-108)]^2}{36} = \frac{(-2)^2}{36} = 0 \qquad (f{=}1) \quad (9.22)$$

Similar calculations are made to obtain ANOVA Table 9.16.

The average of the squares of deviations from the target, V_T, is

$$V_T = \frac{19{,}164}{18} \times 10^{-8} = 0.000\,010\,65 \qquad (9.23)$$

Standard deviation is equal to σ=0.00326 second, or $\pm 3\sigma$=0.0098 second. It is within the specification of 0.04±0.01 second. If variability is to be further reduced, the causes of variation must be reduced.

Table 9.16. ANOVA Table ($\times 10^4$)

Factor		f	S	V	ρ(%)
m		1	3,254	3,254	16.98
A	l	1	243	243	1.26
	res	1	0	0°	
B	l	1	15,194	15,194	79.28
	res	1	67	67	0.34
C	l	1	80	80	0.41
	res	1	0	0°	
D	l	1	127	127	0.66
	res	1	1	1°	
E	l	1	127	127	0.66
	res	1	1	1°	
F	l	1	65	65	0.33
	res	1	0	0°	
e		5	7	1.4	
(e)		(10)	(9)	(0.9)	(0.08)
T		18	19,164		100.00

From the ANOVA table, the effect of factor B, variability due to oil plunger speed during dry cycle, is significant. It is caused by the oil viscosity in the oil press circuit which changes by temperature change. After a die casting machine is selected, plunger speed, V, is measured, and then the speed is controlled by a valve adjustment. If the variability of speed can be reduced by the control to 1/2 in terms of standard deviation, the contribution of the linear term of B becomes $(1/2)^2$ and the quadratic term becomes $(1/2)^4$. In such a case, the mean of the squares of deviations from the target becomes:

$$
\begin{aligned}
V_{T(\text{new})} &= V_T \left[\rho_m + \rho_{Al} + \left(\frac{1}{2}\right)^2 \rho_{Bl} + \left(\frac{1}{2}\right)^4 \rho_{Bq} \right. \\
&\quad \left. + \rho_{Cl} + \rho_{Dl} + \rho_{El} + \rho_{Fl} + \rho_e\right] \times \frac{1}{100} \\
&= \frac{19{,}164}{18} \times 10^{-8} \times [0.169\,8 + 0.012\,6 + \left(\frac{1}{2}\right)^2 \times 0.792\,8 + \left(\frac{1}{2}\right)^4 \\
&\quad \times 0.0034 + 0.004\,1 + 0.006\,6 + 0.006\,6 + 0.0033 + 0.000\,8] \\
&= 1{,}065 \times 10^{-8} \times 0.402\,1 = 0.000\,004\,282 \qquad (9.24)
\end{aligned}
$$

It becomes 1/2.5 in variance and 1/1.58 in standard deviation.

9.2.6 Discussion of Economic Aspects

The filling time is the best at 0.04 seconds. Inner defects occur when it is shorter than 0.03 seconds, and the fluidity of molten metal becomes bad when it is longer than 0.05 seconds. When defects occur, the loss is ¥500 per shot.

The loss due to the results from Equation (9.23) is:

$$L = \frac{A}{\Delta} V_T = \frac{500}{0.01^2} \times 0.000\,010\,65 = 53.25 \text{ (¥/shot)} \tag{9.25}$$

After the speed variability is reduced by 1/2 in the standard deviation by adjusting the speed control valve, the variance of filling time becomes 1/2.5, and its loss is given by:

$$L_{(\text{new})} = \frac{L}{2.5} = 21.3 \text{ (¥/shot)} \tag{9.26}$$

The cost per shot to reduce the standard deviation of speed must be lower than:

$$53.25 - 21.3 = 31.95 \text{ (¥/shot)} \tag{9.27}$$

9.3 CASE WHEN A SIGNAL FACTOR IS ASSIGNED

9.3.1 Layout of Inner and Outer Arrays

The objective function of a product is to assure the stroke of a power cylinder up to 1mm by the input current up to 1 ampare.

There are 9 control factors *A* through *I* for design. The model, which is complicated is not shown here. The stroke can be approximately calculated when *A* through *I* and the input current, *M*, are given.

Table 9.17 shows the 3 levels of control factors *A* through *I* which were set within engineering constraint. The 2nd levels are initial design conditions.

Inner noise factors were assigned in the outer array with current as the signal factor, *M*, as shown in Table 9.18. Noise factor levels were set by the following equation:

$$\begin{cases} \text{1st level:} & m - \sqrt{3/2}\,\sigma \\ \text{2nd level:} & m \\ \text{3rd level:} & m + \sqrt{3/2}\,\sigma \end{cases} \tag{9.28}$$

Where m is the mid value.

Table 9.17. Inner Factor Levels

Factor \ Level	1	2	3
A	6×0.5	6	6×2
B	0.05×0.5	0.05	0.05×2
C	$0.070\,55 \times 0.5$	0.070 55	$0.070\,55 \times 2$
D	16.5×0.5	16.5	16.5×2
E	$0.000\,475 \times 0.5$	0.000 475	$0.000\,475 \times 2$
F	$1.199\,3 \times 0.5$	1.199 3	$1.199\,3 \times 2$
G	0.046×0.5	0.046	0.046×2
H	4.8×0.5	4.8	4.8×2
I	150×0.8	150	150×1.2

Table 9.18. Outer Factor Levels

Factor \ Level	1	2	3
A'	Nominal $\times$ 0.9	Nominal	Nominal $\times$ 1.1
B'	Nominal $\times$ 0.1	Nominal	Nominal $\times$ 1.9
C'	Nominal $\times$ 0.999 3	Nominal	Nominal $\times$ 1.000 7
D'	Nominal $\times$ 0.998 8	Nominal	Nominal $\times$ 1.001 2
E'	Nominal $\times$ 0.5	Nominal	Nominal $\times$ 1.5
F'	Nominal $\times$ 0.999 8	Nominal	Nominal $\times$ 1.000 2
G'	Nominal $\times$ 0.88	Nominal	Nominal $\times$ 1.12
H'	Nominal $\times$ 0.987 5	Nominal	Nominal $\times$ 1.012 5
I'	Nominal $\times$ 0.95	Nominal	Nominal $\times$ 1.05
M	0.1	0.55	1.0

An L_{36} orthogonal array was used for both the inner and outer arrays. Stroke was calculated, and the S/N Ratio for the linear equation was used. Proportional equation was not used since it was clear that no stroke is generated until a certain current value.

9.3.2 ANOVA for S/N Ratios

To calculate the S/N Ratio for each of the inner conditions, level totals of the signal factor, denoted by M_1, M_2 and M_3, are calculated,

$$S_\beta = \frac{(-M_1 + M_3)^2}{12 \times 2} \qquad (f=1) \quad (9.29)$$

$$S_e = (\text{total of squares of 36 data points}) - \frac{(\text{Total})^2}{36} - S_\beta \qquad (f=34) \quad (9.30)$$

$$V_e = \frac{S_e}{34} \tag{9.31}$$

$$\eta = 10 \log \frac{\frac{1}{r}(S_\beta - V_e)}{V_e} \text{ (db)} \tag{9.32}$$

where

$$r = 12Sh^2 = 12 \times 2 \times 0.45^2 \tag{9.33}$$

The S/N Ratios of No.1 through No.36 of the inner array are shown in Table 9.19 in decibel units. Table 9.20 is the supplementary table and Table 9.21 is the ANOVA table.

The estimates of larger effects are as follows:

$\overline{A}_1 = 28.365 \quad \overline{A}_2 = 24.786 \quad \overline{A}_3 = 18.035$

$\overline{B}_1 = 27.016 \quad \overline{B}_2 = 24.664 \quad \overline{B}_3 = 19.505$

$\overline{H}_1 = 18.388 \quad \overline{H}_2 = 24.137 \quad \overline{H}_3 = 28.661$

The process average is calculated using $A_1B_1H_3$.

$$\begin{aligned} \hat{\mu} &= 28.365 + 27.016 + 28.661 - 2 \times \frac{854.226\,2}{36} \\ &= 36.586 \text{ (db)} \end{aligned} \tag{9.34}$$

Current condition, $A_2B_2H_2$ is:

$$\begin{aligned} \hat{\mu}_{\text{current}} &= 24.786 + 24.664 + 24.137 - 2 \times 23.728 \\ &= 26.131 \text{ (db)} \end{aligned} \tag{9.35}$$

The gain is about 10 decibels.

Table 9.19. S/N Ratio

Factor / Column No.	*A* 1	*B* 2	*C* 3	*D* 4	*E* 5	*F* 6	*G* 7	*H* 8	*I* 9	*e* 10	*e* 11	*e* 12	*e* 13	S/N Ratio (db)
1	1	1	1	1	1	1	1	1	1	1	1	1	1	30.204
2	2	2	2	2	2	2	2	2	2	2	2	2	1	26.713
3	3	3	3	3	3	3	3	3	3	3	3	3	1	19.908
4	1	1	1	1	2	2	2	2	3	3	3	3	1	31.688
5	2	2	2	2	3	3	3	3	1	1	1	1	1	30.375
6	3	3	3	3	1	1	1	1	2	2	2	2	1	6.157
7	1	1	2	3	1	2	3	3	1	2	2	3	1	32.138
8	2	2	3	1	2	3	1	1	2	3	3	1	1	20.644
9	3	3	1	2	3	1	2	2	3	1	1	2	1	12.824
10	1	1	3	2	1	3	2	3	2	1	3	2	1	32.004
11	2	2	1	3	2	1	3	1	3	2	1	3	1	16.831
12	3	3	2	1	3	2	1	2	1	3	2	1	1	13.410
13	1	2	3	1	3	2	1	3	3	2	1	2	2	31.497
14	2	3	1	2	1	3	2	1	1	3	2	3	2	12.837
15	3	1	2	3	2	1	3	2	2	1	3	1	2	21.121
16	1	2	3	2	1	1	3	2	3	3	2	1	2	30.443
17	2	3	1	3	2	2	1	3	1	1	3	2	2	26.422
18	3	1	2	1	3	3	2	1	2	2	1	3	2	16.877
19	1	2	1	3	3	3	1	2	2	1	2	3	2	29.824
20	2	3	2	1	1	1	2	3	3	2	3	1	2	26.629
21	3	1	3	2	2	2	3	1	1	3	1	2	2	13.010
22	1	2	2	3	3	1	2	1	1	3	3	2	2	26.476
23	2	3	3	1	1	2	3	2	2	1	1	3	2	19.779
24	3	1	1	2	2	3	1	3	3	2	2	1	2	29.945
25	1	3	2	1	2	3	3	1	3	1	2	2	3	19.877
26	2	1	3	2	3	1	1	2	1	2	3	3	3	30.211
27	3	2	1	3	1	2	2	3	2	3	1	1	3	26.761
28	1	3	2	2	2	1	1	3	2	3	1	3	3	29.787
29	2	1	3	3	3	2	2	1	3	1	2	1	3	25.107
30	3	2	1	1	1	3	3	2	1	2	3	2	3	17.022
31	1	3	3	3	2	3	2	2	1	2	1	1	3	26.602
32	2	1	1	1	3	1	3	3	2	3	2	2	3	31.871
33	3	2	2	2	1	2	1	1	3	1	3	3	3	12.796
34	1	3	1	2	3	2	3	1	2	2	3	1	3	19.835
35	2	1	2	3	1	3	1	2	3	3	1	2	3	30.012
36	3	2	3	1	2	1	2	3	1	1	2	3	3	26.589

Table 9.20. Level Totals

Factor \ Level	1	2	3
A	340.375	297.431	216.420
B	324.188	295.971	234.067
C	286.064	286.211	281.951
D	286.087	280.780	287.359
E	276.782	289.229	288.215
F	289.143	279.156	285.927
G	290.909	291.107	272.651
H	220.651	289.649	343.926
I	285.296	281.373	287.557

Table 9.21. ANOVA Table for S/N Ratio

Factor	f	S	V	ρ(%)
A	2	660.321 5	330.160 8	35.7
B	2	354.130 6	177.090 4	18.9
C	2	0.974 7	0.487 4°	
D	2	2.028 9	1.014 4°	
E	2	7.962 5	3.981 2°	
F	2	4.330 0	2.165 0°	
G	2	19.635 0	9.817 5°	
H	2	636.212 4	318.106 2	34.4
I	2	1.632 0	0.816 0°	
e	17	129.975 5	7.645 6	
(*e*)	(29)	(166.538 5)	(5.742 7)	
T	35	1,817.253 0		

9.3.3 Selection of an Adjustment Factor and Optimum Conditions for the Objective Function

The objective function of this product is that the stroke varies from 0 to 3mm when the current changes from 0.1A to 1A. In other words, the ideal sensitivity is about 10.5 decibels. The analysis of sensitivity is made as follows. Sensitivity corresponds to the numerator of Equation (9.35).

$$10 \log \beta^2 = 10 \log \frac{1}{r}(S_\beta - V_e) \tag{9.36}$$

Table 9.22 shows the sensitivity in decibel. Table 9.23 shows the ANOVA table.

Table 9.22. Sensitivity in Decibel

No. \ Factor · Column	A 1	B 2	C 3	D 4	E 5	F 6	G 7	H 8	I 9	e 10	e 11	e 12	e 13	S/N Ratio (db)
1	1	1	1	1	1	1	1	1	1	1	1	1	1	13.304
2	2	2	2	2	2	2	2	2	2	2	2	2	1	17.085
3	3	3	3	3	3	3	3	3	3	3	3	3	1	20.386
4	1	1	1	1	2	2	2	2	3	3	3	3	1	15.907
5	2	2	2	2	3	3	3	3	1	1	1	1	1	25.375
6	3	3	3	3	1	1	1	1	2	2	2	2	1	3.849
7	1	1	2	3	1	2	3	3	1	2	2	3	1	31.472
8	2	2	3	1	2	3	1	1	2	3	3	1	1	-1.669
9	3	3	1	2	3	1	2	2	3	1	1	2	1	18.592
10	1	1	3	2	1	3	2	3	2	1	3	2	1	17.494
11	2	2	1	3	2	1	3	1	3	2	1	3	1	18.750
12	3	3	2	1	3	2	1	2	1	3	2	1	1	10.829
13	1	2	3	1	3	2	1	3	3	2	1	2	2	9.888
14	2	3	1	2	1	3	2	1	1	3	2	3	2	16.222
15	3	1	2	3	2	1	3	2	2	1	3	1	2	21.162
16	1	2	3	2	1	1	3	2	3	3	2	1	2	9.705
17	2	3	1	3	2	2	1	3	1	1	3	2	2	37.367
18	3	1	2	1	3	3	2	1	2	2	1	3	2	2.341
19	1	2	1	3	3	3	1	2	2	1	2	3	2	29.531
20	2	3	2	1	1	1	2	3	3	2	3	1	2	15.692
21	3	1	3	2	2	2	3	1	1	3	1	2	2	0.084
22	1	2	2	3	3	1	2	1	1	3	3	2	2	18.889
23	2	3	3	1	1	2	3	2	2	1	1	3	2	3.837
24	3	1	1	2	2	3	1	3	3	2	2	1	2	27.940
25	1	3	2	1	2	3	3	1	3	1	2	2	3	2.307
26	2	1	3	2	3	1	1	2	1	2	3	3	3	13.254
27	3	2	1	3	1	2	2	3	2	3	1	1	3	35.186
28	1	3	2	2	2	1	1	3	2	3	1	3	3	23.500
29	2	1	3	3	3	2	2	1	3	1	2	1	3	8.784
30	3	2	1	1	1	3	3	2	1	2	3	2	3	16.448
31	1	3	3	3	2	3	2	2	1	2	1	1	3	19.240
32	2	1	1	1	3	1	3	3	2	3	2	2	3	23.488
33	3	2	2	2	1	2	1	1	3	1	3	3	3	6.374
34	1	3	1	2	3	2	3	1	2	2	3	1	3	15.909
35	2	1	2	3	1	3	1	2	3	3	1	2	3	21.908
36	3	2	3	1	2	1	2	3	1	1	2	3	3	12.928

Table 9.23. ANOVA Table on Sensitivity

Factor	f	S	V	ρ(%)
A	2	44.113 5	22.056 8	1.3
B	2	5.715 0	2.857 5°	
C	2	949.092 1	474.460	29.4
D	2	832.096 0	416.048 0	25.8
E	2	1.392 9	0.696 5°	
F	2	1.181 6	0.590 8°	
G	2	4.041 2	2.020 6°	
H	2	1,285.679 6	642.839 7	39.9
I	2	64.894 5	32.447 3	1.9
e	17	25.878 1	1.522 2	
(e)	(25)	(38.208 0)	(1.528 4)	(1.7)
T	35	3,214.084 8		100.0

The level averages of factors which are significant to sensitivity, A,C, D, H and I, are shown in Table 9.24.

Table 9.24. Level Averages of Significant Factors on Sensitivity

Unit: db

Factor \ Level	1	2	3
A	17.263	16.674	14.676
C	22.387	16.412	9.815
D	10.441	15.962	22.210
H	8.762	16.458	23.394
I	17.951	15.977	14.686

The sensitivity at the optimum condition, $A_1B_1(C_2)(D_2)(E_2)(F_2)(G_2)H_3(I_2)$, is calculated using the factors which are significant on sensitivity, A, C, D, H and I.

$$\hat{\mu} = 17.263 + 16.412 + 15.962 + 23.394 + 15.977 - 4 \times \frac{583.363\,6}{36}$$

$$= 24.190 \text{ (db)} \tag{9.37}$$

This is larger than the ideal sensitivity. Therefore, adjustment is made by selecting the levels of C, D and I which are significant on sensitivity but less effective on S/N Ratio. $C_3D_1I_3$ was selected. The estimate of $A_1C_3H_3I_3$ is:

$$\mu_s = 17.263 + 9.815 + 10.441 + 23.394 + 14.686 - 4 \times \frac{583.363\,6}{36}$$

$$= 10.78 \text{ (db)} \tag{9.38}$$

9.3.4 Tolerance Design

Next, the effects of noise factors *A* through *I* are studied. The level setting was shown in Table 9.18. These are assigned to an L_{36} array and stroke was calculated as shown in Table 9.25.

Table 9.25. Stroke Around the Optimum Condition

Factor · Column	*A*	*B*	*C*	*D*	*E*	*F*	*G*	*H*	*I*	*M*				Stroke
No.	1	2	3	4	5	6	7	8	9	10	11	12	13	(mm)
1	1	1	1	1	1	1	1	1	1	1	1	1	1	0.28
2	2	2	2	2	2	2	2	2	2	2	2	2	1	1.64
3	3	3	3	3	3	3	3	3	3	3	3	3	1	2.90
4	1	1	1	1	2	2	2	2	3	3	3	3	1	2.92
5	2	2	2	2	3	3	3	3	1	1	1	1	1	0.25
6	3	3	3	3	1	1	1	1	2	2	2	2	1	1.60
7	1	1	2	3	1	2	3	3	1	2	2	3	1	1.78
8	2	2	3	1	2	3	1	1	2	3	3	1	1	3.01
9	3	3	1	2	3	1	2	2	3	1	1	2	1	0.20
10	1	1	3	2	1	3	2	3	2	1	3	2	1	0.27
11	2	2	1	3	2	1	3	1	3	2	1	3	1	1.54
12	3	3	2	1	3	2	1	2	1	3	2	1	1	3.17
13	1	2	3	1	3	2	1	3	3	2	1	2	2	1.59
14	2	3	1	2	1	3	2	1	1	3	2	3	2	3.14
15	3	1	2	3	2	1	3	2	2	1	3	1	2	0.25
16	1	2	3	2	1	1	3	2	3	3	2	1	2	2.90
17	2	3	1	3	2	2	1	3	1	1	3	2	2	0.24
18	3	1	2	1	3	3	2	1	2	2	1	3	2	1.63
19	1	2	1	3	3	3	1	2	2	1	2	3	2	0.25
20	2	3	2	1	1	1	2	3	3	2	3	1	2	1.56
21	3	1	3	2	2	2	3	1	1	3	1	2	2	3.17
22	1	2	2	3	3	1	2	1	1	3	3	2	2	3.17
23	2	3	3	1	1	2	3	2	2	1	1	3	2	0.21
24	3	1	1	2	2	3	1	3	3	2	2	1	2	1.61
25	1	3	2	1	2	3	3	1	3	1	2	2	3	0.21
26	2	1	3	2	3	1	1	2	1	2	3	3	3	1.76
27	3	2	1	3	1	2	2	3	2	3	1	1	3	3.08
28	1	3	2	2	2	1	1	3	2	3	1	3	3	3.08
29	2	1	3	3	3	2	2	1	3	1	2	1	3	0.24
30	3	2	1	1	1	3	3	2	1	2	3	2	3	1.71
31	1	3	3	3	2	3	2	2	1	2	1	1	3	1.71
32	2	1	1	1	3	1	3	3	2	3	2	2	3	3.10
33	3	2	2	2	1	2	1	1	3	1	3	3	3	0.22
34	1	3	1	2	3	2	3	1	2	2	3	1	3	1.60
35	2	1	2	3	1	3	1	2	3	3	1	2	3	2.93
36	3	2	3	1	2	1	2	3	1	1	2	3	3	0.25

From Table 9.25, the average stroke at the signal factor levels are calculated.

$\overline{M}_1 = 0.24 \quad \overline{M}_2 = 1.65 \quad \overline{M}_3 = 3.05$ (mm)

From the above results, it is seen that the objective function can be fairly well characterized.

The S/N Ratio is calculated as follows:

$$S_T = 0.28^2 + 1.64^2 + \cdots + 0.25^2 - \frac{(0.28 + 1.64 + \cdots + 0.25)^2}{36}$$

$$= 47.528\,3 \qquad (f{=}35) \quad (9.39)$$

$$S_\beta = \frac{(-M_1 + M_3)^2}{12 \times 2} = 47.376\,6 \qquad (f{=}1) \quad (9.40)$$

$$S_e = 0.151\,7 \qquad (f{=}34) \quad (9.41)$$

$$V_e = \frac{0.151\,7}{34} = 0.004462 \qquad (9.42)$$

$$\eta = 10 \log \frac{\frac{1}{r}(S_\beta - V_e)}{V_e} = 2{,}184.5 = 33.4 \text{ (db)} \qquad (9.43)$$

where $r = 12 \times 2h^2 = 24 \times (0.45)^2 = 4.86$

The magnitude of the error, σ^2, when the objective slope, is equal to 3.12, is calculated by:

$$\eta = 10 \log \frac{3.12^2}{\sigma^2} = 33.4 \qquad (9.44)$$

From this,

$$\sigma^2 = 0.004\,56 \qquad (9.45)$$

The slope was calculated by:

$$\beta = \frac{-M_1 + M_3}{12 \times 2 \times 0.45} = 3.12 \qquad (9.46)$$

To improve the error variance, σ^2=0.00456, is to tighten the inner noise. The ANOVA table which decomposed the inner noise is shown in Table 9.26 where the effect of the signal factor was excluded.

From the table, it is known that the inner effect of I is large. The variability of stroke could be reduced if the variability of I is reduced. However, it was reported that a variability reduction in I is difficult.

Table 9.26. ANOVA Table

Factor		f	S	V	$\rho(\%)$
A	l	1	0.000 058	0.000 058	
	q	1	0.001 545	0.001 545	
B	l	1	0.004 157	0.004 157	
	q	1	0.001 474	0.001 474	
C	l	1	0.000 191	0.000 191	
	q	1	0.003 502	0.003 502	
D	l	1	0.000 113	0.000 113	
	q	1	0.001 482	0.001 482	
E	l	1	0.001 627	0.001 627	
	q	1	0.001 359	0.001 359	
F	l	1	0.000 104	0.000 104	
	q	1	0.002 960	0.002 960	
G	l	1	0.000 426	0.000 426	
	q	1	0.001 425	0.001 425	
H	l	1	0.000 668	0.000 668	
	q	1	0.000 558	0.000 558	
I	l	1	0.137 929	0.137 929	64.96
	q	1	0.000 013	0.000 013	
	e	15	0.049 333	0.003 289	
	(e)	(32)	(0.0709960)	(0.002 219)	35.04
T		33	0.208 928	0.006 331	100.0

EXERCISES

Question (9.1). The relationship between weight, M, and displacement, y, of a weighing scale is given by:

$$y = \beta M$$

where β is the reciprocal of spring constant, K. It is desirable that β be equal to 2.0. β is given by:

$$\beta = \frac{1}{K} = \frac{1}{\dfrac{ABC^3}{3D^2(E - \dfrac{2E^2}{D} + \dfrac{4E^3}{3D^2})} + \dfrac{FG^4}{8H^3I}}$$

where A: longitudinal elastic coefficient of plate spring

B: width of plate spring

C: thickness of plate spring

D: length of plate spring

E: movable length of plate spring
F: crossway elastic coefficient of coil spring
G: coil spring strand diameter
H: average diameter of coil
I: effective number of turns of coil spring

The conditions of current design are:

$A = 20000$ (kgf/mm^2)
$B = 36$ (mm)
$C = 0.2$ (mm)
$D = 50$ (mm)
$E = 2$ (mm)
$F = 8000$ (kgf/mm^2)
$G = 1$ (mm)
$H = 8$ (mm)
$I = 23$ (T)

Three noise levels were set as follows:

(design condition) $-\sqrt{3/2}\,\sigma$
(design condition)
(design condition) $+\sqrt{3/2}\,\sigma$

where σ is the standard deviation of material currently using.

Factor \ Level	1	2	3
A'	Nominal- 3%	Nominal	Nominal+ 3%
B'	Nominal - 1	Nominal	Nominal +1
C'	Nominal - 5	Nominal	Nominal +5
D'	Nominal - 1	Nominal	Nominal +1
E'	Nominal - 5	Nominal	Nominal +5
F'	Nominal - 3	Nominal	Nominal +3
G'	Nominal - 2	Nominal	Nominal +2
H'	Nominal - 5	Nominal	Nominal +5
I'	Nominal - 5	Nominal	Nominal +5

A' through I' were assigned to columns 1 through 9 of an L_{36} orthogonal array, and β was calculated. In the table, a target value of β=2 has been subtracted from each calculated result.

Factor · Column / No.	A	B	C	D	E	F	G	H	I					β - 2.0
	1	2	3	4	5	6	7	8	9	10	11	12	13	
1	1	1	1	1	1	1	1	1	1	1	1	1	1	
2	2	2	2	2	2	2	2	2	2	2	2	2	1	-0.01
3	3	3	3	3	3	3	3	3	3	3	3	3	1	-0.18
4	1	1	1	1	2	2	2	2	3	3	3	3	1	0.28
5	2	2	2	2	3	3	3	3	1	1	1	1	1	0.01
6	3	3	3	3	1	1	1	1	2	2	2	2	1	-0.28
7	1	1	2	3	1	2	3	3	1	2	2	3	1	0.22
8	2	2	3	1	2	3	1	1	2	3	3	1	1	-0.39
9	3	3	1	2	3	1	2	2	3	1	1	2	1	0.21
10	1	1	3	2	1	3	2	3	2	1	3	2	1	-0.12
11	2	2	1	3	2	1	3	1	3	2	1	3	1	0.32
12	3	3	2	1	3	2	1	2	1	3	2	1	1	-0.22
13	1	2	3	1	3	2	1	3	3	2	1	2	2	-0.16
14	2	3	1	2	1	3	2	1	1	3	2	3	2	-0.01
15	3	1	2	3	2	1	3	2	2	1	3	1	2	0.11
16	1	2	3	2	1	1	3	2	3	3	2	1	2	-0.26
17	2	3	1	3	2	2	1	3	1	1	3	2	2	0.49
18	3	1	2	1	3	3	2	1	2	2	1	3	2	-0.17
19	1	2	1	3	3	3	1	2	2	1	2	3	2	0.74
20	2	3	2	1	1	1	2	3	3	2	3	1	2	-0.26
21	3	1	3	2	2	2	3	1	1	3	1	2	2	-0.35
22	1	2	2	3	3	1	2	1	1	3	3	2	2	0.27
23	2	3	3	1	1	2	3	2	2	1	1	3	2	-0.56
24	3	1	1	2	2	3	1	3	3	2	2	1	2	0.40
25	1	3	2	1	2	3	3	1	3	1	2	2	3	-0.28
26	2	1	3	2	3	1	1	2	1	2	3	3	3	-0.03
27	3	2	1	3	1	2	2	3	2	3	1	1	3	0.33
28	1	3	2	2	2	1	1	3	2	3	1	3	3	0.12
29	2	1	3	3	3	2	2	1	3	1	2	1	3	0.04
30	3	2	1	1	1	3	3	2	1	2	3	2	3	-0.19
31	1	3	3	3	2	3	2	2	1	2	1	1	3	-0.11
32	2	1	1	1	3	1	3	3	2	3	2	2	3	0.24
33	3	2	2	2	1	2	1	1	3	1	3	3	3	-0.13
34	1	3	1	2	3	2	3	1	2	2	3	1	3	0.17
35	2	1	2	3	1	3	1	2	3	3	1	2	3	0.25
36	3	2	3	1	2	1	2	3	1	1	2	3	3	-0.43

(a) Calculate the results of condition No.1 and construct an ANOVA table for tolerance design. The following table is a supplementary table from 35 data points, which does not include the result of No.1.

Level \ Factor	*A*	*B*	*C*	*D*	*E*	*F*	*G*	*H*	*I*
1	0.87	0.87	2.98	-2.14	-1.01	0.01	0.79	-0.81	-0.35
2	0.09	0.10	-0.09	0.00	0.15	0.10	0.02	0.21	0.18
3	-0.90	-0.91	-2.83	2.20	0.92	-0.05	-0.75	0.66	0.23
	0.06	0.06	0.06	0.06	0.06	0.06	0.06	0.06	0.06

(b) When the material with 1/2 of current standard deviation is used, how much the average of $(\beta\text{-}2.0)^2$ (square of deviations from the target) would be?

Question (9.2). In order to improve the quality of spring discussed in previous exercise, parameter design was conducted. The levels of control factors *A* through *I* were set as follows (the 2nd level is current level):

Factor \ Level	1	2	3
A	17,000	20,000	22,000
B	30	36	40
C	0.1	0.2	0.3
D	40	50	60
E	1	2	3
F	6,000	8,000	9,000
G	0.5	1	1.5
H	6	8	10
I	15	23	30

Noise factors were set at the same levels as shown in previous exercise.

Control factors were assigned to an L_{36} array and noise factors to an L_{36} as the outer array. β (reciprocal of spring constant, *K*) was calculated from 36×36 combinations. From each of the 36 inner array conditions, the following S/N Ratio was calculated from 36 data points in the outer array:

$$10 \log \eta = 10 \log \frac{\frac{1}{36}(S_m - V_e)}{V_e} \text{ (db)}$$

The results are shown in the following table:

Factor / Column / No.	*A* 1	*B* 2	*C* 3	*D* 4	*E* 5	*F* 6	*G* 7	*H* 8	*I* 9	10	11	12	13	S/N Ratio (db)
1	1	1	1	1	1	1	1	1	1	1	1	1	1	18.4
2	2	2	2	2	2	2	2	2	2	2	2	2	1	
3	3	3	3	3	3	3	3	3	3	3	3	3	1	19.0
4	1	1	1	1	2	2	2	2	3	3	3	3	1	19.8
5	2	2	2	2	3	3	3	3	1	1	1	1	1	19.7
6	3	3	3	3	1	1	1	1	2	2	2	2	1	17.5
7	1	1	2	3	1	2	3	3	1	2	2	3	1	19.9
8	2	2	3	1	2	3	1	1	2	3	3	1	1	17.5
9	3	3	1	2	3	1	2	2	3	1	1	2	1	19.9
10	1	1	3	2	1	3	2	3	2	1	3	2	1	17.6
11	2	2	1	3	2	1	3	1	3	2	1	3	1	17.0
12	3	3	2	1	3	2	1	2	1	3	2	1	1	17.6
13	1	2	3	1	3	2	1	3	3	2	1	2	2	17.5
14	2	3	1	2	1	3	2	1	1	3	2	3	2	18.5
15	3	1	2	3	2	1	3	2	2	1	3	1	2	19.8
16	1	2	3	2	1	1	3	2	3	3	2	1	2	18.3
17	2	3	1	3	2	2	1	3	1	1	3	2	2	18.3
18	3	1	2	1	3	3	2	1	2	2	1	3	2	19.9
19	1	2	1	3	3	3	1	2	2	1	2	3	2	19.3
20	2	3	2	1	1	1	2	3	3	2	3	1	2	17.6
21	3	1	3	2	2	2	3	1	1	3	1	2	2	19.8
22	1	2	2	3	3	1	2	1	1	3	3	2	2	19.6
23	2	3	3	1	1	2	3	2	2	1	1	3	2	18.2
24	3	1	1	2	2	3	1	3	3	2	2	1	2	17.9
25	1	3	2	1	2	3	3	1	3	1	2	2	3	19.6
26	2	1	3	2	3	1	1	2	1	2	3	3	3	17.6
27	3	2	1	3	1	2	2	3	2	3	1	1	3	19.9
28	1	3	2	2	2	1	1	3	2	3	1	3	3	17.5
29	2	1	3	3	3	2	2	1	3	1	2	1	3	19.2
30	3	2	1	1	1	3	3	2	1	2	3	2	3	18.1
31	1	3	3	3	2	3	2	2	1	2	1	1	3	18.6
32	2	1	1	1	3	1	3	3	2	3	2	2	3	18.4
33	3	2	2	2	1	2	1	1	3	1	3	3	3	17.5
34	1	3	1	2	3	2	3	1	2	2	3	1	3	16.8
35	2	1	2	3	1	3	1	2	3	3	1	2	3	17.5
36	3	2	3	1	2	1	2	3	1	1	2	3	3	17.6

(a) Calculate a S/N Ratio from the results in previous exercise.
Note: When calculating the S/N Ratio, be aware that 2 has been subtracted.

(b) The following table is a supplementary table of 35 data points (excluding the S/N Ratio of No.2). Conduct an analysis of variance of the S/N Ratios. Should the nominal values of the current design be changed?

Level \ Factor	*A*	*B*	*C*	*D*	*E*	*F*	*G*	*H*	*I*
1	222.9	225.8	222.3	220.2	219.0	219.2	214.1	221.3	223.7
2	199.5	202.0	206.2	201.1	203.4	204.5	208.2	204.7	202.4
3	224.5	219.1	218.4	225.6	224.5	223.2	224.6	220.9	220.8

CHAPTER 10

LAYOUT TECHNIQUES WHEN MIXED-TYPE ORTHOGONAL ARRAYS CANNOT BE USED

When mixed type orthogonal arrays are used, factors with different numbers of levels can be assigned by using the techniques explained in Chapter 6. In the case when different factors will be assigned under different levels of a factor, it is necessary to utilize interaction columns. In this chapter, the application of the use of general orthogonal arrays with their linear graphs is explained.

10.1 2^n, 3^n SERIES ORTHOGONAL ARRAYS AND THEIR LINEAR GRAPHS

In experimental design, orthogonal arrays such as L_8, L_{16} or L_{27} are frequently used. As explained in Chapter 5, the interaction of two columns is described by a particular column when these orthogonal arrays are used. Various kinds of orthogonal arrays are shown in Table 4 in the Appendix. For 2^n and 3^n series, interaction tables for the allocation of the interactions between two columns, are also presented along with linear graphs which describe the interactions in graphical form for the convenience of application. These are very useful and effective. Among many linear graphs, only standard graphs are shown in the tables.

In an L_8 application, there are two linear graphs. The two linear graphs (a) and (b) in Figure 10.1 indicate the following:

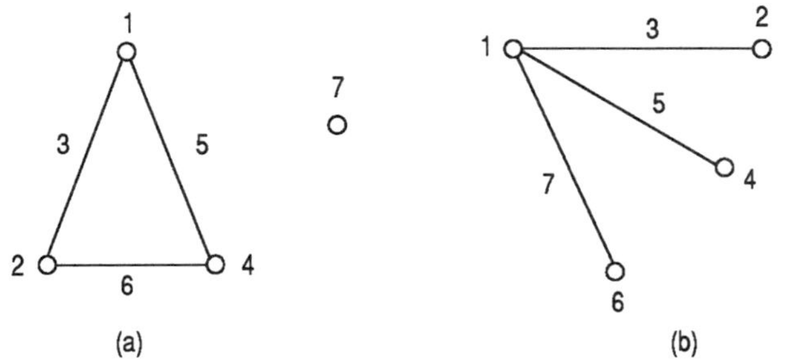

Figure 10.1. Linear Graphs in Orthogonal Array L_8

(1) Linear Graph (a)

The interaction of columns 1 and 2 goes to column 3, and the interaction of columns 1 and 4 goes to column 5. Column 7 is separated from the triangle. When there are four 2-level factors namely *A*, *B*, *C* and *D* which are assigned as:

A to column 1

B to column 2

C to column 3

D to column 7,

then the interactions from these columns are calculated as:

interaction $A \times B$ in column 3

interaction $A \times C$ in column 5

interaction $B \times C$ in column 6

Such a layout is easily made using linear graph (a).

(2) Linear Graph (b)

The linear graph (b) is used when the interactions between a particular factor and some other factors are important. For example, consider an experiment involving two material types, A_1 and A_2, and two annealing methods, B_1 and B_2. To find the best conditions of temperature, *C*, and time, *D*, two levels of these factors are also studied. For each level of *B*, a different furnace will be used, and therefore other conditions may not be the same, so the optimum temperature or time could be different. For material type, on the other hand, a good material will be good no matter what annealing conditions are applied. This implies we need to study its main effect only. Therefore, the effect that we need to study in this experiment are: main effects *A*, *B*, *C*, *D*, and interactions $B \times C$ and $B \times D$.

The required linear graph is shown in Figure 10.2.

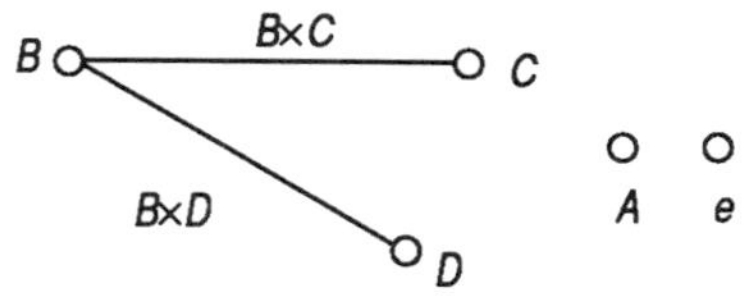

Figure 10.2. Linear Graph (required information)

The layout is easily made with the use of linear graph (b). Two out of the three radical lines are used to assign interactions. The surplus line segment (connecting columns 1 and 6) is removed, and then changed to an independent dot as shown in Figure 10.3. *A* may then be assigned to one of the two dots. The layout is shown in Table 10.1.

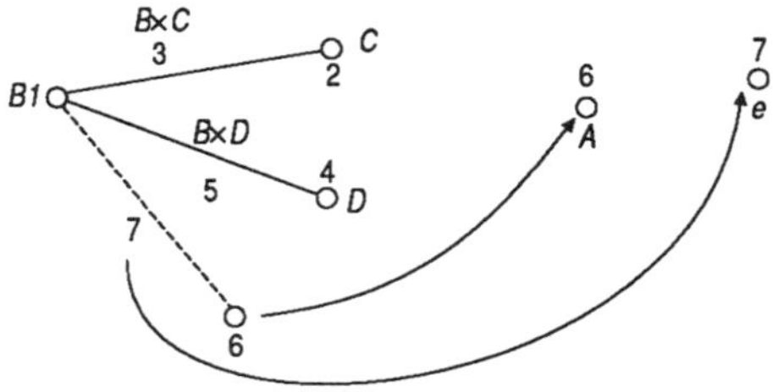

Figure 10.3. Result of Layout

Table 10.1. Table of Layout

Factor · Column / No.	*B*	*C*	*B* × *C*	*D*	*B* × *D*	*A*	*e*
	1	2	3	4	5	6	7
1	1	1	1	1	1	1	1
2	1	1	1	2	2	2	2
3	1	2	2	1	1	2	2
4	1	2	2	2	2	1	1
5	2	1	2	1	2	1	2
6	2	1	2	2	1	2	1
7	2	2	1	1	2	2	1
8	2	2	1	2	1	1	2

The assignment may also be made from linear graph (a). To use linear graph (a), the line segment connecting columns 2 and 4 must be removed, and then changed to an independent dot for the assignment of *A* (see Figure 10.4).

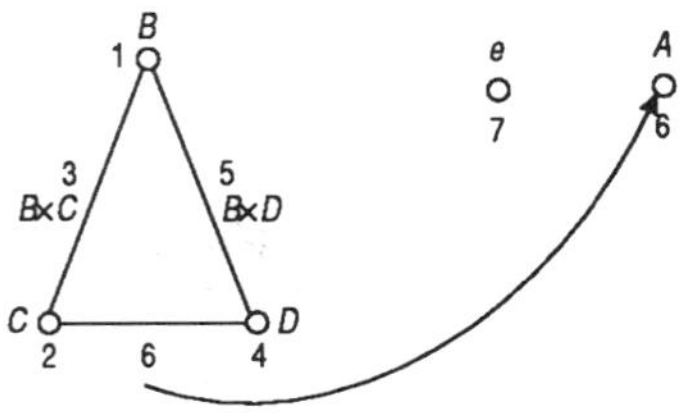

Figure 10.4. Layout Using Linear Graph (a)

10.2 MULTI-LEVEL ARRANGEMENT

10.2.1 Preparation of a 4-level Column

A 4-level factor can be assigned by preparing a 4-level column from orthogonal array L_8.

In an experiment, *A* is a 4-level factor, *B*, *C* and *D* are 2-levels factors. To obtain their main effects, it is necessary to prepare a 4-level column in an L_8. For the preparation of a 4-level column in a 2-level series orthogonal array, it is necessary to use a line segment in the linear graph. For example, consider a line segment in the triangle of Linear Graph (a); a line connected by columns 1 and 2. Remove these three columns from L_8 and insert a 4-level column as shown in Table 10.2 (b). That is to substitute 1, 2, 3 and 4 for the four combinations (1 1), (1 2), (1 3) and (1 4) which are combined by 1 and 2 of columns 1 and 2, then erase those three columns. In other words, a line segment is required to prepare a 4-level column. The arrangement is shown in Tables 10.2 (a), (b) and Figure 10.5.

Table 10.2. Preparation of Orthogonal Array $L_8(4 \times 2^4)$ from Orthogonal Array $L_8\ (2^7)$

(a) $L_8(2^7)$

No. \ Column	1	2	3	4	5	6	7
1	1	1	1(1)	1	1	1	1
2	1	1	1(1)	2	2	2	2
3	1	2	2(2)	1	1	2	2
4	1	2	2(2)	2	2	1	1
5	2	1	2(3)	1	2	1	2
6	2	1	2(3)	2	1	2	1
7	2	2	1(4)	1	2	2	1
8	2	2	1(4)	2	1	1	2

→

(b) $L_8(4 \times 2^4)$

No. \ Column	123	4	5	6	7
1	1	1	1	1	1
2	1	2	2	2	2
3	2	1	1	2	2
4	2	2	2	1	1
5	3	1	2	1	2
6	3	2	1	2	1
7	4	1	2	2	1
8	4	2	1	1	2

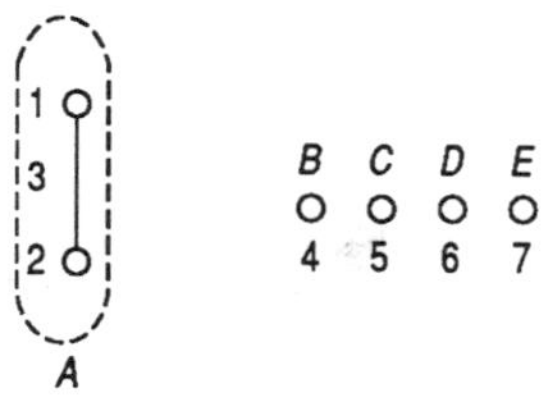

Figure 10.5. Example of the Layout With a 4-level Factor

10.2.2 Preparation of an 8-level Column

An 8-level column with 7 degrees of freedom is made from seven 2-level columns (3 apices, 3 base lines and 1 vertical line). An 8-level column is prepared from a closed triangle (plus an interaction between an apex and its opposite base. An L_{16} is used for the explanation (see Figure 10.6).

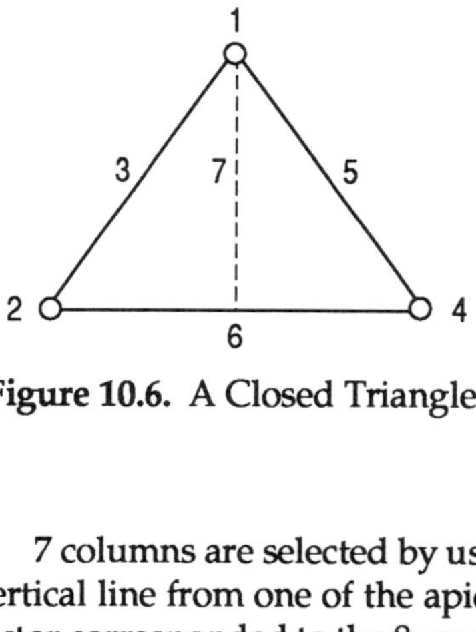

Figure 10.6. A Closed Triangle

Table 10.3. Corresponding 8 Levels

Col. 1	Col. 2	Col. 4	*A*
1	1	1	→1
1	1	2	→2
1	2	1	→3
1	2	2	→4
2	1	1	→5
2	1	2	→6
2	2	1	→7
2	2	2	→8

7 columns are selected by using one of the triangles in an L_{16} plus a vertical line from one of the apices to its opposite base line. An 8-level factor corresponded to the 8 combinations is formed by the 3 apices 1, 2 and 4 of the triangle (see Table 10.3).

From the L_{16}, it is necessary to delete columns 3, 5, 6 and 7 besides deleting columns 1, 2 and 4. After such an arrangement is made, the $L_{16}(2^{15})$ is transformed to an $L_{16}(8x2^8)$. The new orthogonal array, $L_{16}(8x2^8)$, is shown in Table 10.4.

Table 10.4. Orthogonal Array $L_{16}(8 \times 2^8)$ From $L_{16}(2^{15})$

No. \ Original Cols.	1~7	8	9	10	11	12	13	14	15
	1	2	3	4	5	6	7	8	9
1	1	1	1	1	1	1	1	1	1
2	1	2	2	2	2	2	2	2	2
3	2	1	1	1	1	2	2	2	2
4	2	2	2	2	2	1	1	1	1
5	3	1	1	2	2	1	1	2	2
6	3	2	2	1	1	2	2	1	1
7	4	1	1	2	2	2	2	1	1
8	4	2	2	1	1	1	1	2	2
9	5	1	2	1	2	1	2	1	2
10	5	2	1	2	1	2	1	2	1
11	6	1	2	1	2	2	1	2	1
12	6	2	1	2	1	1	2	1	2
13	7	1	2	2	1	1	2	2	1
14	7	2	1	1	2	2	1	1	2
15	8	1	2	2	1	2	1	1	2
16	8	2	1	1	2	1	2	2	1

10.2.3 Preparation of a 9-Level Column From a 3-level Series

A 9-level column can be made from a line segment in a 3-level series linear graph. For example, from a line segment in $L_{27}(3^{13})$, insert 1, 2, 3, 4, 5, 6, 7, 8 and 9 to the combinations of 1, 2 and 3 from

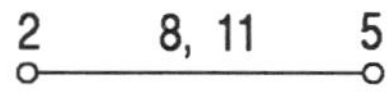

columns 2 and 5, or (11), (12), (13), (21), (22), (23), (31), (32) and(33), then erase columns 2, 5, 8 and 11. A 9-level column with 8 degrees of freedom is thus prepared from four 3-level columns, 2, 5, 8 and 11, with 2 degrees of freedom each.

10.3 Pseudo-Factor Design

10.3.1 What Is Pseudo-factor Design?

Pseudo-factor design is a method that is useful for assigning various type of factors to a single orthogonal array. Pseudo-factor design comprises a method called the trans-factor technique, used in cases where one wishes to experiment on different groups of factors having different levels of a certain factor, and includes the idle-column technique which is used to assign 3-level, 5-level or 7-level factors to a 2-level orthogonal array, or to assign 7-level factors to a 3-level orthogonal array.

When there are one or more "block factors" of a given block size, it is also possible to make the assignment using pseudo-factor design, but this technique is not explained in this book.

10.3.2 Pseudo-Factor Design (1): Trans-Factor Technique

A simple example is used to explain this technique. In the process of tailoring suits, two methods (A) are used to insert the liner are compared:

A_1 = by sewing A_2 = by gluing

The factors to be considered when sewing (A_1) include

B': type of thread

C': sewing method

The factors when gluing include

B'': type of glue

C'': gluing method

This example demonstrates the factors which are studied are different at different levels of A. Beside the above factors, there are common factors D and E with 2-levels each.

In the analysis, $A\times B$, which is the interaction between A and B, is decomposed to

Source	Degrees of Freedom
A	1
B	1
$A\times B$	1
T	3

The purpose of such decomposition is to expect that the interaction is non-existent or not significant.

If the effect of B is quite different at A_1 and A_2, then the effect of B should be compared at different levels of A as shown below:

Source	Degrees of Freedom
A	1
B (at A_1)	1
B (at A_2)	1
T	3

Since the effect of B is determined at different level of A, then factor B may be different at different level of A.

For the tailoring example, the factor B at A_1 is the 2-level factor B', the type of thread, and at A_2, is the 2-level factor, B'', the type of glue. This is not a problem as long as the decomposition is made as follows:

Source	Degrees of Freedom
A	1
B' (at A_1)	1
B'' (at A_2)	1
T	3

Factor B becomes B' at A_1, and becomes at B'' at A_2. Such a factor is called a pseudo-factor, and such assignment is called trans-factor technique.

In order to study the effects of B at A_1 and A_2, it is necessary to secure the interaction between A and pseudo-factor B. The assignment is shown in Figure 10.7.

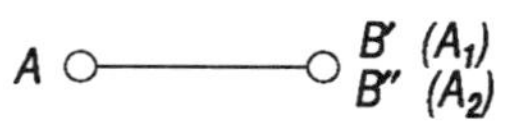

Figure 10.7. Preparation of Trans-factor

For the assignment of C' and C'', the interaction between A and pseudo-factor C is secured. For A, B and D, the linear graph shown in Figure 10.8 is secured.

For factors C and D which are common for $A1$ and $A2$, an independent column is used respectively. Using standard linear graph (1) of the orthogonal array L_8 in Appendix 4, the assignment is shown in Figure 10.9. The actual assignment is shown in Table 10.5. It is understood that the interaction columns secured are not used to assign any factor levels.

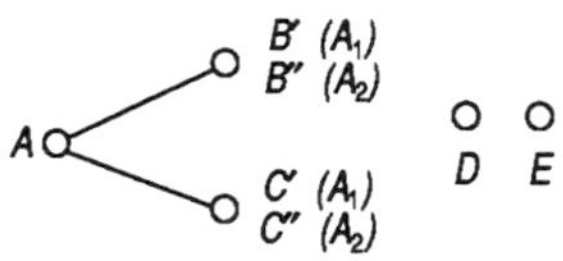

Figure 10.8. Required Linear Graph

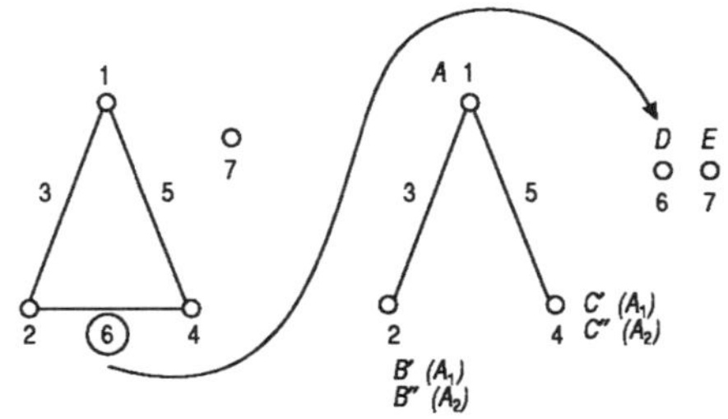

Figure 10.9. Modification of Standard Graph

Table 10.5. Assignment of Trans-factor Technique

No. \ Factor · Column	A 1	B′B″ 2 3	C′C″ 4 5	D 6	E 7
		B′	C′		
1	1	1	1	1	1
2	1	1	2	2	2
3	1	2	1	2	2
4	1	2	2	1	1
		B″	C″		
5	2	1	1	1	2
6	2	1	2	2	1
7	2	2	1	2	1
8	2	2	2	1	2

Since such assignment requires interaction columns, mixed type orthogonal arrays cannot be used, with the exception case of an L_{18} where the interaction between columns 1 and 2 can be assigned without sacrificing any columns.

10.4 MACHINE PROCESSING EXPERIMENT USING TRANS-FACTOR TECHNIQUE

10.4.1 Factors and Levels

In order to study cutting precision in a minute dimensional cutting process, the tool, blade attaching condition, tool shape, and cutting condition are investigated. The types of tools under consideration include super-hard tool, sintered diamond tool and natural diamond tool. The natural diamond tool has been lapped, and hence there is no way to change the blade attachment method or tool shape. Therefore, the blade attachment and tool shape cannot be studied for a natural diamond tool. These factors are reserved for the super-hard tool and sintered diamond tool.

The factors associated with the tool are classified into two groups:

A_1 = super-hard tool and sintered diamond tool which require a change of blade attachment method etc.

A_2 = natural diamond tool

To compare super-hard tool and sintered diamond tool, B is set as a factor only at A_1.

B_1 = super-hard tool

B_2 = sintered diamond tool

Besides B, there are other factors which are studied only under A_1:

C: tool lapping stone	4 levels
D: tool raking angle	2 levels
E: tool approach angle	2 levels

Cutting conditions are common factors for both A_1 and A_2:

F: cutting speed	2 levels
G: feed	2 levels

The factors listed above are inner factors. For the outer factors, the depth of cut is used as the signal factor, M, with 3-levels.

$M_1 = 5 \quad M_2 = 25 \quad M_3 = 50 \quad (\mu m)$

For noise factors, environmental condition, I, and measured direction, J, were studied.

I_1 = with air conditioning I_2 = without air conditioning

J_1 = X direction J_2 = y direction

The material used in the experiment was aluminum.

10.4.2 Layout and S/N Ratio

B through E, the factors only for A_1, are assigned using a trans-factor technique. In order to change factors at different levels of A, it is necessary to have the columns of interactions with A. In this example, we need to secure the interactions between A and B, A and C, A and D, plus A and E. B is a 4-level factor, and a multi-level arrangement is applied. The information required in this example is expressed in Figure 10.10, the required linear graph.

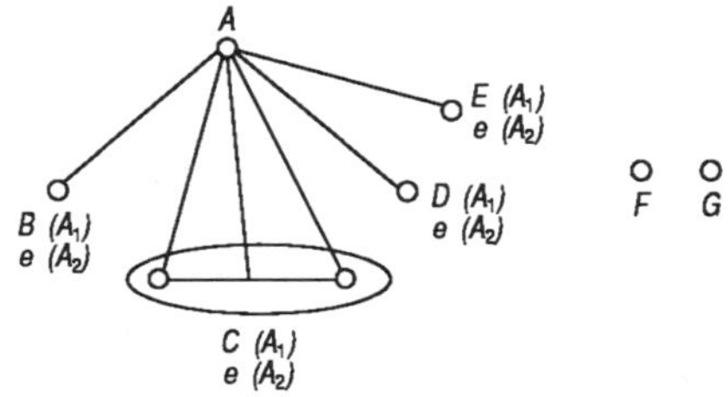

Figure 10.10. Required Linear Graph

From the graph, it is apparent that the required number of degrees of freedom is 15; therefore, an L_{16} orthogonal array is used. The linear graph of L_{16} in appendix 4, (1)a is modified, as shown in Figure 10.11 and Table 10.6.

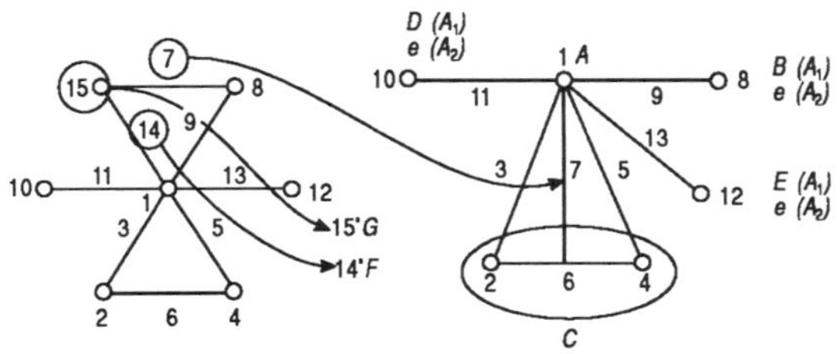

Figure 10.11. Modification of Standard Graph

Table 10.6. Layout of Experiment

No. \ Factor · Column	A 1	C 246357	B 8 9	D 10 11	E 12 13	F 14	G 15	S/N Ratio η (db)
1	1	1	1	1	1	1	1	15.2
2	1	1	2	2	2	2	2	15.2
3	1	2	1	1	2	2	2	16.6
4	1	2	2	2	1	1	1	18.3
5	1	3	1	2	1	2	2	16.6
6	1	3	2	1	2	1	1	19.7
7	1	4	1	2	2	1	1	19.7
8	1	4	2	1	1	2	2	23.7
		e	*e*	*e*	*e*			
9	2	1	1	1	1	1	2	23.4
10	2	1	2	2	2	2	1	22.6
11	2	2	1	1	2	2	1	26.8
12	2	2	2	2	1	1	2	23.9
13	2	3	1	2	1	2	1	23.2
14	2	3	2	1	2	1	2	21.1
15	2	4	1	2	2	1	2	23.5
16	2	4	2	1	1	2	1	22.7

10.4.3 S/N Ratios of Results from Outer Array

For each experiment number in the inner array, the depth of cutting test pieces was changed, and the actual amount of cut measured in the x and y directions under I_1 (with air conditioning) and I_2 (without air conditioning) was measured. The experimental results for inner condition No.1 is shown in Table 10.7.

Table 10.7. Actual Amount of Cut

	M_1		M_2		M_3	
	I_1	I_2	I_1	I_2	I_1	I_2
X	0.3	0.25	2.1	2.1	4.6	4.65
Y	0.2	0.3	2.1	2.25	4.65	4.5
	1.05		8.55		18.4	

The S/N Ratio of the proportional equation is calculated:

$$S_T = 0.3^2 + 0.25^2 + \cdots + 4.5^2 = 103.23 \qquad (f=12) \quad (10.1)$$

$$S_\beta = \frac{(0.5 \times 1.05 + 2.5 \times 8.55 + 5 \times 18.4)^2}{4 \times (0.5^2 + 2.5^2 + 5^2)}$$

$$= 102.96 \qquad (f=1) \quad (10.2)$$

$$S_e = 0.027 \qquad (f=11) \quad (10.3)$$

$$V_e = 0.024\,5 \qquad (10.4)$$

$$\eta = 10 \log \frac{102.96 - 0.0245}{4(0.5^2 + 2.5^2 + 5^2) \times 0.024\,5}$$

$$= 15.2 \text{ (db)} \qquad (10.5)$$

Similar calculations were made and the S/N Ratios in Table 10.6 were obtained.

10.4.4 Analysis of Variance for the Trans-factor Technique

As a supplementary table, level totals are calculated as shown in Table 10.7. Since B through E are the factors for A_1 only, the level totals are calculated from the experiment number under A_1. In the tailoring process example of the previous section, different factors existed for A_1 and A_2. But in this example, none of the factors, corresponding for B through E under A_2. From the results under A_2, the error can be calculated. For checking the calculation, level totals of B, C, D and E when A is at level 2 are calculated.

Table 10.8. Supplementary Table

Factor / Level	For A_1 Only						
	A	C	B	D	E	F	G
1	145.2	30.4	68.1	75.2	73.8	164.8	168.2
2	187.2	34.9	76.9	69.8	71.2	167.4	164.0
3	—	36.3	—	—	—	—	—
4	—	43.4	—	—	—	—	—

Factor / Level	For A_2 Only				Total
	246	8	10	12	
1	46.0	96.9	94.0	93.2	
2	50.7	90.3	93.2	94.0	
3	44.3	—	—	—	
4	46.2	—	—	—	

Variations are calculated as follows:

$$S_T = 15.2^2 + 15.2^2 + \cdots + 23.5^2 - \frac{332.2^2}{16} = 187.42 \qquad (f=15) \quad (10.6)$$

$$S_A = \frac{(A_1 - A_2)^2}{16} = \frac{(145.0 - 187.2)^2}{16} = 111.30 \qquad (f=1) \quad (10.7)$$

Next, S_B, S_C, S_D and S_E are calculated from the data under A_1.

$$S_B = \frac{(B_1 - B_2)^2}{8} = \frac{(68.1 - 76.9)^2}{8} = 9.68 \qquad (f=1) \quad (10.8)$$

$$S_C = \frac{C_1{}^2 + C_2{}^2 + C_3{}^2 + C_4{}^2}{\text{No. of replicationsof } C_i} - \frac{A_1{}^2}{8}$$

$$= \frac{30.4^2 + 34.9^2 + 36.3^2 + 43.4^2}{2} - \frac{145.0^2}{8}$$

$$= 43.59 \qquad (f=1) \quad (10.9)$$

$$S_D = \frac{(D_1 - D_2)^2}{8} = 3.65 \qquad (f=1) \quad (10.10)$$

$$S_E = \frac{(E_1 - E_2)^2}{8} = 0.84 \qquad (f=1) \quad (10.11)$$

S_F and S_G are calculated from all data.

$$S_F = \frac{(F_1 - F_2)^2}{16} = \frac{(164.8 - 167.4)^2}{16} = 0.42 \qquad (f=1) \quad (10.12)$$

$$S_G = 1.10 \qquad (f=1) \quad (10.13)$$

S_e is calculated by subtraction from S_T.

$$S_e = S_T - S_A - S_B - S_C - S_D - S_E - S_F - S_G$$
$$= 16.84 \qquad (f=6) \quad (10.14)$$

This must be equal to the total of the four variation terms under A_2. It is checked as follows:

$$S_{col.2\text{-}4\text{-}6} = \frac{46.0^2 + 50.7^2 + 44.3^2 + 46.2^2}{2} - \frac{187.2^2}{8}$$
$$= 11.23 \qquad (f=3) \quad (10.15)$$

$$S_{col.8} = \frac{(96.9 - 90.3)^2}{8} = 5.44 \qquad (f=1) \quad (10.16)$$

$$S_{col.10} = \frac{(93.2 - 94.0)^2}{8} = 0.08 \qquad (f=1) \quad (10.17)$$

$$S_{col.12} = 0.08 \qquad (f=1) \quad (10.18)$$

From the above,

$$S_e = S_{col.2\text{-}4\text{-}6} + S_{col.8} + S_{col.10} + S_{col.12} = 16.83 \qquad (f=6) \quad (10.19)$$

Table 10.9 is the analysis of variance table.

Table 10.9. ANOVA Table

Factor	f	S	V	ρ(%)
A	1	111.30	111.30	58.2
B	1	9.68	9.68	3.9
C	3	43.59	14.53	19.6
D	1	3.65	3.65°	
E	1	0.84	0.84°	
F	1	0.42	0.42°	
G	1	1.10	1.10°	
e	6	16.84	2.807	
(e)	(10)	(22.85)	(2.285)	(18.3)
T	15	187.42		100.0

The pooled error, (e), is calculated by putting the factors which are either smaller or similar to the error together with the error. Degrees of contribution is calculated as follows:

$$\rho_A = \frac{S_A - f_A V_{(e)}}{S_T} \times 100 = \frac{111.30 - 2.285}{187.42} \times 100$$
$$= 58.2\ (\%) \qquad (10.20)$$

$$\rho_B = \frac{9.68 - 2.285}{187.42} \times 100 = 3.9\ (\%) \qquad (10.21)$$

$$\rho_C = \frac{S_C - f_C V_{(e)}}{S_T} \times 100$$
$$= \frac{43.59 - 3 \times 2.285}{187.42} \times 100 = 19.6\ (\%) \qquad (10.22)$$

$$\rho_{(e)} = \frac{S_{(e)} + (f_A + f_B + f_C) V_{(e)}}{S_T} \times 100$$
$$= \frac{22.85 + 5 \times 2.285}{187.42} \times 100 = 18.3\ (\%) \qquad (10.23)$$

10.4.5 Estimation

A, B and C are the factors of the large effects.

$$\overline{A}_1 = \frac{145.0}{8} = 18.125 \qquad (10.24)$$

$$\overline{A}_2 = \frac{187.2}{8} = 23.4 \qquad (10.25)$$

$$\overline{B}_1 = \frac{68.1}{4} = 17.025 \qquad (10.26)$$

$$\overline{B}_2 = \frac{76.9}{4} = 19.225 \qquad (10.27)$$

$$\overline{C}_1 = \frac{30.4}{2} = 15.2 \qquad (10.28)$$

$$\overline{C}_2 = \frac{34.9}{2} = 17.42 \qquad (10.29)$$

$$\overline{C}_3 = \frac{36.3}{2} = 18.15 \qquad (10.30)$$

$$\overline{C}_4 = \frac{43.4}{4} = 21.7 \qquad (10.31)$$

By comparing $\overline{A}_1$ and $\overline{A}_2$, it seems that A_2 is better. However, they must be compared based on their respective optimum conditions. A_2 is used for comparison since there is no factors under the level. For A_1, its optimum condition is $B_2C_4D_1E_1$. Among the factors under this condition, B and C have large effects; its estimate is:

$$\hat{\mu}(A_1B_2C_4) = \overline{A}_1 + (\overline{B}_2 - \overline{A}_1) + (\overline{C}_4 - \overline{A}_1) = \overline{B}_2 + \overline{C}_4 - \overline{A}_1$$
$$= 19.225 + 21.7 - 18.125 = 22.8\ \text{(db)} \qquad (10.32)$$

Comparing the two conditions, A_2 is slightly better than $A_1B_2C_4$, but the difference is very small. In other words, by selecting a better grinding stone, sintered diamond has equivalent process ability with natural diamond. Calculation of confident limits is not explained here. For the calculation, [see References 1) and 2)].

10.5 Pseudo-Factor Design (2): Idle Column Technique

In Section 10.3, the layout for the case when the type of factors under different levels of another factor are different was discussed. In this section, the layout technique for assigning many 3-level factors to a 2-level orthogonal arrays explained. For the case when there are factors with more number of levels, the reader should refer to Section 10.7.

Let B_1, B_2 and B_3 denote the 3 levels of factor B. The levels of pseudo-factor B' are B_1 and B_2 at A_1 and B_2 and B_3 at A_2.If the interaction column between factor A and pseudo-factor B', namely, column $A \times B'$, is deleted from the orthogonal array, then the main effect of B' can be determined from levels A_1 and A_2, respectively. That is, the difference of B_1 and B_2 from A_1 and the difference of B_2 and B_3 from A_2 will be calculated. B_2 is common in A_1 and A_2, and is called the contrast level. The degrees of freedom are distributed as follows:

Source	Degrees of Freedom
A	1
B_1B_2(from A_1)	1
B_2B_3 (from A_2)	1
T	3

One half of the difference between B_1 and B_3 is included in A, the difference between A_1 and A_2. The difference between B_1 and B_2 is determined from the data at A_1, and the difference between B_2 and B_3 can be determined from the data at A_2. From this result, the difference between B_1 and B_3 can be calculated, and thus, the effect of A can be determined by adjusting the effect of B. However, we should try to avoid adjustment as much as possible.

For this reason, the column belonging to factor A and the interaction column with A are not assigned with any factor. Such an empty column is called an "idle column". Of course the idle column may be assigned with a factor for which there is no room for elsewhere. It is a good idea to utilize the idle column by assigning the factor such as a blocking factor, which is useless.

Suppose *A*, *B* are 3-level and *C*, *D* are 2-level factors. The layout using idle-column technique is made as shown in Figure 10.21.

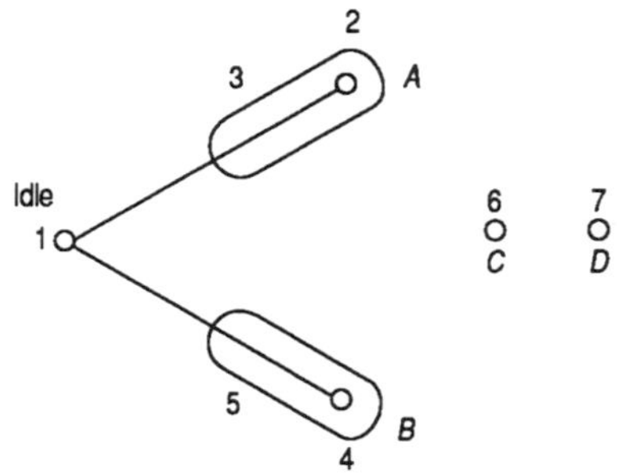

Figure 10.12. Linear Graph for Idle Column Technique

Table 10.10 shows the assignment corresponding to an orthogonal array. Distribution of the degrees of freedom is shown in Table 10.11.

Table 10.10. Layout by Idle Column

Factor · Column / No.	Idle	*A*	B	C	D
	1	2 3	4 5	6	7
1	1	1	1	1	1
2	1	1	2	2	2
3	1	2	1	2	2
4	1	2	2	1	1
5	2	2	2	1	2
6	2	2	3	2	1
7	2	3	2	2	1
8	2	3	3	1	2

Table 10.11. Factors and Distribution of Degrees of Freedom

Source	Degrees of Freedom
Idle	1
(A_1A_2	1
A_2A_3	1
(B_1B_2	1
B_2B_3	1
C	1
D	1
e	0
T	7

The analysis is described next.

10.6 Numerical Example for Idle-Column Technique

(1) Layout and Results

The specification of bearing wear of a component part is 0.05 mm or less. Since current products do not satisfy the specification, the following factors were studied:

A(anti-abrasion agent): A_1(current) A_2(SPO) A_3(NTO)
B(bearing material): B_1(steel) B_2(bronze) B_3(aluminum)
C(rpm): C_1(6000) C_2(2000)
D(load): D_1(300gf) D_2(500gf)

Table 10.12 shows the layout using the idle-column technique. Comparing this layout with the case when L_9 is used, there is only one experiment less than using L_9, indicating that there is not too much merit of using idle-column technique. However, when there are seven 3-level factors, it requires an L_{27} experiments when L_{27} is used. By applying the idle-column techniques, an L_{16} can be used as shown in Figure 10.13.

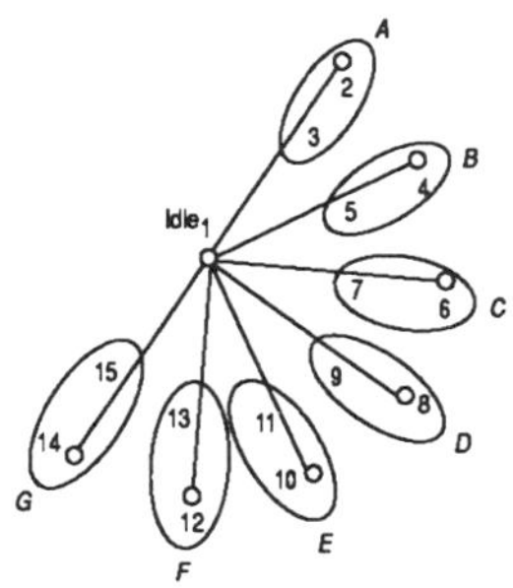

Figure 10.13. Linear Graph of 3^7 Experiment Using L_{16}

From each experiment, 3 test pieces were used to measure the amount of wear. Table 10.12 shows the S/N Ratios in decibels of a smaller-the-better characteristic given by

$$\eta = -10 \log \frac{y_1^{\,2} + y_2^{\,2} + y_3^{\,2}}{3} \qquad (10.33)$$

Table 10.12. Layout and Results [Using $L_8(2^7)$]

Factor / No.	Idle 1	A 2 3	B 4 5	C 6	D 7	Amount of Wear (mm)			$\sigma^2 = \frac{1}{3}\Sigma y^2$	-10 log σ^2 (db)
1	1	1	1	1	1	0.16	0.14	0.15	0.022 6	16.5
2	1	1	2	2	2	0.18	0.13	0.20	0.029 8	15.2
3	1	2	1	2	2	0.04	0.05	0.02	0.001 5	28.2
4	1	2	2	1	1	0.12	0.18	0.17	0.025 2	16.0
5	2	3	3	1	2	0.06	0.04	0.01	0.001 8	27.4
6	2	3	2	2	1	0.02	0.08	0.05	0.003 1	25.1
7	2	1	3	2	1	0.21	0.17	0.14	0.030 9	15.1
8	2	1	2	1	2	0.11	0.18	0.17	0.024 5	16.1

(2) Analysis of Variance of S/N Ratio

From the results in Table 10.12, level totals are calculated as shown in Table 10.13.

Table 10.13. Level Totals

Level \ Factor	*A*		*B*		Total
(Idle)$_1$	A_1	31.7	B_1	44.7	75.9
	A_2	44.2	B_2	31.2	
(Idle)$_2$	A_1	31.2	B_2	41.2	83.7
	A_3	52.5	B_3	42.5	
Total	159.6		159.6		

Level \ Factor	*C*	*D*
1	76.0	72.7
2	83.6	86.9
Total	159.6	159.6

Variations are calculated as follows:

$$S_T = 16.5^2 + 15.2^2 + \cdots + 16.1^2 - \frac{159.6^2}{8}$$

$$= 238.50 \qquad (f{=}7) \quad (10.34)$$

$$S_{\text{Idle}} = \frac{(75.9 - 83.7)^2}{8} = 7.61 \qquad (f=1) \quad (10.35)$$

$$S_{A_1A_2} = \frac{(31.7 - 44.2)^2}{4} = 39.06 \qquad (f=1) \quad (10.36)$$

$$S_{A_1A_3} = \frac{(31.2 - 52.5)^2}{4} = 113.42 \qquad (f=1) \quad (10.37)$$

$$S_{B_1B_2} = \frac{(44.7 - 31.2)^2}{4} = 45.56 \qquad (f=1) \quad (10.38)$$

$$S_{B_2B_3} = \frac{(41.2 - 42.5)^2}{4} = 0.42 \qquad (f=1) \quad (10.39)$$

$$S_C = \frac{(76.0 - 83.6)^2}{8} = 7.22 \qquad (f=1) \quad (10.40)$$

$$S_D = \frac{(72.7 - 86.9)^2}{8} = 25.21 \qquad (f=1) \quad (10.41)$$

Table 10.14 is the analysis of variance table.

Table 10.14. ANOVA Table

Factor	f	S	V
Idle	1	7.61	7.61°
A A_1A_2	1	39.06	39.06
A A_1A_3	1	113.42	113.42
B B_1B_2	1	45.56	45.56
B B_2B_3	1	0.42	0.42°
C	1	7.22	7.22°
D	1	25.21	25.21
(e, pooling o)	(3)	(15.25)	(5.083)
T	7	238.50	

Often in the analysis of variance for the S/N Ratio, there are no degrees of freedom for the error, or it is very small. In such a case, the smaller factorial effects are pooled and the significance is determined by individual judgment. In this example, A_1A_2, A_1A_3, B_1B_2 and D are determined as significant.

(3) Estimation

When an idle column technique is used, the correction is made sometimes by using the contrast level as a base. For example, the contrast level of A, A_1, is estimated as the average of 1 and 2 of the idle column.

$$\overline{A_1} = \frac{31.7 + 31.2}{4} = 15.72 \tag{10.42}$$

A_2 and A_3 are estimated by:

$$\overline{A_2} = \frac{44.2}{2} - \frac{1}{2}(31.7 - 31.2) = 21.8 \tag{10.43}$$

$$\overline{A_3} = \frac{52.5}{2} + \frac{1}{2}(31.7 - 31.2) = 26.5 \tag{10.44}$$

For factor B, only B_1B_2 is significant, and B_2B_3 is not. Therefore, $\overline{B_2}$ and $\overline{B_3}$, which are calculated from the 2nd level of the idle column, are considered as the estimates of B_2. These are calculated together with $\overline{B_2}$, the average calculated from the 1st level of the idle column, to be the estimate of B_2.

$$\begin{aligned}\overline{B_2} &= \frac{1}{2}\left(\frac{31.2}{2} + \frac{42.2 + 42.5}{4}\right) = \frac{1}{2}(15.60 + 20.92) \\ &= 18.26\end{aligned} \tag{10.45}$$

B_1 is estimated by correcting the difference of the contrast level.

$$\begin{aligned}\overline{B_1} &= \frac{44.7}{2} - \frac{1}{2}(15.60 - 20.92) \\ &= 22.35 - (-2.66) = 25.01\end{aligned} \tag{10.46}$$

The estimation of D has nothing to do with the idle column.

$$\overline{D_1} = \frac{72.2}{4} = 18.18 \tag{10.47}$$

$$\overline{D_2} = \frac{86.9}{4} = 21.72 \tag{10.48}$$

The confidence limits is calculated by

$$\pm\sqrt{F \times V_e \times \frac{1}{n_e}} \tag{10.49}$$

where D is the number of units and n_e=1/D. The calculation is not shown.

(4) Optimum Condition and Quality Improvement

Condition $A_3B_1(C_2)D_2$ is considered to give the highest S/N Ratio. The S/N Ratio of this condition is:

$$\eta = \overline{T} + (\overline{A}_3 - \overline{T}) + (\overline{B}_1 - \overline{T}) + (\overline{D}_2 - \overline{T})$$
$$= \overline{A}_3 + \overline{B}_1 + \overline{D}_2 - 2\overline{T}$$
$$= 26.50 + 25.01 + 21.72 - 2 \times \frac{159.6}{8}$$
$$= 33.33 \quad \text{(db)} \tag{10.50}$$

Assuming a current condition of $A_1B_1(C_1)D_1$, its S/N Ratio, denoted by η_0, is:

$$\eta_0 = \overline{A}_1 + \overline{B}_1 + \overline{D}_1 - 2\overline{T}$$
$$= 15.72 + 25.01 + 18.18 - 2 \times 19.95$$
$$= 19.01 \quad \text{(db)} \tag{10.51}$$

The specification of the amount of bearing wear is 0.05mm or less and the loss being out of specification is ¥1,000. Since it is smaller-the-better characteristic, its loss is:

$$L = \frac{A}{\Delta^2}\sigma^2 = \frac{1{,}000}{0.05^2}\sigma^2 \tag{10.52}$$

The mean squared error of the amount of wear, σ^2, under the optimum condition, $A_3B_1D_2$, is:

$$-10 \log \sigma^2 = 33.33 \quad \text{(db)} \tag{10.53}$$

$$\sigma^2 = 0.000464\,5 \tag{10.54}$$

The mean squared error of the amount of wear, $\sigma_0{}^2$, under current condition is:

$$-10 \log \sigma_0{}^2 = 19.01 \quad \text{(db)} \tag{10.55}$$

$$\sigma_0{}^2 = 0.01256 \tag{10.56}$$

The losses of these two conditions are:

$$L_1 = \frac{1{,}000}{0.05^2} \times \sigma^2 = 400{,}000 \times 0.000\,464\,5 = 185.8 \text{ (¥/pc)} \tag{10.57}$$

$$L_2 = 400{,}000 \times 0.01256 = 5{,}024 \text{ (¥/pc)} \tag{10.58}$$

It is seen that a big improvement can be expected.

10.7 The Case When Factors of Various Number of Levels Are Mixed

In this section, the case when there are many multi-level factors is discussed. For example, consider the case when there are 13 types of plasticity, or there are 17 kinds of additives for a crystallization reaction.

These multi-level factors in many cases are qualitative in nature, rather than quantitative. How does one layout such types of experiments? Pseudo-factor design is useful for such applications. As a simple example, the layout of a 7-level factor *A*, 3-level factors *B*, *C*, *D*, and 2-level factors *E* and *F* using L_{16} is discussed.

Consider 4-level pseudo-factor, *A'*, with an idle column as follows:

Idle \ Pseudo-factor	A_1'	A_2'	A_3'	A_4'
1	A_1	A_2	A_3	A_4
2	A_1	A_5	A_6	A_7

Idle \ Pseudo-factor	A_1'	A_2'	A_3'	A_4'
1	A_1	A_2	A_3	A_4
2	A_3	A_4	A_5	A_6

In the above right table, A_1, A_2, A_3 and A_4 are compared under the 1st level of the idle column, A_1, A_5, A_6 and A_7 are compared under the 2nd level of the idle column. From such a layout, all levels can be compared at the common level, A_1. A_1 is the contrast level. A contrast level does not have to be A_1; often a more important level is used. When *A* has 6 levels, the table on the above right may be used where there are two contrast levels.

Using a similar approach, the following pseudo-factors for *B*, *C* and *D* are designed:

Idle \ Pseudo-factor	B_1'	B_2'
1	B_1	B_2
2	B_1	B_3

Idle \ Pseudo-factor	$D1'$	D_2'
1	D_1	D_2
2	D_1	D_3

Of course, B_1, ...,D_1 are contrast levels. Figure 10.14 is the required linear graph.

By connecting columns 2 and 4 in linear graph (4) of L_{16}, the layout shown in Figure 10.15 is made.

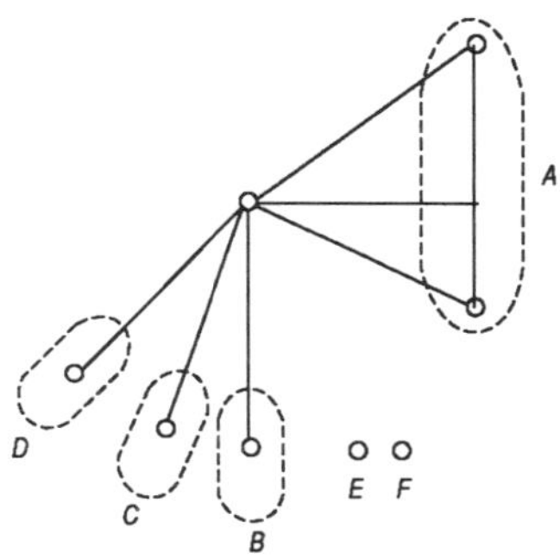

Figure 10.14. Required Linear Graph

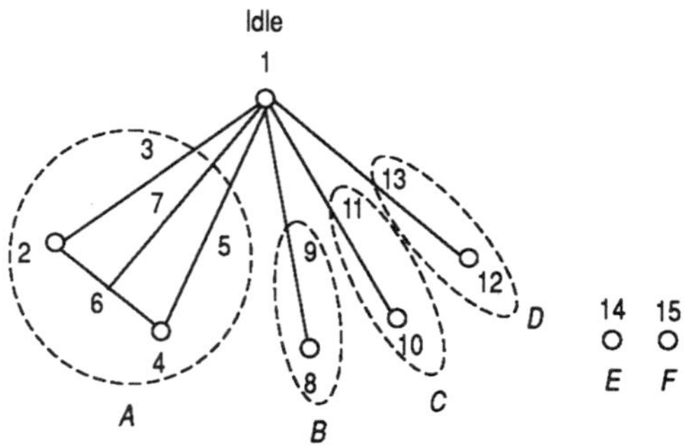

Figure 10.15. Assignment to L_{16}

For the levels of A, correspond A_1, A_2, A_3 and A_4 to the combinations of columns 2, 4 and 6 under the 1st level of the idle column, then correspond A_1, A_5, A_6 and A_7 to the combinations of columns 2, 4 and 6 under the 2nd level of the idle column. The layout is shown in Table 10.15.

Table 10.15. Layout of $7\times3^3\times2^2$ from an L_{16}

No. \ Factor	Idle 1	A 234567	B 8 9	C 10 11	D 12 13	E 14	F 15
1	1	1	1	1	1	1	1
2	1	1	2	2	2	2	2
3	1	2	1	1	2	2	2
4	1	2	2	2	1	1	1
5	1	3	1	2	1	2	2
6	1	3	2	1	2	1	1
7	1	4	1	2	2	1	1
8	1	4	2	1	1	2	2
9	2	1	1	1	1	1	2
10	2	1	3	3	3	2	1
11	2	5	1	1	3	2	1
12	2	5	3	3	1	1	2
13	2	6	1	3	1	2	1
14	2	6	3	1	3	1	2
15	2	7	1	3	3	1	2
16	2	7	3	1	1	2	1

EXERCISES

Question (10.1). In a design of an LSI process, the following 7 factors were considered to be important:

A 2 levels D 4 levels G 2 levels

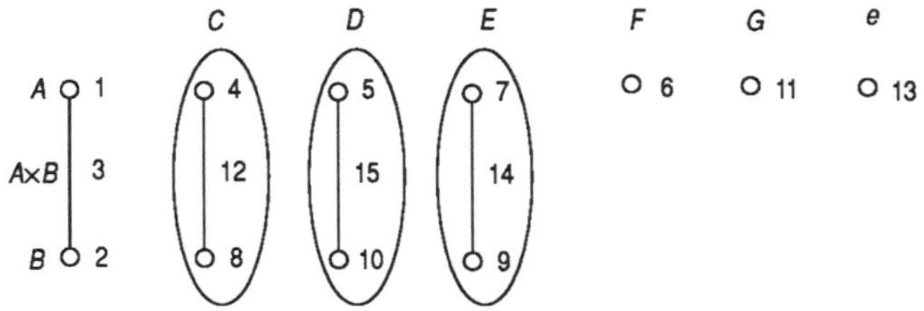

B 2 levels E 4 levels

C 4 levels F 2 levels

The following linear graph was used to assign the experiment to an L_{16} array so that all main effects and interaction $A \times B$ may be obtained.

10 transistors from each wafer of each experiment and V_{CBO} was measured.

(a) 10 V_{CBO} values of Experiment No.16 are as follows:

47.8 47.1 47.6 45.3 45.4 44.1 46.8 45.8 47.8 46.5 (V)

Calculate the S/N Ratio of larger-the-better characteristic in decibel units.

(b) The S/N Ratios in decibels from No.1 to No.15 are shown in the following table. Conduct an analysis of variance and estimate the significant factorial effects.

(c) Determine the optimum condition and estimate its S/N Ratio in decibels and V_{CBO} (V).

Factor · Column No.	*A*	*B*	*A*×*B*	*C*	*D*	*F*	*E*	*G*	*e*	S/N Ratio (db)
	1	2	3	4 8 12	5 10 15	6	7 9 16	11	13	
1										28.1
2										29.5
3										30.9
4										31.2
5										31.0
6										30.9
7										31.4
8										31.4
9										28.3
10										28.9
11										30.0
12										30.7
13										29.8
14										31.7
15										33.0
16										

Question (10.2). The target of a certain electric circuit is 6.6±0.8V; its loss due to becoming out of specification is ¥2,000. To conduct a robust design, the following control factors were assigned to an L_8 orthogonal array using pseudo-factor design.

A: diode	A_1 = use	A_2 = do not use
B: zena diode	B' (Z_1 at A_1)	2 levels
	B'' (Z_2 at A_2)	2 levels
C: resistor	2 levels	
D: transistor (h_{FE})	2 levels	

Noise factors were compounded as follows:

N_1 = worst condition of negative side

N_2 = standard condition

N_3 = worst condition of positive side

Output voltages were measured twice repeatedly.

(a) Calculate sensitivity and S/N Ratios of a nominal-the-best characteristic. Prepare an analysis of variance table.

(b) Determine the optimum condition for S/N Ratio.

(c) Estimate the sensitivity at this condition and compare with the target of 6.6V.

(d) If the output deviates from the target, which factor should be adjusted? Determine the loss due to variability after the adjustment.

Factor · Column / No.	A	B′ B″		C	e	e	D	N_1		N_2		N_3	
	1	2	3	4	5	6	7						
1	1	1	B′	1	1	1	1	6.13	6.42	6.55	6.63	6.82	6.78
2	1	1		2	2	2	2	5.89	6.14	6.17	6.25	6.50	6.38
3	1	2		1	1	2	2	6.52	6.82	6.98	6.78	7.39	7.28
4	1	2		2	2	1	1	5.89	6.48	6.61	6.59	6.89	6.91
5	2	1	B″	1	2	1	2	6.20	5.94	6.31	6.46	6.68	6.85
6	2	1		2	1	2	1	5.88	5.99	6.23	6.31	6.62	6.56
7	2	2		1	2	2	1	6.51	6.25	6.66	6.73	7.02	7.19
8	2	2		2	1	1	2	6.29	6.21	6.57	6.35	6.64	6.98

Question (10.3). The following factors will be used to find the optimum condition for manufacturing a certain valve:

A: material 2 levels
B: dwell time 2 levels
C: heat gun temperature 2 levels
D: bracker pressure 3 levels
E: ultrasonic weld yes/no
F: welding pressure 2 levels (only at E_2)
G: welding time 2 levels (only at E_2)
H: bracker travel time 3 levels

F and G are the factors only under E_2. Assign the factors to an L_{16} orthogonal array to estimate all main effects.

CHAPTER 11

CLASSIFIED DATA

Although the use of continuous variables is recommended whenever possible, sometimes there is no way that one can use continuous variables, and we have to use classified data. Classified data includes classified attributes, classified variables, multi-variables, and multi-attributes. In this chapter, accumulation analysis and minute analysis are introduced as the preferred methods for analyzing classified attributes.

11.1 THE MEANING OF CLASSIFIED DATA: AN ILLUSTRATED EXAMPLE

Consider a crushing process of a certain material wherein the crushed product should neither be too coarse nor too fine. Preferably, the fineness should be between 15 and 50 meshes. In such a case, it is desirable from an economic viewpoint that the yield be 100%. However, such a characteristic, percent yield, is not suitable for efficient determination of the optimum condition.

For example, the current condition is $A_1B_1C_1$. The results of changing A_1 to A_2, B_1 to B_2 and C_1 to C_2 are shown in Table 11.1.

The results show the yield increases 40% when A_1 is changed to A_2, increases 50% when B_1 is changed to B_2, and there is no increase when C is changed. The optimum condition is therefore either $A_2B_2C_1$ or $A_2B_2C_2$.

Table 11.1. Yield Between 15 and 50 Meshes

Condition	%
A_1	40
A_2	80
B_1	40
B_2	90
C_1	40
C_2	40

It is easily forecasted that using any of these two optimum conditions, the yield will probably be close to 100%. But if an experiment were conducted at $A_2B_2C_2$ to obtain an yield under 10%, the engineer would be puzzled. It is indeed possible to have such a result. Here is the reason why. Suppose the distribution of the portion which is coarser than 15 meshes and the portion which is finer than 50 meshes are as shown in Table 11.2.

Table 11.2. Particle Size Distribution(mesh)

	Under 15	15~50	Over 50
A_1	60	40	0
A_2	0	80	20
B_1	60	40	0
B_2	0	90	10
C_1	60	40	0
C_2	0	40	60

The reason why the yield is only 40% is that particle size is too coarse. Factor *C* has no effect on the percentage between 15 to 50 meshes, even though the factor really has the largest effect on particle size distribution. When *C* is changed from C_1 to C_2, the distribution goes toward the finer direction and that made the yield at $A_2B_2C_2$ very low.

This example explains why yield is not a suitable characteristic for the determination of the optimum condition.

Of course the optimum condition would be obtained if all combinations of *A*, *B* and *C* were conducted; only 2^3=8 experiments are required. When there are only three factors, 8 experiments is not a problem, but if the number of levels for *A*, *B* and *C* were either 3 or 4, the total number of experiments would then be 27 or 64.

If there were 3 factors with 10 levels each, then the total number of combinations would be 59,040, etc. By running 10 experiments a day, it would take 20 years to complete.

In order to run the experiments one factor at a time, or a few factors at a time, and still be able to correctly find the optimum condition, the characteristic must be monotonous or additive.

Like the example when we expect the yield to be 100% within a certain range, yield is not an appropriate characteristic. It requires that we observe the percentages on the coarser and the finer sides. If the percentage of the 3 classes are shown, it is obvious that $A_2B_2C_2$ is the worst condition without conducting experiments.

In other words, there is no additivity for the yield of 15 to 50 meshes, but there is additivity if the percentage of 3 classes are shown.

Characteristics such as the percentage within a certain range such as "particle size" in the above example, or condition is expressed as "moderate" or "bad", etc. belong to the type of characteristics which are not additive.

Yield, in the case of a reversible reaction, or over-reaction, has no additivity like the case of particle size. The situation can be easily understood if we think of what kind of person is considered to be important in the world where "balance" or "condition" are important, where color match of dye or paint, machine condition, furnace condition, odor, or physical condition, etc. are to be improved.

The people who are attached importance to have the experience of nearly all combinations. They are experienced people. Without exploring the characteristics having additivity, we can never get out of craftsman's world.

It is difficult in many cases to judge from engineer's specialized field whether or not there is additivity in the characteristic that we are using. Therefore, there is a necessity to find a way to confirm that we are not in a maze. That is, we need to evaluate the additivity of characteristics. To do so, we will make use of orthogonal arrays in the experiment so that the evaluation of additivity becomes possible.

11.2 CLASSIFIED DATA

11.2.1 Classified Attributes

When n times of action or n pieces of product are classified into several classes such as "number of times" or "number of pieces," such results belong to classified attributes.

Classified attributes are classified into the following types:

(1) Graded Data

When the appearance of a product is given by points, then we are using a continuous variable. However, when a grade is given of 5 good, 20 medium and 8 poor, it belongs to graded data.

Whenever we grade using labels such as "excellent", "good", "normal" and "poor", we are generating graded data.

(2) Scale-out data

In a life test of animals or products, the test is terminated before the end of life. For example, if 10 products were subjected to a 1,000 hour life test, and 3 units still functioned after 1,000 hour test, and if the results are classified into:

I: Did not function from the beginning.
II: Life ended between 1 to 100 hours.
III: Life ended between 101 to 500 hours.
IV: Life ended between 501 to 1000 hours
V: Life ends after 1000 hours.

the data can be treated like graded data. Generally, part of scale-out data belongs to continuous variables and part is either above of below a certain value. When a range of continuous variable is classified into about 3 classes, or life test data is classified into 5 classes, these can be treated as classified attributes. In the case of life test, the time factor, k, is set at K_1=0th hour, K_2=100th hour, K_3=200th hour, K_4=300th hour,...., K_{11}=1000th hour, and if the 10 products functioned at each level of K, 1 is given, otherwise 0 is given. This is the case of graded data with 2 classes.

(3) Gauge Value

In quantitative analysis, data are classified into 3 classes:

-(no trace), +(trace), ++(more than trace)

Originally such results belong to variables, but the results are classified into 3 classes by the magnitude. The results such as "passed go gauge", "passed go gauge but stopped by no-go gauge" and "passed no-go gauge" belong to this type.

When we want to measure the temperature that a plastic starts to become brittle, it is cumbersome to measure the exact temperature. Instead, test is made at certain temperatures such as whether it became brittle at 0°C or at -10°C. In this case, it belongs to gauge value.

Many sensory tested results belong to this type.

(4) Ranked Data

The appearance of products can be put in order to get ranked data.

More than one data point can be at the same rank. If the appearance of products from several companies are compared, then the product of our company may be used as a base and classified as follows:

I: Worse than our company
II: Slightly worse than our company
III: Same as our company
IV: Slightly better than our company
V: Better than our company and treated as graded data.

In many cases, figures or graphs may be ranked by observation.

(5) Pure Classified Data

This is the type of data which is classified into classes, but there is no rank, such as types of defective or classified shape, etc.

Among those types, (1), (2) and (3) can be analyzed by the accumulation analysis. Type (5) is analyzed by frequency analysis.

Regular Frequency			Cumulative Frequency		
Excellent	Good	Normal	I(E)	II(E+G)	III(E+G+H)
2	3	1	2	5	6

When accumulation analysis is used, data must be treated as follows. For example, graded data is classified into 3 classes: excellent, good and normal, and the number of data points in each class is 2, 3 and 1.

Classes I and II are analyzed, and no analysis is made for class III. The number of classes used for analysis is (No. of classes of regular frequency - 1).

In the case of frequency analysis, regular frequency data is used; therefore, the number of classes used for analysis is equal to the number of classes of the regular frequency data.

11.2.2 Classified Variables

With classified variables, data in each class is not expressed by a number (integer) as in the case of classified attributes, but by rate (percent). Here are some examples:

(1) Distribution when the total number of units is unknown, such as particle size, polymerization or fiber length.

For example, particle size is classified into

I: Particles which do not pass through 25 mesh screen
II: Particles which pass through 25 mesh screen but do not pass through 50 mesh screen

III: Particles which pass through 50 mesh screen but do not pass through 100 mesh screen

IV: Particles which pass through 100 mesh screen

(2) Continuous curve

When we want to control continuously measured values such as the thickness of steel plate or film, tool temperature of viscosity in a certain range,the whole range is divided into several classes, and we use the rate in each class as shown in Figure 11.1.

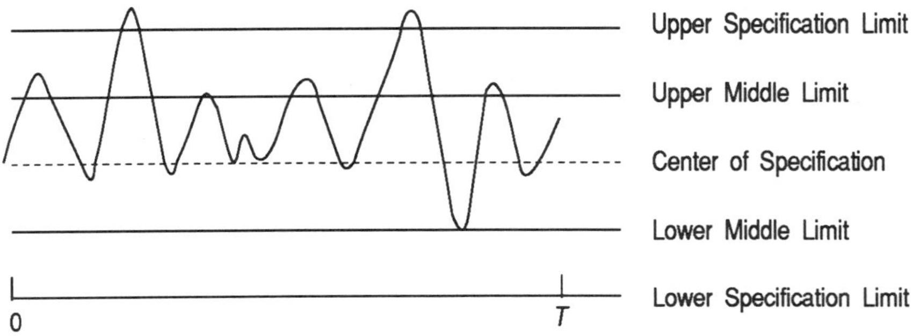

Upper specification limit:	Troubles occur or becomes defective beyond this limit.
Upper middle limit:	Desirable upper limit. Determined from data analysis results or by managerial objective.
Lower middle limit:	Desirable lower limit. Determined data analysis results or by managerial objective.
Lower specification limit:	Troubles occur or becomes defective.

Figure 11.1. Data of Continuous Curve

Observing the curve between points 0 and T, it is classified into 5 classes:

I: Portion exceeded upper specification limit: 5%
II: Portion between upper specification limit and upper middle limit: 15%
III: Portion between upper middle limit and lower middle limit (desirable portion): 80%
IV: Portion below lower middle limit: 0%
V: Portion below lower specification limit: 0%

If the interval [0,T] is equally divided into 100, and the number of points in classes I, II, III, IV and V are counted, such data is analyzed as gauge values in classified attributes.

If the original graphs are available, then samples are taken as continuous variables, and a nominal-the-best S/N Ratio should be calculated.

(3) Pure Classified Variables

Percentage is given to classes which have no rank. Such cases are rare in the engineering field.

(1) and (2) are analyzed by accumulation analysis; the number of classes analyzed is equal to (the initial number of classes - 1).

(3) is analyzed by frequency analysis; the number of classes analyzed is equal to the initial number of classes.

11.2.3 Multi-attributes Example

A group of data consisting of a number (integer) classified into several classes showing the number of large, medium and small defects or large, medium and small troubles belong to this type. The number of units in each class can be counted. When the type of defects cannot be ranked with classification large, medium and small, but classified by the types of defect, frequency analysis is used for analysis.

11.2.4 Multi-variables Example

Dynamite is used to dig up rocks or ore. The materials are weighed, and their particle distribution is classified into 3 classes:

(1) Particle sizes are too large. Further size reduction is necessary.

(2) Right size.

(3) Particle sizes are too small. Either sintering is necessary, or it may be difficult to transfer.

Accumulation analysis is used; the number of classes for the analysis is 3.

11.3 Classified Data and S/N Ratio

11.3.1 Classification Into Two Classes

Quite often, outcomes are classified into two classes—e.g., products are classified as "good" and "bad" or function is classified into "correct" and "fail."

A function is tested, and "1" and "0" are used to express correct and fail outcomes, respectively, out of n time trials. The results will be denoted by $y_1, y_2, \ldots, y_n$. If the number of correct actions is denoted by r, its reliability, p, is given by

$$p = \frac{y_1 + y_2 + \cdots + y_n}{n} = \frac{r}{n} \tag{11.1}$$

The total sum of squares, S_T, is

$$S_T = {y_1}^2 + \cdots + {y_n}^2 = r \qquad (f=n) \quad (11.2)$$

The variation of average (p) is denoted by S_m and is given by

$$S_m = np^2 = n\left(\frac{r}{n}\right)^2 = \frac{r^2}{n} = \frac{(np)^2}{n} \qquad (f=1) \quad (11.3)$$

The error variation, S_e, is given by

$$S_e = S_T - S_m = r - \frac{r^2}{n} = np - \frac{(np)^2}{n} = np(1-p) \qquad (f=n\text{-}1) \quad (11.4)$$

Letting the pure reliability be denoted by p', its pure variation, S_m', is equal to np'^2. The expectation of S_m is given by

$$E(S_m) = S_m' + \sigma^2 = np'^2 + \sigma^2 \tag{11.5}$$

as seen in Chapter 2. The expectation of S_e is

$$E(S_e) = (n-1)\sigma^2 \tag{11.6}$$

$$\therefore E(V_e) = \sigma^2 \tag{11.7}$$

The S/N Ratio equivalent to $10 \log m^2/\sigma^2$ is given by

$$\eta = 10 \log \frac{(\rho')^2}{\sigma^2} = 10 \log \frac{\frac{1}{n}(S_m - V_e)}{V_e} = 10 \log \frac{n-1}{n}\left[\frac{S_m - V_e}{(n-1)V_e}\right]$$

$$= 10 \log \frac{n-1}{n}\left[\frac{S_m - \frac{S_e}{n-1}}{S_e}\right] = 10 \log \frac{n-1}{n}\left[\frac{\frac{(np)^2}{n} - \frac{np(1-p)}{n-1}}{np(1-p)}\right]$$

$$\approx 10 \log \frac{p}{1-p} = -10 \log\left(\frac{1}{p} - 1\right) \text{ (db)} \tag{11.8}$$

This is called the "omega" transformation. To ease the transformation, the values of η(db) corresponding to p(%) are shown in Appendix Table 3.

11.3.2 An Efficient Improvement of the Reliability Test Using the Omega Transformation: Vending Machine Example

The reliability of automatic vending machines (the probability of the commodity correctly coming out when coins are put in) is estimated as 99.2%. In order to improve reliability, five countermeasures A, B, C, D and E were used to conduct an experiment using orthogonal array L_9.

A: component part	A_1 = current	A_2 = new
B: component part	B_1 = current	B_2 = new
C: circuit	C_1 = current	C_2 = new1
	C_3 = new2	
D: power consumption	D_1 = -50%	D_2 = current
	D_3 = +50%	
E: dust prevention	E_1 = current	E_2 = new device

In order to improve the efficiency of the test, the power supply was lowered to 1/3 of the actual condition, and the test was conducted under vibration and dusty conditions.

A combination design was used for factors A and B, and level E_1 was dummy treated.

Table 11.3. Layout and Results

No. \ Factor	(AB) 1	C 2	D 3	E 4	Results (No. of Correct Actions)	η (db)
1	11	1	1	1	57	1.224
2	11	2	2	2	94	11.950
3	11	3	3	1′	60	1.761
4	12	1	2	1′	58	1.402
5	12	2	3	1	98	16.902
6	12	3	1	2	60	1.761
7	21	1	3	2	83	6.886
8	21	2	1	1′	88	8.653
9	21	3	2	1	62	2.126
Total					660	52.665

Table 11.4. Supplementary Table

Level \ Factor	(AB)	C	D	E
1	14.935	9.512	11.638	20.252
2	20.065	37.505	15.478	20.597
3	17.665	5.648	25.549	11.816 (E_1')

Table 11.5. ANOVA Table

Factor	f	S	V	ρ(%)
(AB)	2	4.392 2	2.196 1°	
A	(1)	(1.242 2)	(1.242 2)	
B	(1)	(4.386 2)	(4.386 2)	
C	2	201.489 7	100.744 8	74.4
D	2	34.409 6	17.204 8	9.3
E	1	4.626 9	4.626 9°	
e	1	11.861 0	11.861 0°	
(e)	(4)	(20.880 1)	(5.220 0)	(16.3)
T	8	256.779 4		100.0

The right column of Table 11.3 gives decibel value by the omega transformation. An analysis of variance was conducted using the decibel values. The optimum condition is $A_2B_2C_2D_3E_2$. The average is estimated using the larger effects, C and D.

$$\hat{\mu} = \overline{C}_2 + \overline{D}_3 - \overline{T} = \frac{37.505 + 25.549}{3} - \frac{52.665}{9} = 15.166 \quad \text{(db)} \tag{11.9}$$

If the effects of A and B are large, the estimate of A_2B_2, which was not conducted in the experiment, is estimated by

$$\begin{aligned} \overline{A_2B_2} &= \overline{A_1B_1} + (\overline{A_2B_1} - \overline{A_1B_1}) + (\overline{A_1B_2} - \overline{A_1B_1}) \\ &= \overline{A_2B_1} + \overline{A_1B_2} - \overline{A_1B_1} \end{aligned} \tag{11.10}$$

Since the current condition is $A_1B_1C_1D_2E_1$, its average is

$$\hat{\mu}_{(\text{current})} = \overline{C}_1\overline{D}_2 - \overline{T} = \frac{9.512 + 15.478}{3} - \frac{52.665}{9} = 2.478 \quad \text{(db)} \tag{11.11}$$

The gain is

$$15.166 - 2.478 = 12.688 \quad \text{(db)} \tag{11.12}$$

Current reliability is 99.2%. We want to know the reliability at the optimum condition. Its decibel value is 20.934. With a gain of 12.688 decibels, the reliability of the optimum condition, p_m, is calculated by

$$-10 \log \left(\frac{1}{p_m} - 1 \right) 20.934 + 12.688 = 33.622 \tag{11.13}$$

$$\therefore p_m = 99.957(\%) \tag{11.14}$$

The omega transformation is useful to predict operating conditions during manufacturing, and in the market from the results obtained in the laboratory. The efficiency of experimentation on reliability can thus be improved.

11.3.3 The Case When There Are Two Types of Inputs

The experimental results from the automatic vending machines' reliability test are classified into two kinds: good and bad. However, errors in judgement sometimes do occur. In such cases when there are two types of faults, e.g., 1 is judged as 0 and 0 is judged as 1, results are analyzed by the standard S/N Ratio. For this, see Chapter 13 in Quality Engineering Series, Vol.1, *Taguchi Methods: Research and Development.*

11.3.4 The Case With 3 Classes Or More

When classification categories are "good," "normal" and "bad," they are expressed as 0, 1 and 2 respectively, and a smaller-the-better S/N Ratio may be used.

For this see Chapter 8 of Quality Engineering Series Vol.1, *Taguchi Methods: Research and Development.*

11.4 CLASSIFIED ATTRIBUTES AND ACCUMULATION ANALYSIS

Here, accumulation analysis and minute analysis are described for the purpose of comparing variability or distribution.

11.4.1 What Is Accumulation Analysis?

When data are classified into several classes, a distribution can be observed. It is a good way to know variability by observing distribution. Since quality improvement starts with the reduction of variation, distributions must be compared. In the case of continuous variables, the distribution is not used for the improvement of variability. Instead, quality improvement has been achieved through the improvement of S/N Ratio, which is the overall index consisting of the mean and variance.

Accumulation analysis and minute analysis are the analysis methods which are used to compare the mean and variance together. Therefore, such analysis methods meet the objective of quality improvement.

They are described next:

11.4.2 Variation In Accumulation Analysis

Three component parts used for assembly were produced under three manufacturing conditions, A_1, A_2 and A_3 respectively. The conditions at during assembling were classified into:

(1) tight
(2) good
(3) loose

The results are shown in Table 11.6. Such data are called regular frequency.

Table 11.6. Classified Data

Level \ Class	(1)	(2)	(3)	
A_1	3	2	0	5
A_2	1	3	1	5
A_3	0	1	4	5

There is an analysis method called accumulation analysis, which is to observe the effect of A on the average and the type of distribution. As described in Section 11.1, we should not observe only class (2) which is "good," but class (1), "tight," and class (3), "loose," as well.

When there is order in classes, the cumulative frequency is prepared. The selected order does not necessarily mean "good order" or "bad order," but the order of the characteristic itself:

I: the number of (1)
II: the numbers of (1) + (2)
III: the numbers of (1) + (2) + (3)

The cumulative frequencies are shown in Table 11.7.

Table 11.7. Cumulative Frequency

	I	II	III
A_1	3	5	5
A_2	1	4	5
A_3	0	1	5
	4	10	15

Letting an observation in class I be denoted by "1," and one that is not in class I be "0," the five results for A_1, A_2 and A_3 are rewritten as shown in Table 11.8:

Table 11.8. Results of Class I

A_1	1	1	1	0	0	3
A_2	1	0	0	0	0	1
A_3	0	0	0	0	0	0

All results consist of 0 and 1. The variation is calculated as in the case of continuous variables. Total variation, S_T(I), variation of A, S_A(I), and error variation, S_e(I), are given by

$$S_T(I) = 1^2 + 1^2 + \cdots + 0^2 - \frac{(1 + 1 + \cdots + 0)^2}{15}$$

$$= 4 - \frac{4^2}{15} \qquad (f\!=\!14) \quad (11.15)$$

$$S_A(I) = \frac{(\text{Total of } A_1)^2 + (\text{Total of } A_2)^2 + (\text{Total of } A_3)^2}{\text{No. of replications of } A_i}$$

$$- \frac{(\text{Grand total})^2}{\text{Total no. of replications}}$$

$$= \frac{3^2 + 1^2 + 0^2}{5} - \frac{4^2}{15} \qquad (f\!=\!2) \quad (11.16)$$

$$S_e(I) = S_T(I) - S_A(I) \qquad (f\!=\!12) \quad (11.17)$$

Next, letting an observation in class II be "1" and one that is not in class II be "0," Table 11.9 is obtained:

Table 11.9. Results of Class II

A_1	1	1	1	1	1	5
A_2	1	1	1	1	0	4
A_3	1	0	0	0	0	1

From these results, the total variation, S_T(II), variation of A, S_A(II), and error variation, S_e(II), are given by

$$S_T(II) = 10 - \frac{10^2}{15} \qquad (f\!=\!14) \quad (11.18)$$

$$S_A(II) = \frac{5^2 + 4^2 + 1^2}{5} - \frac{10^2}{15} \qquad (f\!=\!2) \quad (11.19)$$

$$S_e(II) = S_T(II) - S_A(II) \qquad (f\!=\!12) \quad (11.20)$$

The results of class III are identical for A_1, A_2 and A_3. Therefore, no analysis will be made.

The variation is calculated from each class and added together. The total variation of class I, S_T(I), is rewritten as

$$S_T(I) = 4 - \frac{4^2}{15} = \frac{4 \times 15 - 4^2}{15}$$

$$= 15 \times \frac{4}{15} \times (1 - \frac{4}{15}) = 15 \times P_I(1 - P_I) \tag{11.21}$$

where P_I is the rate in class I or 4/15.

Similarly,

$$S_T(II) = 15 \times \frac{10}{15} \times (1 - \frac{10}{15})\ 15 \times P_{II}(1 - P_{II})$$

P_{II}: the rate in Class II (11.22)

where P_{II} is the rate in class II.

This suggests that the total variation, S_T, is equal to n times $P(1\text{-}P)$, depending on P, the rate in each class. In other words, variation is maximum when P is equal to 0.5 and becomes smaller when P becomes closer either to 0 or 1.

Similar to the total variation, other factorial effects are also evaluated by their relative magnitude depending on the magnitude of P. However, it is not appropriate to add the variation of different classes with the same weight. It means that an appropriate measure is required.

Since the total variation is evaluated by $P(1\text{-}P)$ when the rate of occurrence in a class is P, the reciprocal of $P(1\text{-}P)$, or $1/P(1\text{-}P)$, is used as the weight for the overall variation of all classes.

For the variation of class I,

$$W_1 = \frac{1}{P_I(1-P_I)} = \frac{1}{\frac{4}{15}(1 - \frac{4}{15})}$$

$$= \frac{15^2}{4 \times (15 - 4)} = 5.11 \tag{11.23}$$

is used as the multiplier, and for the calculation of class II,

$$W_2 = \frac{1}{P_{II}(1 - P_{II})} = \frac{1}{\frac{10}{15}(1 - \frac{10}{15})}$$

$$= \frac{15^2}{10 \times (15 - 10)} = 4.50 \tag{11.24}$$

is used as the multiplier. Then they are added together.

Letting the overall total variation be S_T, the overall variation of A be S_A and the overall error variation be S_e,

$$S_T = S_T(I)W_1 + S_T(II)W_2$$
$$= 15 \times \frac{4}{15}(1 - \frac{4}{15})W_1 + 15 \times \frac{10}{15}(1 - \frac{10}{15})\ W_2$$
$$= 15 + 15 = 30 \qquad (f=14\times2=28) \quad (11.25)$$

The total variation after subtracting the general mean is given by

$$(\text{No. of measurements}) \times (\text{No. of classes} - 1) \qquad (11.26)$$

$$S_A = S_A(I)W_1 + S_A(II)W_2$$
$$= \frac{(3^2 + 1^2 + 0^2)W_1 + (5^2 + 4^2 + 1^2)W_2}{5} - CF$$
$$= 48.02 - 35.45 = 12.57 \qquad (f=2\times2=4) \quad (11.27)$$

where

$$CF = \frac{4^2 \times W_1 + 10^2 \times W_2}{15} = 35.45 \qquad (11.28)$$

$$S_e = S_T - S_A = 30 - 12.57 = 17.43 \qquad (f=24) \quad (11.29)$$

Table 11.10 shows the ANOVA table.

Table 11.10 . ANOVA Table

Factor	f	S	V	ρ(%)
A	4	12.57	3.14	32.2
e	24	17.43	0.726	67.8
T	28	30.00		100.0

Since the effect of A is significant, the next step is to determine which level of A is the best. Since there is only one factor, A, the percentage of each class is calculated and the distribution is compared using graphs.

In the case when there is a rank in the order from good to bad, the optimum condition is determined by selecting the class with the highest "rate of good." But as in this case when the second class is the best, the determination is not that straightforward. The highest rate of good in this case is A_2. However, the tightness or looseness is adjustable by the diameter. The average can be changed using an adjusting factor. If this is the case, then A_3 is better than A_2.

Table 11.11. Estimation

	Cumulative Percent			Regular Percent		
	I	II	III	(1)	(2)	(3)
A_1	60	100	100	60	40	0
A_2	20	80	100	20	60	20
A_3	0	20	100	0	20	80

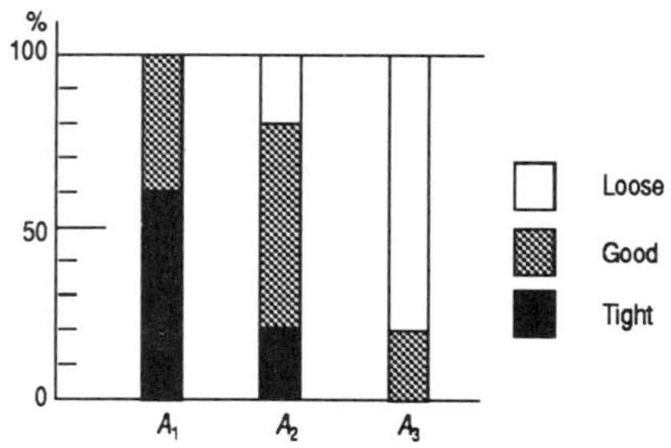

Figure 11.2. Factorial Effect

Without adjusting, A_3 is 80% loose. If the average is changed, the rate of good can be increased. However, such judgment is not quantitative. It is recommended that the data be transformed into continuous variables and the S/N Ratio be used for the analysis. The following is an example with many factors:

11.4.3 Accumulation Analysis Using Orthogonal Array

(1) Layout and Results

In order to improve anti-corrosion of a chrome plated part, the following five factors were studied:

A = Pre-finishing	2 levels
B = Plating solution	2 levels
C = Plating time	3 levels
D = Finishing method	3 levels
E = Type of additive	2 levels

An orthogonal array L_9 was used, and the results were classified into good, normal and poor. A combination design was used to assign A and B, and E was dummy treated.

When data is classified in an order such as "good," "normal" and "poor," and points are awarded as,

Classification	Points
Good	0
Normal	1
Poor	2

and a smaller-the-better S/N Ratio may be used. In this example, accumulation analysis is used to simultaneously analyze the mean and variance through the distribution.

Table 11.12. Layout and Results

No. \ Factor	(*AB*) 1	*C* 2	*D* 3	*E* 4	Results: Good	Normal	Poor	Cumulative Frequency: I	II	III
1	11	1	1	1	0	0	3	0	0	3
2	11	2	2	1′	0	0	3	0	0	3
3	11	3	3	2	1	2	0	1	2	3
4	21	1	2	2	0	0	3	0	0	3
5	21	2	3	1	0	2	1	0	2	3
6	21	3	1	1′	2	1	0	2	3	3
7	12	1	3	1′	0	2	1	0	2	3
8	12	2	1	2	3	0	0	3	3	3
9	12	3	2	1	1	2	0	1	3	3
					7	9	11	7	16	27

First, a supplementary table is prepared from the cumulative frequencies to produce Table 11.13:

Table 11.13. Supplementary Table

	I	II	III		I	II	III
$(AB)_1$	1	3	9	D_1	5	6	9
$(AB)_2$	2	5	9	D_2	1	3	9
$(AB)_3$	4	8	9	D_3	1	7	9
C_1	0	2	9	E_1	1	5	9
C_2	3	5	9	E_1'	2	5	9
C_3	4	9	9	E_2	4	6	9
				Total	7	16	27

(2) Calculation of Variation

First, weights W_1 and W_2 are calculated. The totals for classes I and II are 7 and 16, respectively, and the grand total is 27.

$$W_1 = \frac{27^2}{7 \times (27-7)} = 5.21 \tag{11.30}$$

$$W_2 = \frac{27^2}{16 \times (27-16)} = 4.14 \tag{11.31}$$

Total variation is calculated by (total number of measurements) × (number of classes analyzed).

$$S_T = 27 \times 2 = 54 \qquad (f{=}52) \tag{11.32}$$

The total degrees of freedom, f=52, is calculated by 26×2.

$$CF = \frac{7^2 \times W_1 + 16^2 \times W_2}{27} = 48.71 \qquad (f{=}1{\times}2{=}2) \tag{11.33}$$

$S_{(AB)}$ is given by

$$S_{(AB)} = \frac{(1^2+2^2+4^2) \times W_1 + (3^2+5^2+8^2) \times W_2}{9} - CF$$
$$= 8.53 \qquad (f{=}2{\times}2{=}4) \tag{11.34}$$

Next, S_A and S_B are calculated. A is calculated from $(AB)_1$ and $(AB)_2$, and B is from $(AB)_1$ and $(AB)_3$. Since both are 2 levels, the equation for 2 levels is used.

$$S_A = \frac{(A_1B_1 - A_2B_1)^2}{(\text{Total no. of measurements in } A_1B_1 \text{ and } A_2B_1)} \tag{11.35}$$

Using this equation to calculate classes I and II, then summarize them to obtain the overall sum of squares of A.

$$S_A = \frac{(1-2)^2 \times W_1 + (3-5)^2 \times W_2}{18} = 1.21 \qquad (f{=}1{\times}2{=}2) \tag{11.36}$$

Similarly,

$$S_B = \frac{(1-4)^2 \times W_1 + (3-8)^2 \times W_2}{18} = 8.36 \qquad (f{=}1{\times}2{=}2) \tag{11.37}$$

Next, S_C and S_D are calculated.

$$S_C = \frac{(0^2+3^2+4^2) \times W_1 + (2^2+5^2+9^2) \times W_2}{9} - CF = 16.36$$
$$(f{=}2{\times}2{=}4) \tag{11.38}$$

$$S_D = \frac{(5^2+1^2+1^2) \times W_1 + (6^2+3^2+7^2) \times W_2}{9} - CF = 10.16$$
$$(f{=}2{\times}2{=}4) \tag{11.39}$$

Since E is dummy treated,

$$S_E = \frac{(E_1 + E_1')^2}{\text{Total of no. of replications of } E_1 \text{ and } E_1'} + \frac{{E_2}^2}{\text{No. of replications of } E_2} - CF \qquad (11.40)$$

This is summarized using the weights.

$$S_E = \left[\frac{(1+2)^2}{18} + \frac{4^2}{9}\right] \times W_1 + \left[\frac{(5+5)^2}{18} + \frac{6^2}{9}\right] \times W_2 - CF$$
$$= 2.72 \qquad (f{=}1{\times}2{=}2) \quad (11.41)$$

S_{e1} is calculated from the difference between E_1 and E_1'.

$$S_{e1} = \frac{(1-2)^2 \times W_1 + (5-5)^2 \times W_2}{18} = 0.29 \qquad (f{=}2) \quad (11.42)$$

The total of the above calculated will be equal to the variation between experiments, S_{T1}.

$$S_{T1} = \frac{1}{3}[(0^2 + 0^2 + 1^2 + \cdots + 1^2) \times W_1 + (0^2 + 0^2 + 3^2 + \cdots + 3^2) \times W_2] - CF$$
$$= 38.06 \qquad (f{=}8{\times}2{=}16) \quad (11.43)$$

Up to this section, most of the variation between experiments is equal to the total variation. In this example, the total variation, S_T, is 54; its degrees of freedom is equal to 52. The difference is because there are 3 repetitions in each of the nine experiments. In the case of continuous variables as seen in previous chapters, there are repetitions, too. However, one S/N Ratio is calculated from the repetitions of each inner condition. For the analysis of S/N Ratio, the number or repetitions is equal to one.

In the 3 observations in class I of experiment No.1, there are no 1's. Its variation is

$$0 - \frac{0^2}{3} = 0 \qquad (11.44)$$

This means there is no variation in 3 observations. The variation of No.2 to No.9 are calculated and added together.

$$S_{e2}(I) = (0 - \frac{0^2}{3}) + (0 - \frac{0^2}{3}) + (1 - \frac{12}{3}) + \cdots + (1 - \frac{1^2}{3})$$
$$= 7 - \left(\frac{0^2 + 0^2 + \cdots + 1^2}{3}\right) \qquad (11.45)$$

A similar calculation is made for class II.

$$S_{e2}(II) = 16 - \frac{0^2 + 0^2 + \cdots + 3^2}{3} \tag{11.46}$$

Summarizing those with the weight,

$$S_{e2} = S_T - S_{T1} = 15.94 \qquad (f=18\times 2=36) \tag{11.47}$$

This is the variation between repetitions. To obtain the degrees of freedom for each class, 1 is subtracted from 3, the number of repetitions, and multiplied by 9, the total number of experiments, to get 18. It is multiplied by 2, the number of classes analyzed, to get 18×2=36.

(3) Analysis of Variance

Table 11.14 shows the ANOVA table.

Table 11.14. ANOVA Table

Factor	f	S	V
(AB)	4	8.53	2.13
(A)	(2)	(1.21)	(0.60)°
(B)	(2)	(8.36)	(4.18)
C	4	16.36	4.09
D	4	10.16	2.54
E	2	2.72	1.36
e_1	2	0.29	0.14°
e_2	36	15.94	0.443
(e)	(40)	(17.44)	(0.436)
T	52	54.00	

When there are two types of error, e_1 and e_2, judgment is made as described in Chapter 5, comparing e_1 and e_2. In this case, e_1 is smaller than e_2, so pooling is made with e_2. A and e_1 are pooled with e_2 to get (e).

If the effect of A can be neglected, the combined factor (AB) can be considered as the result of the experiment of B as if B were dummy treated. Therefore, the variation of B is calculated as

$$S_B = \frac{(B_1 + B_1')^2}{18} + \frac{{B_2}^2}{9} - CF \tag{11.48}$$

Since A is pooled with the error, it is easier to calculate as follows:

$$S_B = S_{(AB)} - S_A = 7.32 \qquad (f=1\times 2=2) \tag{11.49}$$

The ANOVA table is rearranged to get Table 11.15.

Degrees of contribution are calculated as follows:

$$\rho_B = \frac{S_B - f_B V_e}{S_T} \times 100 = 11.9 \ (\%) \tag{11.50}$$

Table 11.15.

Factor	f	S	V	ρ(%)
B	2	7.32	3.66	11.9
C	4	16.36	4.09	27.1
D	4	10.16	2.54	15.6
E	2	2.72	1.36	3.4
e	40	17.44	0.436	42.0
T	52	54.00		100.0

$$\rho_C = \frac{S_C - f_C V_e}{S_T} \times 100 = 27.1 \ (\%) \tag{11.51}$$

$$\rho_D = \frac{S_D - f_D V_e}{S_T} \times 100 = 15.6 \ (\%) \tag{11.52}$$

$$\rho_E = \frac{S_E - f_E V_e}{S_T} \times 100 = 3.4 \ (\%) \tag{11.53}$$

$$\rho_e = \frac{S_e + (f_B + f_C + f_D + f_E) V_e}{S_T} \times 100 = 42.0 \ (\%) \tag{11.54}$$

(4) Estimate

Large effects, B, C, D and E are estimated in percent as shown in Table 11.16.

Table 11.16. Estimate

	Cumulative Frequency			%			%		
	I	II	III	I	II	III	Good	Normal	Poor
B_1	3	8	18	16.4	44.4	100.0	16.7	27.7	55.6
B_2	4	8	9	44.4	88.9	100.0	44.4	44.5	11.1
C_1	0	2	9	0	22.2	100.0	0.0	22.2	77.6
C_2	3	5	9	33.3	55.6	100.0	33.3	22.2	44.4
C_3	4	9	9	44.4	100.0	100.0	44.4	55.6	0.0
D_1	5	6	9	55.6	66.7	100.0	55.6	11.1	33.3
D_2	1	3	9	11.1	33.3	100.0	11.1	22.2	66.7
D_3	1	7	9	11.1	77.8	100.0	11.1	66.7	22.2
E_1	3	10	18	16.7	55.6	100.0	16.7	38.9	44.4
E_2	4	6	9	44.4	66.7	100.0	44.4	22.2	33.3
Total	7	16	27	25.9	59.3	100.0	25.9	33.4	40.7

(5) Optimum Condition and Estimation by Omega Transformation

The optimum condition is determined as $B_2C_3D_1E_2$ by placing the importance on class I. Sometimes, the selection of the best level is difficult. For example, even though D_1 has a higher rate of "poor" outcomes than D_3, D_1 was selected because the rate of "poor" can be significantly reduce by selecting C_3.

If the estimate were made using the regular method for rate,

$$\hat{\mu} = \overline{B}_2 + \overline{C}_3 + \overline{D}_1 + \overline{E}_2 - 3\overline{T} \tag{11.55}$$

then the percentage in class I would exceed 100%, which does not make sense. For the characteristics with limits from 0 to 100%, using "rate" or "fraction defective" is not an appropriate characteristic.

$$\hat{\mu} = 44.4 + 44.4 + 55.6 + 44.4 - 3 \times 25.9 = 111.1 \tag{11.56}$$

Let's consider fraction defective, for example. When the fraction defective is as low as 2%, it is not easy to reduce it to about 1%. However, if the fraction defective is about 30%, reducing it by 1% is not very difficult at all. That is the difficulty of reducing a rate by 1% in the case of from 2% to 1% and in the case of from 30% to 29% is entirely different. As shown in Figure 11.3, the amount of difficulty to reduce the rate from 2% to 1% is A and the same from 30% to 29% is B. In other words, the effect of reducing the rate from 2% to 1 cannot be considered to be the same as the effect of reducing the rate from 30% to 29%. The omega transformation is used to transform the curve in Figure 11.3 to the one in Figure 11.4 so that the effect becomes uniform everywhere in the range, 0 to 100%. [This is the S/NRatio shown in Equation (11.8) for the case when data consists of only 0 and 1]. In Figure 11.4,

$$p' = -10 \log \left(\frac{1}{p} - 1\right) \tag{11.57}$$

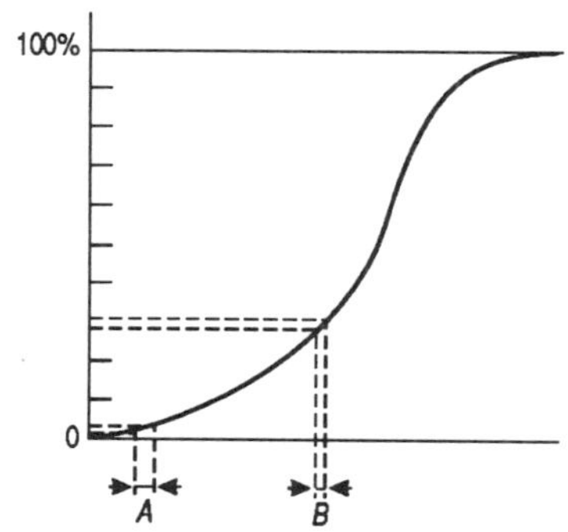

Figure 11.3. Effects of a Characteristic With Limits

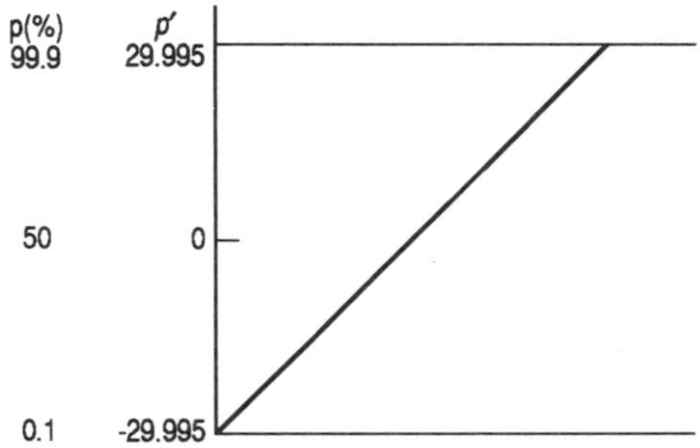

Figure 11.4. Effects After Omega Conversion

By the omega transformation, the effect of reducing a rate from 2% to 1% is equivalent to reducing -16.901 to -19.995 on the omega scale, which about 15 times bigger than the effect of reducing the rate from 30% to 29% (reducing -3.679 to -3.888). Thus, comparisons can be made on the same scale.

When the rates are fairy equal, such a transformation may not be necessary.

Using the additivity on omega scale, the decibel value of $\hat{\mu}$ is given by

$$(\text{db of } \hat{\mu}) = (\text{db of } \overline{B}_2) + (\text{db of } \overline{C}_3) + (\text{db of } \overline{D}_1) + (\text{db of } \overline{E}_2) - 3 \times (\text{db of } \overline{T}) \tag{11.58}$$

The calculation is made for classes I and II, and the cumulative frequency must be used instead of regular frequency.

Table 11.7 presents the decibel values for $\overline{B}_2$, $\overline{C}_3$, $\overline{D}_1$, $\overline{E}_2$ and $\overline{T}$ with the aid of Appendix 3.

Table 11.17. Estimation by Omega Transformation

	I		II	
	%	db	%	db
B_2	44.4	-0.976	88.9	9.036
C_3	44.4	-0.976	100.0	∞
D_1	55.6	0.977	66.7	3.017
E_2	44.4	-0.976	66.7	3.017
T	25.9	-4.564	59.3	1.635
$\hat{\mu} = \overline{B}_2 + \overline{C}_3 + \overline{D}_1 + \overline{E}_2 - 3\overline{T}$	93.7	11.741	100.0	∞

From Table 11.7,

$$\text{db of } \hat{\mu}(I) = -0.976 - 0.976 - 0.977 - 0.976 - 4 \times (-4.564) = 11.741 \qquad (11.59)$$

$$\therefore \hat{\mu}(I) = 93.7 \ (\%)$$

$$\text{db of } \hat{\mu}(II) = 9.036 + \infty + 3.017 + 3.017 - 3 \times 1.635 = \infty$$

$$\therefore \mu(II) = 100 \ (\%) \qquad (11.60)$$

The rates of occurrence of classes I and II and III are 93.7%, 100% and 100% respectively, and the rates of occurrence of "good," "normal" and "poor" are 93.7%, 6.3% and 0% respectively.

Confidence interval calculations will not be discussed here. [See references 1) and 2).]

11.5 Other Cases of Classified Data

11.5.1 Classified Variables

Classified variables can be considered as dividing an object into a small unit (in this case, a surface area is the aggregate of minute areas), and each of such minute unit belong to one of classes as is the case of classified attributes.

The calculation of minute units is treated the same as the case of classified attributes. However, the total variation is not calculated from minute units. Total variation of the highest unit is the unit in the data (lower order unit).

No numerical example is shown here. [See Literatures 1) and 2).]

11.5.2 Multi-attributes and Multi-variables

In the cases of multi-attributes and multi-variables, the total of all classes is either the number of samples or equal to 100%.

It is recommended whenever one would want to model the number of arrivals or the number of traffic accidents within a certain period of time for example, that the Poisson distribution to approximate this. The Poisson distribution is derived based on the assumption that individual occurrences have a small probability of occurrence, and are independent to each other. Note that for the Poisson distribution, the mean is equal to the variance.

In the case of multi-attribute data, the means of classified classes can be quite different from each other, and therefore, their variation must also be quite different. Therefore, the variation of different classes should not be added without further consideration.

The characteristics to be treated as multi-attributes do not necessarily follow Poisson distribution; therefore, we cannot always consider that variance is equal to the mean. However, there is a tendency that when the mean increases, the variance increases. In multi-attributes and multi-variables, we consider that the variance is proportional to the mean, and the following weight is used for the "i"th class:

$$W_i = \frac{1}{\overline{T}_i} \tag{11.61}$$

where $\overline{T}_i$ is the mean of "i"th class.

When there are k classes, the total variation of k classes is summarized. The degrees of freedom becomes k times the degrees of freedom of each class.

11.5.3 The Case With No Order

In the case when there is no order between classified classes, frequency analysis is used instead of accumulation analysis. In accumulation analysis, the cumulative frequency of each class is calculated, and the weight of each class is used to summarize the variation of the original data in each class. In frequency analysis, the cumulative frequency is not calculated. Instead, the variation of the original data in each class is summarized.

The calculation of the weight is the same as in the case of accumulation analysis. With classified attributes or classified variables, the weight is calculated using p, the rate of occurrence in each class.

$$W = \frac{1}{p(1-p)} \tag{11.62}$$

In multi-attributes and multi-variables, the weight is given by

$$W = \frac{1}{\sigma^2} \propto \frac{1}{\text{(Average no. of occurrences of individual class)}} \tag{11.63}$$

The method of calculation is the same as the case of accumulation analysis.

11.6 S/N Ratio for Accumulation Analysis and Frequency Analysis

There are many machines that automatically classify a mixture of coins or paper currency. At the development stage, we recommend using the S/N Ratio of physical or electrical characteristics using continuous variables to improve machine performance. At the final product testing stage, actual coins or paper money will be used.

For example, we want to avoid the coins from other countries being mixed in with ¥100 coins. The test will be conducted with several different kinds of coinage. In the test, a signal factor with k levels (M_1, M_2, ...,M_k) is provided to observe the result of classification.

As another example, the weather forecasting capability of a weather bureau will be compared to that of a neuralgia patient with respect to ability to predict the weather in 24 hours. S/N Ratio will be used to analyze experimental data:

(1) Factor, Levels and Characteristic

Factor to compare S/N Ratio: A

A_1 = The Meteorological Agency
A_2 = Neuralgia patient

Signal factor M: Actual weather

M_1 = Clear
M_2 =Partially cloudy
M_3 = Cloudy
M_4 = Cloudy, occasional rain
M_5 = Rain

Characteristic: 5 class classified attributes

Results of 2 week forecast and actual weather are shown in Table 11.18.

Table 11.18.

A_1 The Meteorological Agency

Judgment / Actual	M_1	M_2	M_3	M_4	M_5	Total
M_1	3	2	0	0	0	5
M_2	0	3	0	0	0	3
M_3	1	0	1	2	0	4
M_4	0	0	0	2	0	2
M_5	0	0	0	0	0	0
Total	4	5	1	4	0	14

A_2 Neuralgia Patient

Judgment / Actual	M_1	M_2	M_3	M_4	M_5	Total
M_1	3	0	2	0	0	5
M_2	2	1	0	0	0	3
M_3	0	0	2	2	0	4
M_4	0	0	1	0	1	2
M_5	0	0	0	0	0	0
Total	5	1	5	2	1	14

Since there is order from M_1 to M_5, accumulation analysis is used. The cumulative frequencies are shown in Table 11.19.

Table 11.19. Cumulative Frequency

A_1						A_2					
Signal Factor \ Class	I	II	III	IV	V	Signal Factor \ Class	I	II	III	IV	V
M_1	3	5	5	5	5	M_1	3	3	5	5	5
M_2	0	3	3	3	3	M_2	2	3	3	3	3
M_3	1	1	2	3	3	M_3	0	0	2	4	4
M_4	0	0	0	2	2	M_4	0	0	1	1	2
M_5	0	0	0	0	0	M_5	0	0	0	0	0
Total	4	9	10	14	14	Total	5	6	11	13	14

Class V shows the number of repetitions, which are not equal. 5 levels were provided for M, but there is no data for level M_5; therefore, the signal factor has only 4 levels. There were no judged results corresponding to M_5 in A_1, but there were for A_2. For this reason, the number of classes analyzed is 3 for A_1, and 4 for A_2.

(2) Calculation of Variation

[The case of A_1]

Weights are calculated.

$$W_1 = \frac{14^2}{4 \times (14-4)} = 4.9 \quad (11.64)$$

$$W_2 = \frac{14^2}{9 \times (14-9)} = 4.36 \quad (11.65)$$

$$W_3 = \frac{14^2}{10 \times (14-10)} = 4.9 \quad (11.66)$$

Total variation, S_T, is

$$S_T = 14 \times 3 = 42 \quad (f=13\times3=39) \quad (11.67)$$

Letting the number of repetitions of M_1, M_2, M_3, and M_4 be r_1, r_2, r_3 and r_4, respectively, the variation of M or S_M is given by

$$S_M = \frac{{M_1}^2}{r_1} + \frac{{M_2}^2}{r_2} + \frac{{M_3}^2}{r_3} + \frac{{M_4}^2}{r_4} - \frac{(M_1 + \cdots + M_4)^2}{r_1 + \cdots + r_4} \quad (11.68)$$

S_M is summarized using the weights.

$$S_M = \left(\frac{3^2}{5} + \frac{0^2}{3} + \frac{1^2}{4} + \frac{0^2}{2} - \frac{4^2}{14}\right) \times W_1$$
$$+ \left(\frac{5^2}{5} + \frac{3^2}{3} + \frac{1^2}{4} + \frac{0^2}{2} - \frac{9^2}{14}\right) \times W_2$$
$$+ \left(\frac{5^2}{5} + \frac{3^2}{3} + \frac{2^2}{4} + \frac{0^2}{2} - \frac{10^2}{14}\right) \times W_3$$
$$= 1.407 \times 4.9 + 2.464 \times 4.36 + 1.85 \times 4.9$$
$$= 26.702 \qquad (f{=}3{\times}3{=}9) \quad (11.69)$$

$$S_e = S_T - S_M = 29.298 \qquad (f{=}30) \quad (11.70)$$

[The case of A_2]

There are 4 classes to analyze.

$$\begin{cases} W_1 = \dfrac{14^2}{5 \times (14-5)} = 4.36 \\ W_2 = \dfrac{14^2}{6 \times (14-6)} = 4.08 \\ W_3 = \dfrac{14^2}{11 \times (14-11)} = 5.94 \\ W_4 = \dfrac{14^2}{13 \times (14-13)} = 15.08 \end{cases} \qquad (11.71)$$

$$S_T = 14 \times 4 = 56 \qquad (f{=}13{\times}4{=}42) \quad (11.72)$$

$$S_M = \left(\frac{3^2}{5} + \frac{2^2}{3} + \frac{0^2}{4} + \frac{0^2}{2} - \frac{5^2}{14}\right) \times W_1$$

$$+ \left(\frac{3^2}{5} + \frac{3^2}{3} + \frac{0^2}{4} + \frac{0^2}{2} - \frac{6^2}{14}\right) \times W_2$$

$$+ \left(\frac{5^2}{5} + \frac{3^2}{3} + \frac{2^2}{4} + \frac{1^2}{2} - \frac{11^2}{14}\right) \times W_3$$

$$+ \left(\frac{5^2}{5} + \frac{3^2}{3} + \frac{4^2}{4} + \frac{1^2}{2} - \frac{13^2}{14}\right) \times W_4$$

$$= 1.348 \times 4.36 + 2.229 \times 4.08 + 0.857 \times 5.94 + 0.428 \times 15.08$$

$$= 26.516 \qquad (f{=}3{\times}4{=}12) \qquad (11.73)$$

$$S_e = S_T - S_M = 56 - 26.516 = 29.483 \qquad (f{=}40) \qquad (11.74)$$

(3) ANOVA Table and S/N Ratio

Tables 11.20 and 11.21 show the ANOVA tables of A_1 and A_2.

Table 11.20. ANOVA Table of A_1

Factor	f	S	V	$\rho(\%)$
M	9	26.702	2.967	52.6
e	30	15.298	0.510	47.4
T	39	42.000		100.0

Table 11.21. ANOVA Table of A_2

Factor	f	S	V	$\rho(\%)$
M	12	26.516	2.210	31.6
e	40	29.483	0.737	68.4
T	52	56.000		100.0

The S/N Ratio is given by

$$\eta = 10 \log \frac{\rho_M}{\rho_e} \qquad (11.75)$$

$$\eta(A_1) = 10 \log \frac{52.6}{47.4} = 0.45 \quad \text{(db)} \qquad (11.76)$$

$$\eta(A_2) = 10 \log \frac{31.6}{68.6} = -3.37 \quad \text{(db)} \qquad (11.77)$$

The results reveal that the Weather Bureau is better by 3.82 decibels. Note that if there had been many factors assigned to an orthogonal array, the S/N Ratios of the conditions of the inner array are analyzed. In this example, the S/N Ratio is calculated using accumulation analysis. The S/N Ratio of the proportional equation might also have been used by assigning points from 1 to 5 to the signal factor and the output characteristic.

When there is no slope, the following S/N Ratio is often used:

$$\eta = 10 \log \frac{\frac{1}{r}(V_M - V_e)}{V_e} \tag{11.78}$$

There is little difference between the S/N Ratio using this equation and the one using percent contribution, as shown in this example.

If Equation (11.78) is used, the following calculation is made to obtain the harmonic mean:

$$\frac{1}{r} = \frac{1}{4}\left(\frac{1}{5} + \frac{1}{3} + \frac{1}{4} + \frac{1}{2}\right) = 0.321 \tag{11.79}$$

$$\bar{r} = 3.115 \tag{11.80}$$

It is the reciprocal of the mean of the reciprocals of the repetitions. In this case, it means that the number of repetitions is equivalent to 3.115.

$$\eta(A_1) = 10 \log \frac{\frac{1}{r}(V_M - V_e)}{V_e} = 10 \log \frac{\frac{1}{3.115}(2.967 - 0.510)}{0.510}$$

$$= 1.89 \quad \text{(db)} \tag{11.81}$$

$$\eta(A_2) = 10 \log \frac{\frac{1}{3.115}(2.210 - 0.737)}{0.737} = -1.93 \quad \text{(db)} \tag{11.82}$$

The difference between A_1 and A_2 is 3.82 decibels.

The S/N Ratio of classified attributes is not as meaningful as the case of continuous variables; it is mainly used for comparison purposes. Therefore, it is good enough to just using the ratio of percent contribution.

11.7 Minute Analysis

11.7.1 What Is Minute Analysis?

All kinds of data can be expressed by "yes" or "no" for co-ordinate variables. For example, in the case of a 1,000 hour life test, the time factor is denoted by K and the levels of K could be K_1=0th hour, K_2=100th hour, K_3=200th hour,...., K_{11}=1,000th hour, etc. 5 unit products are tested. At each level of K, "1" is given for alive, and "0" is given for dead to obtain Table 11.22.

Table 11.22. Minute Data for Life Test

No. \ Hour	K_1	K_2	K_3	K_4	K_5	K_6	K_7	K_8	K_9	K_{10}	K_{11}	Total
1	1	1	0	0	0	0	0	0	0	0	0	2
2	1	1	1	1	1	0	0	0	0	0	0	5
3	1	1	1	1	1	1	1	1	1	0	0	9
4	1	1	1	1	1	1	1	1	1	1	1	11
5	1	1	1	1	1	1	1	1	1	1	1	11
Total	5	5	4	4	4	3	3	3	3	2	2	38

In the table, the No.1 product was alive at the K_2=100th hour, but had failed by the K_3=200th hour. This is graded data for 2 classes. More information can be obtained from such data in data analysis. In the table, the total in the longitudinal direction is just the number of survivors, and the total in the horizontal direction shows the life. But Nos. 4 and 5 products were alive at the 1,000th hour; it shows the hours of survival instead of the time of failure.

Data expressed this way is called "minute data," and the analysis method for such data is called minute analysis. Using such method, a simple continuous variable, such as strength, can be expressed as

K_1 = minimum strength 150 (kgf/cm^2)
K_2 = 152
K_3 = 154
.
.
.
K_{21} = maximum strength 190

where K is the co-ordinate variable. For example, 6 data points with recorded bonding strengths of 152, 155, 160, 186, 189 and 190 (kgf/cm^2) are expressed with a coordinate width of 2 kgf/cm^2 are shown in Table 11.23.

Table 11.23. Minute Data of Strength

No.	150 K_1	152 K_2	154 K_3	156 K_4	158 K_5	160 K_6	162 K_7	···	184 K_{18}	186 K_{19}	188 K_{20}	190 K_{21}	Total
1	1	0	0	0	0	0	0		0	0	0	0	1
2	1	1	1	0	0	0	0		0	0	0	0	3
3	1	1	1	1	1	1	0	···	0	0	0	0	6
4	1	1	1	1	1	1	1		1	1	0	0	19
5	1	1	1	1	1	1	1		1	1	1	0	20
6	1	1	1	1	1	1	1		1	1	1	1	21
Total	6	5	5	4	4	4	3	···	3	3	2	1	70

In Table 11.23, the totals in the longitudinal direction are just cumulative histogram data. The totals in the horizontal direction are the strength after subtracting a working mean of 149, which is then rounded by 2 kg/cm^2. K is called a co-ordinate factor. It is recommended that the number of levels of a coordinate factor be between 5 or 6 and 20 classes, not to exceed 50.

Often, ω(omega) is used to express the coordinate, K. Omega is considered as a factor, and conduct the analysis of 0 and 1 data. This is called minute analysis.

Accumulation analysis as described before is the method for overall observation of the mean and variability. In comparison, in minute analysis, there is the advantage that the mean and the distribution are separated. In the case of a nominal-the-best type characteristic, the factor which is used to reduce variability, and the one to adjust the mean can be found.

11.7.2 Life Test of Fluorescent Lamps Example

(1) Layout and Results

In a manufacturing process for fluorescent lamps, five 2-level factors and an interaction, $A \times B$, were assigned to an orthogonal array L_8. The results after stress test were analyzed. The results show alive or death each other day for 20 days with 2 repetitions. For example, result 4 means it was alive the fourth day, but had failed by the sixth day. ◎ marks in Table 11.24 indicate they were alive on the 20th day.

Table 11.24. Layout and Results

No. \ Factor · Column	A 1	B 2	A × B 3	C 4	D 5	E 6	e 7	Life	
1	1	1	1	1	1	1	1	14	◎
2	1	1	1	2	2	2	2	20	10
3	1	2	2	1	1	2	2	8	◎
4	1	2	2	2	2	1	1	18	◎
5	2	1	2	1	2	1	2	◎	◎
6	2	1	2	2	1	2	1	13	◎
7	2	2	1	1	2	2	1	17	◎
8	2	2	1	2	1	1	2	12	14

(2) Preparation of Minute Data and Supplementary Table

Set the levels of ω as follows:

$\omega_1=0 \quad \omega_2=2 \;\ldots\ldots\; \omega_{11}=20$

Minute data are shown in Table 11.25.

ω is put outside similar to the indicative factor as shown in Chapter 5. A supplementary table is prepared in the same way to obtain Table 11.26.

(3) Calculation of Variation and ANOVA Table

The total number of data points is 8×2×11. Analysis is made for orthogonal array L_8 with factor ω of 11 levels outside the array. The analysis is complicated since there are many repetitions.

$$S_m = \frac{148^2}{8\times 2\times 11} = 124.45 \tag{11.83}$$

$$S_T = 148 \tag{11.84}$$

$$S_{T1} = \frac{1}{2}(2^2+2^2+2^2+\cdots+0^2) - S_m = \frac{284}{2} - S_m = 17.55 \tag{11.85}$$

$$S_{T2} = \frac{1}{11}(8^2+11^2+11^2+\cdots+8^2) - S_m = \frac{1{,}436}{11} - S_m = 6.10 \tag{11.86}$$

$$S_{T3} = \frac{1}{2\times 11}(19^2+22^2+\cdots+15^2) - S_m = \frac{2{,}840}{22} - S_m = 4.64 \tag{11.87}$$

$$S_A = \frac{(73-75)^2}{8\times 2\times 11} = 0.02 \tag{11.88}$$

$$S_B = \frac{(81-67)^2}{176} = 1.11 \tag{11.89}$$

$$S_{A\times B} = \frac{(76-72)^2}{176} = 0.09 \tag{11.90}$$

Table 11.25. Minute Data

No.	Day	0	2	4	6	8	10	12	14	16	18	20	Subtotal
	R \ ω	ω_1	ω_2	ω_3	ω_4	ω_5	ω_6	ω_7	ω_8	ω_9	ω_{10}	ω_{11}	
1	R_1	1	1	1	1	1	1	1	1	0	0	0	8
	R_2	1	1	1	1	1	1	1	1	1	1	1	11
	Subtotal	2	2	2	2	2	2	2	2	1	1	1	19
2	R_1	1	1	1	1	1	1	1	1	1	1	1	11
	R_2	1	1	1	1	1	1	1	1	1	1	1	11
	Subtotal	2	2	2	2	2	2	2	2	2	2	2	22
3	R_1	1	1	1	1	1	0	0	0	0	0	0	5
	R_2	1	1	1	1	1	1	0	0	0	0	0	6
	Subtotal	2	2	2	2	2	1	0	0	0	0	0	11
4	R_1	1	1	1	1	1	1	1	1	1	1	0	10
	R_2	1	1	1	1	1	1	1	1	1	1	1	11
	Subtotal	2	2	2	2	2	2	2	2	2	2	1	21
5	R_1	1	1	1	1	1	1	1	1	1	1	1	11
	R_2	1	1	1	1	1	1	1	1	1	1	1	11
	Subtotal	2	2	2	2	2	2	2	2	2	2	2	22
6	R_1	1	1	1	1	1	1	1	0	0	0	0	7
	R_2	1	1	1	1	1	1	1	1	1	1	1	11
	Subtotal	2	2	2	2	2	2	2	1	1	1	1	18
7	R_1	1	1	1	1	1	1	1	1	1	0	0	9
	R_2	1	1	1	1	1	1	1	1	1	1	1	11
	Subtotal	2	2	2	2	2	2	2	2	2	1	1	20
8	R_1	1	1	1	1	1	1	1	0	0	0	0	7
	R_2	1	1	1	1	1	1	1	1	0	0	0	8
	Subtotal	2	2	2	2	2	2	2	1	0	0	0	15

Table 11.26. Supplementary Table

Factor · Level	ω_1	ω_2	ω_3	ω_4	ω_5	ω_6	ω_7	ω_8	ω_9	ω_{10}	ω_{11}	Total
A_1	8	8	8	8	8	7	6	6	5	5	4	73
A_2	8	8	8	8	8	8	8	6	5	4	4	75
B_1	8	8	8	8	8	8	8	7	6	6	6	81
B_2	8	8	8	8	8	7	6	5	4	3	2	67
$(A{\times}B)_1$	8	8	8	8	8	8	8	7	5	4	4	76
$(A{\times}B)_2$	8	8	8	8	8	7	6	5	5	5	4	72
C_1	8	8	8	8	8	7	6	6	5	4	4	72
C_2	8	8	8	8	8	8	8	6	5	5	4	76
D_1	8	8	8	8	8	7	6	4	2	2	2	63
D_2	8	8	8	8	8	8	8	8	8	7	6	85
E_1	8	8	8	8	8	8	8	7	5	5	4	77
E_2	8	8	8	8	8	7	6	5	5	4	4	71
e_1	8	8	8	8	8	8	8	7	6	5	4	78
e_2	8	8	8	8	8	7	6	5	4	4	4	70
Total	16	16	16	16	16	15	14	12	10	9	8	148

$$S_C = \frac{(72-76)^2}{176} = 0.09 \tag{11.91}$$

$$S_D = \frac{(63-85)^2}{176} = 2.75 \tag{11.92}$$

$$S_E = \frac{(77-71)^2}{176} = 0.20 \tag{11.93}$$

$$S_{e1} = \frac{(78-70)^2}{176} = 0.36 \tag{11.94}$$

$$S_{e2} = S_{T2} - S_{T3} = 1.46 \tag{11.95}$$

$$\begin{aligned} S_\omega &= \frac{\omega_1{}^2 + \cdots + \omega_{11}{}^2}{\text{No. of replications}} - S_m \\ &= \frac{1}{16}(16^2 + 16^2 + \cdots + 8^2) - S_m = \frac{2{,}090}{16} - S_m \\ &= 6.18 \qquad (f{=}10) \end{aligned} \tag{11.96}$$

$S_{A\times\omega}$ is normally calculated by

$$S_{A\times\omega} = \frac{A_1\omega_1{}^2 + \cdots + A_2\omega_{11}{}^2}{8} - S_m - S_A - S_\omega \tag{11.97}$$

Since A has 2 levels, it is calculated by

$$S_{A\times\omega}=\frac{(A_1\omega_1-A_2\omega_2)^2+\cdots+(A_1\omega_{11}-A_2\omega_{11})^2}{16}-S_A \tag{11.98}$$

$$S_{A\times\omega}=\frac{1}{16}\left[(8-8)^2+(8-8)^2+\cdots+(4-4)^2\right]-S_A=0.36 \tag{11.99}$$

Similarly,

$$\begin{cases} S_{B\times\omega}=1.26 \\ S_{A\times B\times\omega}=0.54 \\ S_{C\times\omega}=0.22 \\ S_{D\times\omega}=3.38 \\ S_{E\times\omega}=0.42 \\ S_{e_1\times\omega}=S_{e_3}=0.52 \end{cases} \tag{11.100}$$

$$S_{e_4}=S_T-S_m-S_{T_1}-S_{e_2}=4.54 \tag{11.101}$$

The above results are summarized to obtain Table 11.27:

Table 11.27. NOVA Table

	Factor	f	S	V
	m	1	124.45	124.45
	A	1	0.02	0.02°
	B	1	1.11	1.11
	$A\times B$	1	0.09	0.09°
S_{T_3}	C	1	0.09	0.09°
S_{T_2}	D	1	2.75	2.75
	E	1	0.20	0.20°
	e_1	1	0.36	0.36°
S_{T_1}	e_2	8	1.46	0.182
S_T + S_{e_2}	ω	10	6.18	0.618
	$A\times\omega$	10	0.36	0.036°
	$B\times\omega$	10	1.26	0.126
	$A\times B\times\omega$	10	0.54	0.054°
	$C\times\omega$	10	0.28	0.028°
	$D\times\omega$	10	3.38	0.338
	$E\times\omega$	10	0.42	0.042°
	e_3	10	0.52	0.052°
	e_4	80	4.54	0.057
	T	176	148.00	

e_1 and e_2 are considerably larger than e_4. $e_1, e_2, A, A\times B, C$ and E are pooled to obtain (e_1), also $A\times\omega$, $A\times B\times\omega$, $C\times\omega$, $E\times\omega$, e_3 and e_4 are pooled to obtain (e_2). These are rearranged to obtain Table 11.28.

Table 11.28. Rearranged Table

Factor	f	S	V
m	1	124.45	124.45
B	1	1.11	1.11
D	1	2.75	2.75
(e_1)	13	2.22	0.171
ω	10	6.18	0.618
$B\times\omega$	10	1.26	0.126
$D\times\omega$	10	3.38	0.338
(e_2)	130	6.65	0.051
T	176	148.00	

Table 11.29. Estimation of $B\times\omega$ and $D\times\omega$

	ω_1	ω_2	ω_3	ω_4	ω_5	ω_6	ω_7	ω_8	ω_9	ω_{10}	ω_{11}	Total
B_1	8	8	8	8	8	8	8	7	6	6	6	81
B_2	8	8	8	8	8	7	6	5	4	3	2	67
D_1	8	8	8	8	8	7	6	4	2	2	2	63
D_2	8	8	8	8	8	8	8	8	8	7	6	85
Total	16	16	16	16	16	15	14	12	10	9	8	148

(4) Estimation

From Table 11.29, it is known that B_1D_2 is the optimum condition. The rate of survival of B_1D_2 at w_k is

$$\text{db of } \hat{\mu}_k = (\text{db of } \overline{B_1\omega_k}) + (\text{db of } \overline{D_2\omega_k}) - (\text{db of } \overline{\omega_k}) \tag{11.102}$$

From the db value of $\hat{\mu}_k$, $\hat{\mu}_k$ is calculated.

$(\hat{\mu}_1 \sim \hat{\mu}_9)$ is 100%. $\hat{\mu}_{10}$ and $\hat{\mu}_{11}$ are calculated as

$$\begin{aligned}\text{db of } \hat{\mu}_{10} &= (\text{db of } 75.0\%) + (\text{db of } 87.5\%) - (\text{db of } 75.0\%)\\ &= 4.771 + 8.451 - 4.771 = 8.451\end{aligned} \tag{11.103}$$

$$\therefore \hat{\mu}_{10} = 87.5\ (\%) \tag{11.104}$$

Similarly,

$$\hat{\mu}_{11} = 76.5\ (\%) \tag{11.105}$$

The rate of survival is shown in Figure 11.5. Calculation of the confidence limits is not shown here. See Literature 1).

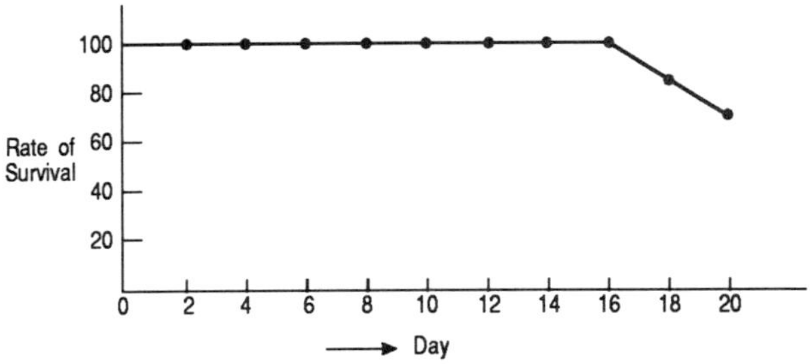

Figure 11.5. Graph of Rate of Survival

EXERCISES

Question (11.1). In order to improve the quality of an ink, the following factors were assigned to an L_{18}. The appearance after etching is classified into

0 small peel-off

Δ medium peel-off

X large peel-off

3 repetitions were taken since there is variation in the thickness. The layout and results are shown below.

No. \ Factor / Column	*A* 1	*B* 2	*C* 3	*D* 4	*E* 5	*F* 6	*G* 7	*H* 8	Appearance
1	1	1	1	1	1	1	1	1	○ Δ X
2	1	1	2	2	2	2	2	2	○ ○ ○
3	1	1	3	3	3	3	3	3	Δ Δ ○
4	1	2	1	1	2	2	3	3	Δ ○ ○
5	1	2	2	2	3	3	1	1	Δ ○ X
6	1	2	3	3	1	1	2	2	X X ○
7	1	3	1	2	1	3	2	3	Δ Δ Δ
8	1	3	2	3	2	1	3	1	X X ○
9	1	3	3	1	3	2	1	2	Δ X X
10	2	1	1	3	3	2	2	1	Δ X X
11	2	1	2	1	1	3	3	2	○ ○ ○
12	2	1	3	2	2	1	1	3	X ○ Δ
13	2	2	1	2	3	1	3	2	○ ○ ○
14	2	2	2	3	1	2	1	3	○ ○ ○
15	2	2	3	1	2	3	2	1	Δ Δ Δ
16	2	3	1	3	2	3	1	2	○ X ○
17	2	3	2	1	3	1	2	3	Δ Δ Δ
18	2	3	3	2	1	2	3	1	Δ Δ Δ

(a) Perform analysis of variance using accumulation analysis.

(b) Find the optimum condition and estimate the process average.

Question (11.2). In the life test experiment of fluorescent lamps given in the text, brightness was also measured. The results after subtracting a working mean 40 are shown below:

No. \ Factor · Column	A 1	B 2	$A \times B$ 3	C 4	D 5	E 6	e 7	Brightness	
1	1	1	1	1	1	1	1	0	5
2	1	1	1	2	2	2	2	3	4
3	1	2	2	1	1	2	2	0	9
4	1	2	2	2	2	1	1	2	1
5	2	1	2	1	2	1	2	2	2
6	2	1	2	2	1	2	1	-4	3
7	2	2	1	1	2	2	1	-6	-1
8	2	2	1	2	1	1	2	2	5

(a) Set the levels of brightness, ω as $\omega_1=-6$, ..., $\omega_{16}=9$, and perform minute analysis.

(b) Estimate the process average, or the optimum condition, using the significant factors.

(c) What would be the optimum condition considering both the life test and brightness?

CHAPTER 12

TREATMENT OF MISSING DATA

Sometimes during experimentation, data can not be obtained as planned because of lost data or equipment troubles. In this chapter, the treatment of missing data is described. The treatment includes the case of nominal-the-best characteristic when the error is too large and the S/N Ratio becomes -∞.

12.1 WHAT ARE MISSING DATA?

Missing data occur in the case when data is supposed to be available, but is not because some troubles occurred in the experiment. However, instances where data is not reported because of the occurrence of extreme conditions at experimental setting should not be treated as missing data. In such cases, it is recommended to record a -∞ decibel for data analysis. Even where data is available, but the error is too large in the case of a nominal-the-best characteristic, the decibel value is indeterminate. In the following equation,

$$\eta = 10 \log \frac{\frac{1}{r}(S_\beta - V_e)}{V_e} \tag{12.1}$$

if $V_e > S_\beta$, the decibel value is indeterminate, and -∞ is put. When there is no error on the other hand, it is indeterminate, too, and +∞ is put for the analysis.

-∞ or +∞ do not belong to the case which is called missing data. However, some treatment must be made so that the inner array can be analyzed.

When all data are equal to zero for smaller-the-better ($-\infty$ decibel) and for larger-the-better ($+\infty$ decibel), these do not belong to the case of missing data. These cases will be illustrated later.

In the case of missing data, sometimes all data of a certain inner array condition are missing, or part of the data are missing. In the former case, the S/N Ratio cannot be calculated; there is missing data for the analysis of the inner array. In the latter case, the S/N Ratio can be calculated in some cases and cannot be calculated in other cases. This case is illustrated in Section 12.5.

First, the estimation of missing data for the inner array analysis is considered. For such purpose, Yates' method is known. The method described in the next section which is called sequential approximation is a universal method. It is also applied for the missing data of the outer array.

12.2 Sequential Approximation

Sequential approximation is a method used to estimate the data which suppose to exist. In the experiment for the amount of wear characteristic in Chapter 5, assume that the data of Experiment No.3 in orthogonal array L_{12} is missing. This is the case of missing data; the S/N Ratio cannot be calculated.

Sequential approximation is made using the following procedure:

(1) Place the average into the missing location; this is called the "zero"th approximation.

(2) Estimate the missing location using larger effects, and call this the first approximation.

(3) Repeat (2) until the estimation converges.

The following is the estimation of experiment No.3 using sequential approximation:

(1) "Zero"th approximation

The average of 11 pieces of data is -27.15 decibels.

(2) ANOVA and estimation of No.3.

Table 12.1. Missing Data in No.3

No. \ Factor · Column	A 1	B 2	C 3	D 4	E 5	F 6	G 7	H 8	I 9	J 10	K 11	Amount of Wear (μm) N_1		N_2		η (db)
1	1	1	1	1	1	1	1	1	1	1	1	39	19	9	10	-27.12
2	1	1	1	1	1	2	2	2	2	2	2	21	20	3	16	-24.42
3	1	1	2	2	2	1	1	1	2	2	2					
4	1	2	1	2	2	1	2	2	1	1	2	42	27	13	24	-29.08
5	1	2	2	1	2	2	1	2	1	2	1	37	35	9	29	-29.44
6	1	2	2	2	1	2	2	1	2	1	1	68	85	38	64	-36.38
7	2	1	2	2	1	1	2	2	1	2	1	21	10	5	2	-21.54
8	2	1	2	1	2	2	2	1	1	1	2	38	23	17	4	-27.55
9	2	1	1	2	2	2	1	2	2	1	1	30	34	30	24	-29.46
10	2	2	2	1	1	1	1	2	2	1	2	72	46	24	40	-33.75
11	2	2	1	2	1	2	1	1	1	2	2	10	6	1	2	-15.47
12	2	2	1	1	2	1	2	1	2	2	1	30	12	8	0	-24.42

Put -27.15 db into No.3, and the ANOVA table is used to find larger effects. Table 12.2 shows the ANOVA table.

Table 12.2. ANOVA Table Using 0th Approximation

Factor	f	S	V
A	1	38.16	38.16
B	1	10.64	10.64°
C	1	55.64	55.64
D	1	4.84	4.84°
E	1	5.91	5.91°
F	1	0.01	0.01°
G	1	0.08	0.08°
H	1	7.68	7.68°
I	1	53.68	53.68
J	1	139.40	139.40
K	1	9.97	9.97°
T	11	326.02	

From Table 12.2, larger effects are A, C, I and J. They are used to estimate the condition of No.3: $A_1(B_1)C_2(D_2E_2F_1G_1H_1)I_2J_2(K_2)$.

Table 12.3. Supplementary Table

Factor	Level Total	Factor	Level Total
A_1	173.59	G_1	162.39
A_2	152.19	G_2	163.39
B_1	157.24	H_1	158.09
B_2	168.54	H_2	167.69
C_1	149.97	I_1	150.20
C_2	175.81	I_2	175.58
D_1	166.70	J_1	183.34
D_2	159.08	J_2	142.44
E_1	158.68	K_1	168.36
E_2	167.10	K_2	157.42
F_1	163.06		
F_2	162.72		

$$\hat{\mu}(\text{No.3}) = \frac{-173.59 + (-175.81) + (-175.58) + (-142.44)}{6} - 3 \times \frac{-325.78}{12} = -29.78 \text{ (db)} \qquad (12.2)$$

The first approximation is -29.78 db.

Using Procedure (2) by putting -29.78 db,

$$\hat{\mu}(\text{No.3}) = -30.88 \text{ (db)} \qquad (12.3)$$

is obtained. Using the same procedure, the fifth approximation is obtained as -31.62 db.

This is close to the fourth approximation of -31.53, and calculation is discontinued. The actual result was -34.02; these are not very close, but the estimate is within the confidence limit of +/-3.80 db.

The above calculation is terminated at the fifth approximation using 4 factors A, C, I and J. Since the effect of E becomes larger and larger, the approximation could be better if E were included in the calculation.

Next, the optimum condition is estimated using the ANOVA table from the fifth approximation.

The optimum condition is $A_2B_1C_1D_2E_1F_2G_2H_1I_1J_2K_2$. Using the effects of A, C, E, I and J,

$$\hat{\mu} = \overline{A}_2 + \overline{C}_1 + \overline{E}_1 + \overline{I}_1 + \overline{J}_2 - 4\overline{T}$$
$$= -25.36 - 25.00 - 26.45 - 25.03 - 24.48 - 4 \times (-27.52)$$
$$= -16.24 \text{ (db)} \qquad (12.4)$$

In Chapter 5, the estimate without using missing data is -15.84 db; both are fairly close to each other, suggesting the method is practical.

Sequential approximation can be used when there is more than one missing experiment. But it cannot be used when all data under a certain level of a certain factor is missing.

Table 12.4. Supplementary Table Using 5th Approximation

	A	*B*	*C*	*D*	*E*	*F*	*G*
1	-178.06	-161.71	-149.97	-166.70	-158.68	-167.53	-166.86
2	-152.19	-168.54	-180.28	-163.55	-171.57	-162.72	-163.39
Total	-330.25						

	H	*I*	*J*	*K*
1	-162.56	-150.20	-183.34	-168.36
2	-167.69	-180.05	-146.91	-161.89

Table 12.5. ANOVA Table Using 5th Approximation

Factor	*f*	*S*	*V*	ρ(%)
A	1	55.77	55.77	15.6
B	1	3.89	3.89°	
C	1	76.56	76.56	21.6
D	1	0.83	0.83°	
E	1	13.85	13.85	3.4
F	1	1.93	1.93°	
G	1	1.00	1.00°	
H	1	2.19	2.19°	
I	1	74.25	74.25	20.9
J	1	110.60	110.60	31.5
K	1	3.49	3.49°	
(*e*)	(6)	(13.33)	(2.22)	(7.0)
T	11	344.36		100.0

12.3 SEQUENTIAL APPROXIMATION WHEN FACTORS ARE CONTINUOUS VARIABLES

When factors are continuous variables, a better approximation can be made using a functional expression.

For example, orthogonal array L_9 was used in the following experiment:

A: Amount of urea (l)	A_1=1.0	A_2=1.4	A_3=1.8
B: Amount of water (ml)	B_1=120	B_2=200	B_3=280
C: Reaction time (min)	C_1=10	C_2=15	C_3=20
D: Amount of solvent (l)	D_1=5	D_2=15	D_3=25

Assume the result (or S/N Ratio) of No.5 is missing. The actual result is 28.5. The average of eight results is 23.6.

Table 12.6. Results Putting 0th Approximation

No. \ Factor · Column	A 1	B 2	C 3	D 4	y
1	1	1	1	1	11.5
2	1	2	2	2	22.7
3	1	3	3	3	22.6
4	2	1	2	3	19.0
5	2	2	3	1	23.6
6	2	3	1	2	24.0
7	3	1	3	2	25.1
8	3	2	1	3	30.3
9	3	3	2	1	33.3

The ANOVA table considering linear and quadratic terms is shown in Table12.7.

Table 12.7. ANOVA Table Using 0th Approximation

Factor		f	S	V
A	l	1	169.60	169.60
	q	1	8.41	8.41
B	l	1	98.42	98.42
	q	1	17.40	17.40
C	l	1	5.04	5.04
	q	1	9.25	9.25
D	l	1	2.04	2.04
	q	1	0.61	0.61
T		8	310.76	

For the first approximation, larger effects A_l, B_l and B_q are used.

When factor x has equal intervals, its orthogonal polynomial equation is given by

$$y = m + a_1(x - \bar{x}) + a_2\left[(x - \bar{x})^2 - \frac{k^2 - 1}{12}h^2\right] \tag{12.5}$$

where m is the average of y's

k is the number of levels

h is level interval

a_1 and a_2 are calculated from the following equation using the coefficients of Orthogonal Polynomial Equation.

$$a_i = \frac{W_1y_1 + W_2y_2 + \cdots + W_ky_k}{r_0\lambda sh^i} \tag{12.6}$$

When k=3,

$$a_1 = \frac{-1 \times y_1 + 0 \times y_2 + 1 \times y_3}{r_0 \times 2 \times h} \tag{12.7}$$

$$a_2 = \frac{1 \times y_1 + 2 \times y_2 + 1 \times y_3}{r_0 \times 2 \times h^2} \tag{12.8}$$

Considering the terms m, A_1, B_l and B_q, y is given by

$$y = m + a_1(A - \bar{A}) + b_1(B - \bar{B}) + b_2\left[(B - \bar{B})^2 - \frac{3^2 - 1}{12} \times h_{13}{}^2\right] \tag{12.9}$$

To calculate m, a_1, b_1 and b_2, level totals are calculated as shown in Table 12.8.

Table 12.8. Level Totals of A and B

Level \ Factor	A	B
1	56.8	55.6
2	66.6	76.6
3	88.7	79.9
Total	212.2	212.1

$$m = \frac{212.2}{9} = 23.6 \tag{12.10}$$

$$a_1 = \frac{-56.8 + 88.7}{3 \times 2 \times 0.4} = 13.29 \tag{12.11}$$

$$b_1 = \frac{-55.6 + 79.9}{3 \times 2 \times 80} = 0.051 \tag{12.12}$$

$$b_2 = \frac{55.6 - 2\times 76.6 + 79.9}{3\times 2\times 80^2} = -0.000\,46 \tag{12.13}$$

y is estimated as

$$y = m + a_1(A - \overline{A}) + b_1(B - \overline{B}) + b_2[(B - \overline{B})^2 - \frac{2}{3}\times 80^2]$$
$$= 23.6 + 13.29(A - 1.4) + 0.051(B - 200)$$
$$- 0.000\,46[(B - 200)^2 - \frac{2}{2}\times 80^2] \tag{12.14}$$

where $\overline{A}$ and $\overline{B}$ are the averages of A and B respectively.

Since the condition of No.5 is $A_2B_2C_3D_1$, put A=1.4 and B=200 into Equation(12.14) to obtain 25.6 as the first approximation. Using this value, ANOVA table is made as shown in Table 12.9.

Table 12.9. ANOVA Table Using 1st Approximation

Factor		f	S	V
A	l	1	169.60	169.60
	q	1	3.82	3.82
B	l	1	98.42	98.42
	q	1	26.16	26.16
C	l	1	9.38	9.38
	q	1	6.60	6.60
D	l	1	0.38	0.38
	q	1	0.09	0.09
T		8	314.45	

Since effects A_q, D_l and D_q becomes smaller, C_l and C_q are added into the equation.

$$y = 23.8 + 13.29(A - 1.4) + 0.051(B - 200)$$
$$-0.000\,57\,[(B - 200)^2 - \frac{2}{3}\times 80^2] + 0.254(C - 15)$$
$$-0.007\,18\,[(C - 15)^2 - \frac{2}{3}\times 5^2] \tag{12.15}$$

Putting A=1.4, B=200 and C=20 into the equation to obtain 26.9 as the second approximation.

Repeating such calculation, the estimate converged to be 28.4 after theeighth approximation, which is used for No.5.

It is cumbersome to conduct analysis of variance every time. In this case, analysis of variance was conducted for the first and the second approximation, and from the third time on, approximation was repeated using A_l, B_l, C_l and C_q.

Table 12.10 shows the ANOVA table using 28.4 as the substituted value.

Table 12.10. ANOVA Table Using 28.4 As The Final Substitution

Factor		*f*	*S*	*V*
A	*l*	1	169.60	169.60
	q	1	0.41	0.41°
B	*l*	1	98.42	98.42
	q	1	41.40	41.40
C	*l*	1	17.68	17.68
	q	1	3.65	3.65
D	*l*	1	0.28	0.28°
	q	1	0.13	0.13°
T		8	331.57	

Larger effects are used to express data, and the remaining effects (o mark) are pooled to be the error. It is recommended that the total degrees of freedom and the one for the error be subtracted by the number of missing data. In this case it is one. Table 12.11 shows the rearranged ANOVA table.

Table 12.11. Rearranged ANOVA Table

Factor	*f*	*S*	*V*
Al	1	169.60	169.60
Bl	1	98.42	98.42
Bq	1	41.40	41.40
Cl	1	17.68	17.68
Cq	1	3.65	3.65
e	2	0.82	0.41
T	7	331.57	

Trial was made using process average estimation approach as shown in the previous section, but a good estimate was not obtained. It seems that when there are not many data points, the approach using function of this section is better.

12.4 THE CASES OF -∞ AND +∞ DECIBELS

When there are -∞ and +∞ data values, there is usually a large variation in S/N Ratios between experiments. If the objective is to find the factors which affect the S/N Ratio, decibel values are classified into a few classes then analyzed by accumulation analysis. Alternatively, sequential approximation may be used. If the latter is to be used, a value which is 3 to 5 decibels either lower or higher than any decibel value may be used instead of the average.

12.5 MISSING DATA OUTSIDE

12.5.1 The Case of Smaller-the-better or Larger-the-better Characteristic

As a principle, when noise factors are not compounded, S/N Ratio is calculated from the existing data. The number of repetitions will be different from one condition to another in the inner array. When error factors are compounded and either the standard condition or the worst condition is missing, the S/N Ratio should not be calculated. Instead, this inner array condition is treated as missing data.

In the case of smaller-the-better or larger-the-better characteristics, if some of the data (not all data) are equal to zero, it is recommended that the zero be replaced by about 1/10 of the smallest value, and then calculate the S/N Ratio.

12.5.2 The Case of Nominal-the-best and Dynamic Characteristics

When noise factors are compounded, it is treated as missing data in principle. When there is no signal factor, the S/N Ratio is calculated from the remaining data. When there is a signal factor, the S/N Ratio is calculated from the remaining data. However, the numbers of repetitions are not equal, and the tables of orthogonal polynomial equation cannot be used even when level intervals are equal. For such a case, the following equations are used.

Table 12.12. The Case With Unequal Number of Repetitions

M_1 (r_1)	M_2 (r_2)	M_3 (r_3)	···	M_1 (r_k)
y_1	y_2	y_3	···	y_k

[Proportional Equation]

S_T = Total of the squares of data $\qquad (f_T = r_1 + \cdots + r_k) \qquad (12.16)$

$$S_\beta = \frac{(M_1y_1 + M_2y_2 + \cdots + M_ky_k)^2}{r_1M_1{}^2 + r_2M_2{}^2 + \cdots + r_kM_k{}^2} \qquad (f=1) \quad (12.17)$$

$$S_e = S_T - S_\beta \qquad (f_e = r_1 + \cdots + r_k - 1) \quad (12.18)$$

$$V_e = \frac{S_e}{f_e} \quad (12.19)$$

$$\eta = 10 \log \frac{\frac{1}{r_1M_1{}^2 + \cdots + r_kM_k{}^2}(S_\beta - V_e)}{V_e} \quad (12.20)$$

[Linear Equation]

S_T = Total of the squares of data $\qquad (f_T = r_1 + \cdots + r_k) \quad (12.21)$

$$S_m = \frac{(\text{Total})^2}{r_1 + \cdots + r_k} \qquad (f=1) \quad (12.22)$$

$$S_\beta = \frac{[(M_1 - \overline{M})y_1 + \cdots + (M_k - \overline{M})y_k]^2}{r_1(M_1 - \overline{M})^2 + \cdots + r_k(M_k - \overline{M})^2} \qquad (f=1) \quad (12.23)$$

$$\text{Where, } \overline{M} = \frac{r_1M_1 + r_2M_2 + \cdots + r_kM_k}{r_1 + r_2 + \cdots + r_k} \quad (12.24)$$

$$S_e = S_T - S_m - S_\beta \qquad (f_e = r_1 + \cdots + r_k - 2) \quad (12.25)$$

$$V_e = \frac{S_e}{f_e} \quad (12.26)$$

$$\eta = 10 \log \frac{\frac{1}{r}(S_\beta - V_e)}{V_e} \quad (12.27)$$

$$\text{Where, } r = r_1(M_1 - \overline{M})^2 + \cdots + r_k(M_k - \overline{M})^2 \quad (12.28)$$

Table 12.13. Data With Unequal Number of Repetitions

M	M_1	M_2	M_3	Total
	44.0	34.2	27.4	
	43.7	31.9	31.8	
	46.6	35.5	27.3	
		42.8	25.5	
		33.7	27.6	
		28.1	34.6	
Total	134.3	206.2	174.2	514.7

Table 12.13 shows the results of experiment for steering wheel angle and turning radius. There are only 3 data points in M_1.

$$S_m = \frac{514.7^2}{15} = 17{,}661.07 \qquad (f{=}1) \quad (12.29)$$

$$S_T = 44.0^2 + 43.7^2 + \cdots + 34.6^2 = 18{,}338.35 \qquad (f{=}15) \quad (12.30)$$

$$\overline{M} = \frac{1}{15}(3 \times 200 + 6 \times 250 + 6 \times 300) = 260 \qquad (12.31)$$

Therefore,

$$M_1 - \overline{M} = -60 \quad M_2 - \overline{M} = -10 \quad M_3 - \overline{M} = 40$$

$$\therefore S_\beta = \frac{(-60 \times 134.3 - 10 \times 206.2 + 40 \times 174.2)^2}{3 \times (-60)^2 + 6 \times (-10)^2 + 6 \times 40^2}$$

$$= \frac{(-3{,}152)^2}{21{,}000} = 473.10 \qquad (f{=}1) \quad (12.32)$$

$$S_e = S_T - S_m - S_\beta = 240.18 \qquad (f{=}13) \quad (12.33)$$

$$V_e = \frac{240.18}{13} = 15.706 \qquad (12.34)$$

$$\therefore \eta = 10 \log \frac{\frac{1}{21{,}000}(472.10 - 15.706)}{15.706} = -28.58 \text{ (db)} \qquad (12.35)$$

In the case when noise factors are compounded but there are repetitions and there is at least one data point in every combination, a S/N Ratio can be calculated as the case of unequal number of repetitions. But it is not described here.

12.6 WHEN EXPERIMENTAL CONDITION WAS MISTAKENLY SET (1)

In the experiment of L_9 described in Section 12.3, suppose that the condition of No.5 was mistakenly conducted under condition $A_2B_2C_3D_2$ instead of $A_2B_2C_3D_1$. In such a case, it may be treated as missing data. However, we might utilize the existing data under the mistakenly conducted condition. For example, No.5 was conducted under condition $A_2B_2C_3D_2$ to get the result shown i n Table 12.14

A: Amount of urea(l)	A_1=1.0	A_2=1.4	A_3=1.8
B: Amount of water(ml)	B_1=120	B_2=200	B_3=280
C: Reaction time(min)	C_1=10	C_2=15	C_3=20
D: Amount of solvent(l)	D_1=5	D_2=15	D_3=25

There are several ways to analyze using mistaken data. If only one data point was mistaken, analysis can be made as is. In this example, Table 12.15 is obtained.

Table 12.14. Layout And Results

No. \ Factor · Column	A 1	B 2	C 3	D 4	y(%)
1	1	1	1	1	11.5
2	1	2	2	2	22.7
3	1	3	3	3	22.6
4	2	1	2	3	19.0
5	2	2	3	②*	29.3
6	2	3	1	2	24.0
7	3	1	3	2	25.1
8	3	2	1	3	30.3
9	3	3	2	1	33.3

* Mistakenly set as 2 instead of 1.

Table 12.15. ANOVA Table Ignoring the Mistake

Factor		f	S	V
A	l	1	169.60	169.60
	q	1	0.04	0.04°
B	l	1	98.42	98.42
	q	1	47.04	47.04
C	l	1	18.03	18.03
	q	1	2.88	2.88
D	l	1	0.38	0.38°
	q	1	0.32	0.32°
(e)		(2)	(0.74)	(0.247)
T		8	340.19	

Since there is no degrees of freedom for the error term, smaller factorial effects A_q, D_l and D_q are pooled together to get the error variance(e).

The mistake of setting experiment No.5 at $A_2B_2C_3D_2$ instead of $A_2B_2C_3D_1$ might affect the variance of D_l and D_q, but even if the experiment were conductedcorrectly, D would not be significant. That means there would be no problem if the existing data were used for analysis. Therefore, significant terms A_l, B_l, B_q, C_l and D_q are put into orthogonal polynomial equation.

$$y = m + a_1(A - \overline{A}) + b_1(B - \overline{B}) + b_2\left[(B - \overline{B})^2 - \frac{3^2 - 1}{12}h_B{}^2\right] + c_1(C - \overline{C}) + c_2\left[(C - \overline{C})^2 - \frac{3^2 - 1}{12}h_C{}^2\right] \tag{12.36}$$

where

$$\hat{m} = \frac{11.5 + 22.7 + \cdots + 33.3}{9} = 24.2 \tag{12.37}$$

$$\hat{a}_1 = \frac{-A_1 + A_3}{\gamma_0 \lambda S h_A} = \frac{-56.8 + 88.7}{3 \times 2 \times 0.4} = 13.292 \tag{12.38}$$

$$\hat{b}_1 = \frac{-B_1 + B_3}{\gamma_0 \lambda S h_B} = \frac{-55.6 + 79.9}{3 \times 2 \times 80} = 0.051 \tag{12.39}$$

$$\hat{b}_2 = \frac{B_1 - 2B_2 + B_3}{\gamma_0 \lambda S h_B{}^2} = \frac{55.6 - 2 \times 82.3 + 79.9}{3 \times 2 \times 80^2} = -0.000\,758 \tag{12.40}$$

$$\hat{c}_1 = \frac{-C_1 + C_3}{\gamma_0 \lambda S h_C} = \frac{-65.8 + 77.0}{3 \times 2 \times 5} = 0.373 \tag{12.41}$$

$$\hat{c}_2 = \frac{C_1 - 2C_2 + C_3}{\gamma_0 \lambda S h_C^{\ 2}} = \frac{65.8 - 2 \times 75.0 + 77.0}{3 \times 2 \times 5^2}$$
$$= -0.048 \qquad (12.42)$$

$$\begin{cases} \overline{A} = 1.4 \\ \overline{B} = 200 \\ \overline{C} = 15 \end{cases} \qquad (12.43)$$

Putting these figures into Equation (12.36),

$$y = 24.2 + 13.292(A - 1.4) + 0.051(B - 200)$$
$$-0.000\,758\,[(B - 200)^2 - \frac{2}{3} \times 80^2]$$
$$+0.373(C - 15) - 0.048\,[(C - 15)^2 - \frac{2}{3} \times 5^2]$$
$$= -47.290 + 13.292A + 0.354\,2B$$
$$-0.000\,758B^2 + 1.813C - 0.048C^2 \qquad (12.44)$$

12.7 WHEN EXPERIMENTAL CONDITION WAS MISTAKENLY SET (2)

In the example of previous section, if the level of a significant factor such as *A*, *B* or *C* are mistakenly set, it is inappropriate to ignore the mistake. For example, the condition of *C* of experiment No.5 was set at 30 minutes instead of 20 minutes and the result was 21.0. One way is to ignore this result and treat it as missing data. Another way is to recognize the condition as 30 minutes and the result of 21.0, and determine the equation by regression analysis from such non-orthogonal data.

To explain the latter approach, the equation is written as

$$y = m + a_1A + a_2A^2 + b_1B + b_2B^2 + c_1C + c_2C^2 + d_1D + d_2^2 \qquad (12.45)$$

From Table 12.16, coefficients $m, a_1, a_2, \cdots, d_2$ are determined.

Table 12.16. Data For Regression Analysis

No.	A	A^2	B	B^2	C	C^2	D	D^2	y
1	1.0	1.0	120	14,400	10	100	5	25	11.5
2	1.0	1.0	200	40,000	15	225	15	225	22.7
3	1.0	1.0	280	78,400	20	400	25	625	22.6
4	1.4	1.96	120	14,400	15	225	25	625	19.0
5	1.4	1.96	200	40,000	30	900	5	25	21.0
6	1.4	1.96	280	78,400	10	100	15	225	24.0
7	1.8	3.24	120	14,400	20	400	15	225	25.1
8	1.8	3.24	200	40,000	10	100	25	625	30.3
9	1.8	3.24	280	78,400	15	225	5	25	33.3

For the determination of coefficients, the method called experimental regression analysis is recommended. See Reference 2.

In this example, the following equation was obtained.

$$y = -43.26 + 11.5A + 0.625A^2 + 0.317B - 0.000\,666B^2 + 1.89C - 0.051\,9C^2 + 0.005D + 0.000\,2D^2 \quad (12.46)$$

This equation may be used to estimate the result of experiment No.5.

If a factor is a non-continuous or attribute type, the linear or quadratic term cannot be determined. It is necessary to compare every level of such an attribute type factor. Suppose factor A is this type for example. The following equation is used.

$$y = m + a_2'A_2' + a_3'A_3' + b_1B + b_2B^2 + \cdots \quad (12.47)$$

Where A_2' and A_3' are:

A_2': Put 1 to the second level of A and 0 to other levels.

A_3': Put 1 to the third level of A and 0 to other levels.

therefore, a_2' and a_3' show (A_2-A_1) and (A_3-A_1) respectively. To determine the regression equation, the data in Table 12.17 are used instead of the ones from Table 12.16.

Table 12.17. Treatment of Attribute Type Factor, A

No.	A_2'	A_3'	B	B^2	C	C^2	D	D^2	y
1	0	0	120	14,400	10	100	5	25	11.5
2	0	0	200	40,000	15	225	15	225	22.7
3	0	0	280	78,400	20	400	25	625	22.6
4	1	0	120	14,400	15	225	25	625	19.0
5	1	0	200	40,000	30	900	5	25	21.0
.	.	.	.	.	.	.	.	.	.
.	.	.	.	.	.	.	.	.	.
.	.	.	.	.	.	.	.	.	.
9	0	1	280	78,400	15	225	5	25	33.3

References

1) Genichi Taguchi (1977): *System of Experimental Design*, 3rd Edition, Vol. 1 and 2, American Supplier Institute, UNIPUB.
2) Genichi Taguchi, Yoshiko Yokoyama (1980): Management Engineering Series, Vol. 5, *Statistical Analysis*, Japanese Standards Association.
3) Genichi Taguchi, et al. (1984): *Parameter Design for New Product Development*, Japanese Standards Association.
4) Genichi Taguchi, Editor, (1984): *Reliability Design Case Studies for New Product Development*, Japanese Standards Association.
5) Genichi Taguchi, Yoshiko Yokoyama (1987): Management Engineering Series, Vol. 18, *Experimental Design*, New Edition, Japanese Standards Association.
6) Seiso Konishi, et al. (1988): Quality Engineering Series, Vol. 3, *Signal-to-Noise Ratio for Quality Evaluation*, Japanese Standards Association, American Supplier Institute.
7) Genichi Taguchi (1988): Quality Engineering Series, Vol. 1, *Research and Development*, Japanese Standards Association, American Supplier Institute.

Appendix A

1. Variation and Sum of Squares
2. Degrees of Freedom
3. Linear Equation and its Variation
4. Orthogonality
5. Expectation
6. F Test
7. Objectives of Using Orthogonal Arrays in Experimentation
8. 3-level Orthogonal Arrays and Mixed-type Orthogonal Arrays
9. Layout Only with Main Effects
10. Discount Coefficient and the Meaning of Over-estimate
11. Equation for Confidence Intervals, Effective Number of Replications
12. Polynomial Expansion and Chebyshev's Orthogonal Polynomial Equation
13. Percent Contribution When e_1 is Significant Against e_2
14. Square of Sensitivity and Decibel
15. Noise Factor Level Setting
16. Effective Number of Replications for Treatment
17. Estimation for Idle Column Method
18. Objective Using Cumulative Data in Accumulation Analysis

1. VARIATION AND SUM OF SQUARES

There are two reasons to use squares. One is from an engineering viewpoint, and another is from an economic viewpoint.

In physics, work is defined by the product of force, F, and time, t.

Work = Ft

This definition may be all right for natural phenomena where efficiency is not the concern. In the real world however, if a person spent a long time working hard but acccomplished nothing, the amount of work should be considered as zero. The amount of work must be measured by the magnitude of its impact to the output (or objective charactersitics).

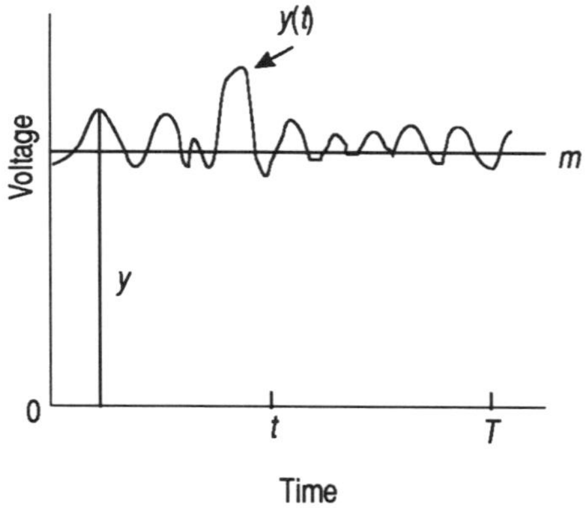

Figure 1. Current Flowing in a Telphone Circuit

Figure 1 shows the output of a telephone circuit. When we pick up a phone, 48 volt direct current flows in the circuit. But direct current does not transfer sound. When we speak to the phone, the longitudinal air wave vibrates the vibrating board in the receiver. Behind the vibrating board, there is carbon powder which works as the resistor of the telephone circuit. When the vibrating board vibrates, carbon powder is compressed or loosened. The voice wave is converted into contact resistance and thus, vibrating current is generated. When time and instantaneous voltage are denoted by t and $y(t)$ respectively, the amount of work done by the current during the time $(0, T)$ is the total amount of work by alternating current:

$$P = \int_0^T [y(t) - \bar{y}]^2 \, dt$$

$$= \int_0^T y^2(t)dt - \frac{1}{T}\left[\int_0^T y(t)dt\right]^2$$

The above equation is rewritten as

$$\int_0^T y^2(t)dt = T\left[\frac{1}{T}\int_0^T y(t)dt\right]^2 + \int_0^T [y(t) - \bar{y}]^2 dt$$

Total of squares $S_T = T \times (\text{direct current voltage})^2$

$= T \times (\text{alternating current voltage})^2$

This is the same form as the following equation:

Total sum of squares S_T = (general mean S_m) + (error variation S_e)

Only by using the form of squares, can the effects of various frequency components or various factorial effects be added together.

Also when square is used, it is proportional to the magnitude of the actual loss caused by variability or error. Therefore, the evaluation of factorial effects in the form of square is appropriate both from economic and engineering viewpoints.

2. Degrees of Freedom

(1) Concept

Degrees of freedom is defined in JIS Z 8101 as:

"The number of independent units in the whole residual, or the whole estimate of factorial effects.

"In sample $(x_1, x_2, ..., x_n)$, the number of degrees of freedom of residual $(x_i - \bar{x})$, i=1, 2, ..., n is (n-1). In a 2-way layout experiment with two factors, A (a levels) and B (b levels), the number of degrees of freedom of the estimate of main effect A, $(\bar{x}_i - \bar{x}..)$, i=1, 2, ..., a, is (a-1)."

In the definition, the number of independent units is the number of unknowns. To estimate the effect of a-level factor A, or $\bar{A}_1, \bar{A}_2, ..., \bar{A}_a$,

$$(\bar{A}_1 - \bar{T}) + (\bar{A}_2 - \bar{T}) + \cdots + (\bar{A}_a - \bar{T}) = 0$$

when $\bar{T}$ is the grand average. If (a-1) units are known, the remaining unit can be determined. In this case, (a-1) is the number of degrees of freedom.

Generally, a factorial effect requires the following condition:

$$(\bar{A}_1 - \bar{T}) + (\bar{A}_2 - \bar{T}) + \cdots + (\bar{A}_a - \bar{T}) = 0$$

Therefore, the number of degrees of freedom of a factor with a-levels is equal to (a-1). Since there is a restriction, the actual number of unknowns is less than the number of unknowns needed to be estimated.

Degrees of freedom can be intuitively understood by the following example. The position of a point which is freely moving the space is determined by three coordinates. Its number of degrees of freedom is equal to three. If the movement is restricted to be on a flat surface or a curved surface, its number of degrees of freedom is equal to two.

(2) Degrees of Freedom in Experimentation

Variation is calculated to measure the variability of deviations. However, the variation of two deviations and the one of ten deviations are different and it is necessary to specify the number of deviations for calculation. In statistics, degrees of freedom is used to specify the number of deviations.

When there are k independent linear equations among n data points, $y_1, y_2, ..., y_n$, the following S_T

$$S_T = y_1^2 + y_2^2 + \cdots + y_n^2$$

is the variation with the number of degrees of freedom equal to (n-k). For example,

(a) When $y_1, y_2, ..., y_n$ are deviations from a target, there is no restrictions, i.e., n units are independent. The number of degrees of freedom of the following S_T

$$S_T = y_1^2 + y_2^2 + \cdots + y_n^2$$

is equal to n.

(b) When $y_1, y_2, ..., y_n$ are deviations from the mean, there is a restriction of

$$y_1 + y_2 + \cdots + y_n = 0$$

The number of degrees of freedom of the following S_T

$$S_T = y_1^2 + y_2^2 + \cdots + y_n^2$$

is equal to (n-1).

Such variations can be algebraically expressed in a quadratic form. The number of degrees of freedom is equal to the rank of the coefficient matrix of quadratic form.

(3) Calculation of Degrees of Freedom

Degrees of freedom in analysis of variance relates to variation. Such degrees of freedom may be found from the number of squares in the equation to calculate variation.

For example, the variation, S_A, of a-level factor is given by

$$S_A = \frac{A_1{}^2 + A_2{}^2 + \cdots + A_a{}^2}{r_0} - \frac{(A_1 + \cdots + A_a)^2}{ar_0}$$

where A_1 is the total of r_0 data points at "i"th level. The number of degrees of freedom of S_A is given by the number of squares in the equation to be (n-1). The variation of interaction $A \times B$, denoted by $S_{A \times B}$, is given by

$$S_{A \times B} = \frac{A_1B_1{}^2 + \cdots + A_aB_b{}^2}{r_0} - \frac{(A_1B_1 + \cdots + A_aB_b)^2}{abr_0} - S_A - S_B$$

Its number of degrees of freedom is calculated as

$$f_{A \times B} = ab - 1 - (a - 1) - (b - 1)$$
$$= (a - 1)(b - 1)$$

It is the product of the numbers of degrees of freedom of A and B. In the equation, A_iB_j, is the total of combination of A at "i"th level and B at "j"th level, r_0 is the number of replications of the combination, and $f_{A \times B}$ is the number of degrees of freedom of the interaction.

For the variation of a linear equation and its number of degrees of freedom, please see the next section.

3. Linear Equation and Its Variation

When n observations are denoted by $y_1, y_2, \ldots, y_n$, the equation of constant coefficients of $y_1, y_2, \ldots, y_n$ in linear terms is called a linear equation.

$$L = c_1y_1 + c_2y_2 + \cdots + c_ny_n$$

Some of $c_1, c_2, \ldots, c_n$ may be zero, but assume not all are equal to zero. **The total of the squares of coefficients $c_1, c_2, \ldots, c_n$** is called the **number of units of linear equation L, and is denoted by D.**

$$D = c_1{}^2 + c_2{}^2 + \cdots + c_n{}^2$$

The square of a linear equation divided by its number of units, D, is the variation with 1 degrees of freedom and consists part of the total variation, S_T.

The variation of the above linear equation, denoted by S_L, is given by

$$S_L = \frac{(c_1y_1 + c_2y_2 + \cdots + c_ny_n)^2}{c_1{}^2 + c_2{}^2 + \cdots + c_n{}^2} \qquad (f\text{=}1)$$

The number of degrees of freedom is explained as follows.
S_L is the variation of linear equation L in quadratic form. Its coefficient matrix, denoted by A, is given by

$$A = \begin{pmatrix} \frac{c_1{}^2}{D} & \frac{c_1c_2}{D} & \frac{c_1c_3}{D} & \cdots & \frac{c_1c_n}{D} \\ \frac{c_2c_1}{D} & \frac{c_2{}^2}{D} & \frac{c_2c_3}{D} & \cdots & \frac{c_2c_n}{D} \\ \vdots & & \ddots & & \\ \frac{c_nc_1}{D} & \frac{c_nc_2}{D} & \cdots & & \frac{c_n{}^2}{D} \end{pmatrix}$$

As described in the section of degrees of freedom, the rank of coefficient matrix is equal to its degrees of freedom. Therefore, the degrees of freedom for the variation of a linear equation is always equal to 1.

A linear equation when

$$c_1 + \cdots + c_n = 0$$

is called contrast or comparison.

The average of n data points

$$\bar{y} = \frac{y_1 + \cdots + y_n}{n}$$

is a linear equation but not a contrast.

Variation of average is expressed by the variation of a linear equation as

$$S_m = \frac{(\bar{y})^2}{\text{Number of units, } D} = \frac{\left(\frac{y_1 + \cdots + y_n}{n}\right)^2}{\left(\frac{1}{n}\right)^2 n}$$

$$= \frac{(y_1 + \cdots + y_n)^2}{n}$$

4. ORTHOGONALITY

(1) What Is Orthogonality?

Two vectors A and B on the same plane are orthogonal means the two vectors literally meet at a right angle. In other words, the projection of B to A is zero and vice versa. That means the two are independent to each other; it is impossible to express A relative to B or vice versa.

Basically, such concept of orthogonality can be extended to the third order or more in the same way. But in a space of "n"th order, right angle is invisible. Vectors A and B

A: $(a_1, a_2, ..., a_n)$, B: $(b_1, b_2, ..., b_n)$

in a space of "n"th order, A and B are orthogonal when

$$a_1b_1 + a_2b_2 + \cdots + a_nb_n = 0 \quad (1)$$

When two vectors on the same plane, A and B, are orthogonal, it is obvious that Equation (1) holds for the case shown in Figure 1.

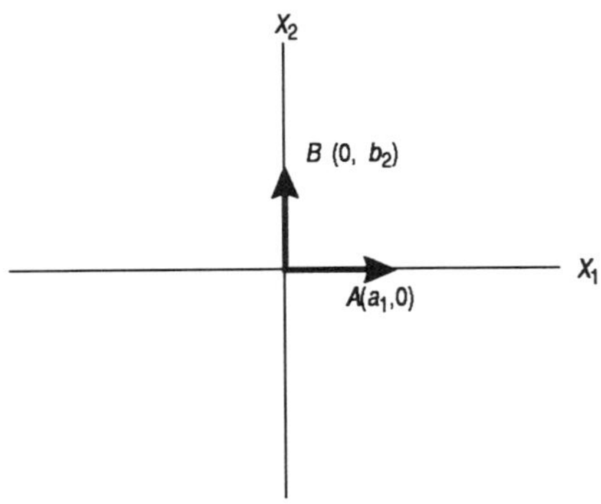

Figure 1. Orthogonal Vectors on a Plane

(2) Orthogonality in Factorial Layout

In experimental design or in data analysis, orthogonality plays a very important role.

The orthogonality in factorial layout means that in any level of a factor, the number of occurrences of any level of any other factor are equal.

For example, the number of all combinations of 2-level factors A and B is 4.

No.1 A_1B_1

No. 2 A_1B_2

No. 3 A_2B_1

No. 4 A_2B_2

No.1 and No.2 are at the first level of A; No.1 is at B_1 and No.2 is at B_2. The contribution of B to A_1 is zero. It is the same for the second level of A, A_2. No.1 and No.3 are at the first level of B, No.2 and No.4 are at the second level of B. In both cases, the contribution of A is zero.

When the contribution of B to A is zero, or the contribution of A to B is zero, A and B are orthogonal. It is the same as the case when two vectors meet at a right angle, the projection of one to the other is zero.

When experiments are conducted under conditions from No.1 to No.4, the results are denoted by y_1, y_2, y_3 and y_4. Since A is not affected by B, the effect of A is determined by comparing the first level of A, or y_1 and y_2, with the second level of A, or y_3 and y_4.

$$y_1 + y_2 - y_3 - y_4 \tag{2}$$

Similarly, the effect of B is calculated as

$$y_1 - y_2 + y_3 - y_4 \tag{3}$$

The coefficients of the above equations are given in Table 1.

Table 1. Coefficients of y's

	A	B
1	1	1
2	1	-1
3	-1	1
4	-1	-1

The product of these coefficients is equal to zero.

$$1 \times 1 + 1 \times (-1) + (-1) \times 1 + (-1) \times (-1) = 0 \tag{4}$$

That means Equation (4) holds true in a space of 4 order.

If instead, the 4 experiments were conducted as

No.1 A_1B_1

No.2 A_1B_2

No.3 A_1B_1

No.4 A_2B_2

The 1st level of A was conducted in No.1, No.2 and No.3. In these three experiments, two experiments were at B_1 and one experiment at B_2. In such a case, we cannot state that B has no contribution. A and B are not orthogonal to each other.

From above, it is clear that in a multi-factor layout where all combinations of factors are totaled, all factors are orthogonal to each other.

The layout in an orthogonal array has the same situation. In any level of any column, all levels of other columns appear the same number of times.

For example, orthogonal array $L_4(2^3)$ is shown in Table 2.

Table 2. Orthogonal Array $L_4(2^3)$

No. \ Column	1	2	3
1	1	1	1
2	1	2	2
3	2	1	2
4	2	2	1

In any column, 1 and 2 occur twice each. In any pair of columns, combinations (1 1), (1 2), (2 1) and (2 2) occur once.

Columns 1 and 2 in Table 2 correspond to A and B in Table 1. If the 2's in Table 2 are substituted by -1, column 3 becomes 1, -1, -1, 1. These coefficients show

$$y_1 - y_2 - y_3 + y_4 \tag{5}$$

It is the difference between the difference of B at A_1, or $(y_1\text{-}y_2)$, and the difference of B at A_2, or $(y_2\text{-}y_4)$, which is the interaction between A and B.

(3) Orthogonality in Data Analysis

Orthogonality is used in orthogonal polynomial equation. It is not described here, but in the equation, each term is independent.

Orthogonality is very valuable in data analysis. The objective of analysis of variance is to decompose the total variation of data into the sum of variations of various sources. To be able to decompose variation, causes must be orthogonal.

For example, equations (2), (3) and (5) are orthogonal to each other. When the variations of (2), (3) and (5) are individually calculated, their sum agrees with the total variation.

$$y_1{}^2 + y_2{}^2 + y_3{}^2 + y_4{}^2 - \frac{(y_1 + y_2 + y_3 + y_4)^2}{4}$$
$$= \frac{(y_1 + y_2 - y_3 - y_4)^2}{4} + \frac{(y_1 - y_2 + y_3 - y_4)^2}{4}$$
$$+ \frac{(y_1 - y_2 - y_3 + y_4)^2}{4} \quad (6)$$

That means

$$S_T = S_A + S_B + S_{A\times B} = S_{\text{col.1}} + S_{\text{col.2}} + S_{\text{col.3}} \quad (7)$$

5. EXPECTATION

Let's consider measuring a certain article many times. Although the mass is constant, usually the measured results change. If the cause of variation cannot be found no matter how hard you tried, some people tend to say that it is a matter of probability. Such a statement is very irresponsible. In statistical terms, such a variable is called a random variable.

If the measured results are plotted as a distribution, there exists a center of distribution. The value showing the central value is called the expectation, denoted by E.

Let m and σ^2 denote, respectively, the average and variance from infinite number of measurements, X.

$$E(X) = m$$
$$V(X) = \sigma^2$$

When we consider the distribution of value X squared, its central value or expectation appears to be m^2, but actually it is known as

$$E(X^2) = m^2 + \sigma^2 \quad (1)$$

For explanation, see the books of statistics. In other words, X^2 estimates $m^2 + \sigma^2$.

When the average of $X_1, X_2, \ldots, X_k$ are denoted by $m_1, m_2, \ldots, m_k$ respectively and the variance is σ^2, the expectation, $E(L)$, and variance, $V(L)$, of linear equation

$$L = c_1X_1 + c_2X_2 + \cdots + c_kX_k$$

are given by

$$E(L) = c_1m_1 + c_2m_2 + \cdots + c_km_k \quad (2)$$

$$V(L) = (c_1{}^2 + c_2{}^2 + \cdots + c_k{}^2)\sigma^2 \quad (3)$$

Next, the expectation of variation A, denoted by S_A, is derived.

Assume factor A has "a" levels with "r" repetitions. Let the total of each level of A be $A_1, A_2, ..., A_n$. The variation of A is given by

$$S_A = \frac{A_1{}^2 + A_2{}^2 + \cdots + A_a{}^2}{r} - \frac{(A_1 + A_2 + \cdots + A_a)^2}{ar} \quad (4)$$

Let the average and variance of the "i"th level be $m+a$ ($a_1+a_2+ ...+a_n=0$) and σ^2 respectively. $E(A_i)$ and $V(A_i)$ are given by

$$E(A_i) = r(m + a_i) \quad (5)$$

$$V(A_i) = r\sigma^2 \quad (6)$$

Therefore,

$$\begin{aligned} &E(A_1 + A_2 + \cdots + A_a) \\ &= r[(m + a_1) + (m + a_2) + \cdots + (m + a_a)] \\ &= arm \end{aligned} \quad (7)$$

$$V(A_1 + A_2 + \cdots + A_a) = ar\sigma^2 \quad (8)$$

From Equation (1),

$$\begin{aligned} &E[(A_1 + A_2 + \cdots + A_a)^2] \\ &= (arm)^2 + ar\sigma^2 \end{aligned} \quad (9)$$

$$\therefore E\left[\frac{(A_1 + A_2 + \cdots + A_a)^2}{ar}\right] = arm^2 + \sigma^2 \quad (10)$$

Also from Equation (5),

$$\begin{aligned} &E(A_i{}^2) = [r(m + a_i)]^2 + r\sigma^2 \\ &\therefore E(A_1{}^2 + A_2{}^2 + \cdots + A_a{}^2) \\ &= r^2[(m + a_1)^2 + (m + a_2)^2 + \cdots + (m + a_a)^2] + ar\sigma^2 \end{aligned} \quad (12)$$

From the above,

$$E(S_A) = E\left[\frac{A_1{}^2 + A_2{}^2 + \cdots + A_a{}^2}{r} - \frac{(A_1 + A_2 + \cdots + A_a)^2}{ar}\right]$$

$$\begin{aligned} &= \frac{[r^2(m + a_1)^2 + (m + a_2)^2 + \cdots + (m + a_n)^2 + ar\sigma^2]}{r} \\ &\quad - (arm^2 + \sigma^2) \\ &= r(a_1{}^2 + a_2{}^2 + \cdots + a_a)^2 + (a - 1)\sigma^2 \\ &= (a - 1)\,(r\sigma_A{}^2 + \sigma^2) \end{aligned} \quad (13)$$

$\sigma_A{}^2$ shows the magnitude of the effect of A and is given by

$$\sigma_A{}^2 = \frac{a_1{}^2 + a_2{}^2 + \cdots + a_a{}^2}{a-1} \tag{14}$$

Equation (15) shows what is estimated by the variance which is calculated from actual data in the process of the analysis of variance.

$$E(V_A) = \frac{E(S_A)}{a-1} = r\sigma_A{}^2 + \sigma^2 \tag{15}$$

Next, the expectation of the signal factor linear effect, denoted by S, is used to calculate the S/N Ratio.

Let $y_1, y_2, \ldots, y_k$ denote observational results at signal factor levels $M_1, M_2, \ldots, M_k$ respectively. The linear coefficient, a, is estimated by

$$\beta = \frac{(M_1 - \overline{M})y_1 + \cdots + (M_k - \overline{M})y_k}{(M_1 - \overline{M})^2 + \cdots + (M_k - \overline{M})^2} \tag{16}$$

Equation (16) is the linear equation with respect to y_i; its variance, $V(\hat{\beta})$, is given by

$$V(\hat{\beta}) = (\text{No. of units of } \hat{\beta}) \times \sigma^2 = \frac{\sigma^2}{r} \tag{17}$$

where

$$r = (M_1 - \overline{M})^2 + \cdots + (M_k - \overline{M})^2$$

σ^2, the variance of y_i, is estimated by

$$\hat{\sigma}^2 = V_e = \frac{S_e}{k-2} = \frac{S_T - S_\beta}{k-2} \tag{18}$$

where

$$S_T = y_1{}^2 + y_2{}^2 + \cdots + y_k{}^2 - \frac{(y_1 + y_2 + \cdots + y_k)^2}{k}$$

$$S_\beta = \frac{(\hat{\beta})^2}{\text{No. of units}}$$
$$= \frac{[(M_1 - \overline{M})y_1 + \cdots + (M_K - \overline{M})y_k]^2}{(M_1 - \overline{M})^2 + \cdots + (M_K - \overline{M})^2}$$

From Equation (1), the expectation of $\hat{\beta}^2$ is

$$E(\hat{\beta}^2) = \beta^2 + V(\hat{\beta}) = \beta^2 + \frac{\sigma^2}{r} \tag{19}$$

Since S_β, the variation of $\hat{\beta}$, is

$$S_\beta = \frac{\hat{\beta}^2}{\text{No. of units}} = r\hat{\beta}^2 \tag{20}$$

The expectation of S_β is

$$E(S_\beta) = E(r\hat{\beta}^2) = r\left(\beta^2 + \frac{\sigma^2}{r}\right) = r\beta^2 + \sigma^2 \tag{21}$$

6. F TEST

When n pieces of data are collected, the contribution of the general mean or the difference between A_1 and A_2 can be calculated. F test is a method developed to answer the question if such a contribution might become to zero if more data were collected.

The question can be asked in a different way. Suppose the truth is that there is neither an effect on the general mean nor an effect from A. But from the result of data collection followed by calculations, there are large contributions. Was it caused by chance?

This method is used to concluded whether or not the variation terms such as S_m or S_A before calculating contributions are larger than the error variance.

If it is definite that S_A is greater than the error variance (not judged from the number but from the true effect), it is meaningful to calculate the contribution.

If instead, neither deviation nor the effect of A exists, the magnitude of S_m or S_A is about the same as the error variance, and the following variance ratio should be in the neighborhood of 1:

$$F_m = \frac{S_m}{V_e} \tag{1}$$

$$F_A = \frac{S_A}{V_e} \tag{2}$$

What value of F should be reached to conclude that there is an effect is answered by the use of the F Table internationally. (F Table was named to commemorate the originator of experimental design, R.A. Fisher of England. For the F Table, see Appendix Table 1.)

If neither the general mean nor the effect of A are existing, F_m or F_A from Equations (1) or (2) hardly becomes larger than the F value at the degrees of freedom of numerator (degrees of freedom of S_m or S_A) being equal to 1 and the degrees of freedom of denominator (degrees of freedom of the error V_e) being equal to (n-2). The F values show such limits. As the probability of occurrence, two values, 5% and 1%, are used.

For example, if n=20, the F value in the F Table at the degrees of freedom of numerator being equal to 1 and the degrees of freedom of denominator being equal to (20-2)=18 is 4.41 for 5% and 8.29 for 1%. It means that if the genreal mean or the effect of A does not exist, either F_m from Equation (1) or F_A from Equation (2) becomes greater than 4.41 with a probability of 5 out of 100 times and becomes greater than 8.28 with a probability of 1 out of 100 times; it must be a miracle.

When the values of F_m or F_A calculated from actual data showed to be marginally greater than 4.41 and therefore concluded that F_m or F_A has an effect, it could be a wrong conclusion at a chance of 5 out of 100.

5% is called the level of significance. When the value of F_m or F_A calculated from actual data is greater than 4.41, the 5% level, and smaller than 8.29, the 1% level, the effect is significant, and customary, an asterisk (*) is put to the F_0, the variance ratio in the analysis of variance table.

When variance ratio F_m or F_A is greater than 8,29, the 1% level, the effect is significant at 1% level and two asterisks (**) are put.

In principle, the contribution is calculated only for those causes (or factorial effects) which are significant. However, those without asterisks should not be neglected if the contribution is large. It is possible that those factorial effects have large impacts. There are two cases when a factor is insignificant; the case when there is truly no effect, and the case when the degrees of freedom is not large enough to be conclusive.

7. Objective of Using Orthogonal Arrays in Experimentation

It is often said that orthogonal arrays are used in experimentation because it is impossible to conduct experiments of all combinations if there are many factors, even if their number of levels is only 2. But that is not the objective of using orthogonal arrays. In quality engineering, it is recommended to use orthogonal array L_{12} to conduct an experiment with 11 factors of 2 levels. The comparison of those 11 factors could be made by the way shown in Table 1, where the number of runs is 12, the same as the case of L_{12}.

Table 1. Layout Using One-Factor-At-A-Time Method

No.	*A*	*B*	*C*	*D*	*E*	*F*	*G*	*H*	*I*	*J*	*K*
1	1	1	1	1	1	1	1	1	1	1	1
2	2	1	1	1	1	1	1	1	1	1	1
3	2	2	1	1	1	1	1	1	1	1	1
4	2	2	2	1	1	1	1	1	1	1	1
5	2	2	2	2	1	1	1	1	1	1	1
6	2	2	2	2	2	1	1	1	1	1	1
7	2	2	2	2	2	2	1	1	1	1	1
8	2	2	2	2	2	2	2	1	1	1	1
9	2	2	2	2	2	2	2	2	1	1	1
10	2	2	2	2	2	2	2	2	2	1	1
11	2	2	2	2	2	2	2	2	2	2	1
12	2	2	2	2	2	2	2	2	2	2	2

Using the method of Table 1, A_1 and A_2 are compared by fixing the conditions of *A*, *C*, ..., *K* at $B_1C_1D_1E_1F_1G_1H_1I_1J_1K_1$ for example. This method has been frequently used, called one-factor-at-a-time method.

Using this method, we need to vary the condition of only one factor each time, and it is much easier to conduct the experiment than the case using an orthogonal array. Nevertheless, orthogonal array experimentation is recommended for the following reason.

By the method in Table 1, two levels A_1 and A_2 are compared while the conditions of other factors, *B*, *C*, ..., *K* are fixed to a certain level such as fixed to the first level. From such an experiment, the difference between A_1 and A_2 can be precisely obtained. Using an orthogonal array, A_1 and A_2 are compared while the levels of other factors are varied, and the average effect (called the main effect) between A_1 and A_2 is obtained. Many factor conditions must be varied from run to run and it takes long to do it, and generally, there will be a larger variability between data. Instead, comparison is made from the average of A_1, or the average of No.1 through No.6, and the average of A_2, or the average of No.7 through No.12.

Although the precision of individual data point becomes worse, comparison is made by the averages; the precision could be better in this sense. However, there is no guarantee that precision is better overall. Then, what is the reason to recommend the use of the cumbersome orthogonal array experiment?

The reason is the importance of discovering the factorial effects with a high reproducibility. In orthogonal experimentation, the difference between two levels A_1 and A_2 is obtained as an average effect while the conditions of other factors varied.

If the impact of A_1 and A_2 to the output is consistent while the conditions of other factors vary, its effect will come out to be large. But if the output becomes either larger or reversed by the conditions of other factors, it will come out to be small.

When an experiment is conducted using an orthogonal array, a consistent effect which is not affected by other factorial conditions will be estimated to be large. It is important to discover the factor whose effect, especially the order of superiority and inferiority is not changed by other factorial conditions. In other words, only reliable factorial effects are discovered and the reliability of factorial effects is thus evaluated.

In one-factor-at-a-time methods on the other hand, the difference between A_1 and A_2 is estimated under the condition of other factors being fixed, and no matter how precisely the effect is estimated and nicely graphed, there is no guarantee at all that such an effect will be consistent when the conditions of the other factors are different.

It is questionable that the results obtained from such a method can be reproduced if the person conducting the experiment is different or raw material is different.

This is a very important point, especially in an experiment for stability. In such an experiment, noise factors are assigned to the outer array. It would be good to conduct the experiment with many noise factors, but it is difficult to consider various product usage conditions or time as a factor in the experiment. But if there are many control factors, we can consider that for one specific control factor, other control factors are disturbing factors, or noise factors. Therefore, even if there are not too many noise factors, if the experiment is conducted with an orthogonal array and with main effects only, a large factor is reliable from the viewpoint of stability.

8. 3-Level Series Orthogonal Arrays and Mixed-Type Orthogonal Arrays

3-level series orthogonal arrays such as L_9 or L_{27} have the following features:

(1) In any column, numbers 1, 2 and 3 occur the same number of times.

(2) In any pair of columns, combinations (1 1), (1 2), (1 3), (2 1), (2 2), (2 3), (3 1), (3 2) and (3 3) occur the same number of times.

(3) The interaction of any pair of columns confound with particular two columns.

Let's consider any two columns to be *i* and *j*. The interaction of their 2-way layout is the variation of three levels denoted by (○, △, X) in a diagonal direction. One component is from upper right to lower left and another is from upper left to lower right as shown in Table 1.

Table 1. Interaction of 3-level Factors

i \ *j*	1	2	3
1	○	X	△
2	△	○	X
3	X	△	○

i \ *j*	1	2	3
1	△	X	○
2	X	○	△
3	○	△	X

In 3-level orthogonal arrays, the 2 columns where the interaction of any 2 columns confounds are the 2 columns as shown in Table 1, which must exist in L_9 or L_{27}. For example, the 2 interaction columns between columns 1 and 2 in L_{27} go to columns 3 and 4. The combinations of 1, 2 and 3 of columns 1 and 2 are shown in Tables 2 and 3: column 3 is from upper right to lower left, and column 4 is from upper left to lower right.

Table 2. Columns 1, 2 and 3

Col. 1 \ Col. 2	1	2	3
1	1	2	3
2	2	3	1
3	3	1	2

Table 3. Columns 1, 2 and 4

Col. 1 \ Col. 2	1	2	3
1	1	2	3
2	3	1	2
3	2	3	1

No columns other than columns 3 and 4 have such a relationship with columns 1 and 2. Similarly, columns 1 and 7 are the interaction columns of columns 5 and 6. Tables 4 and 5 show that the above mentioned relationship is satisfied.

Table 4. Columns 5, 6 and 1

Col. 5 \ Col. 6	1	2	3
1	1	2	3
2	3	1	2
3	2	3	1

Table 3. Columns 5, 6 and 7

Col. 5 \ Col. 6	1	2	3
1	1	3	2
2	3	2	1
3	2	1	3

Orthogonal arrays L_{18} or L_{36} have features (1) and (2) but do not have (3). Therefore, these arrays cannot be used to modify and provide multi-levels or apply pseudo-factor design by using interaction columns.

9. Layout Only With Main Effects

(1) Reproducibility of Results

It is obvious that as long as rational characteristics are used, there is additivity between factorial effects, and it is good enough to just assign main effects in the experiment. However, it is hard to completely guarantee the additivity; in the real world, interactions exist more or less. If so, and if the columns confounded with interactions are not reserved for the identification of interactions, these columns will confound with main effects. As a result, there is no way to tell which effect caused the column significant. In such a case, the experiment assigned only with main effects will fail.

Should we assign interactions then? It is right in a sense, but not desirable for the following two reasons.

First, we do not know which interactions should be assigned. If we try to assign all suspicious interactions because of this reason, it turns out to be impossible to assign them. In practice, there is no way to consider so many interactions. But this is not the real reason for not assigning interactions. The following reason is far more important.

In many cases, the objective of experiments is to find the optimum condition. If there are interactions, those effect may be considered together for the determination of the optimum condition. It is true that the condition so selected is the best within the range of the experiment. But can this condition be applicable outside the range? It might be too much to ask. But it is obvious that the optimum condition would be affected by environmental conditions, deterioration, scale of the manufacturing process, etc.

The existence of interactions between control factors implies that there is a big risk of having significant interactions between outer noise factors. When there are significant interactions, it is questionable if the optimum condition determined by considering interactions is really optimum.

But the optimum condition determined only from main effects has a high possibility to hold the same even if a factor such as environmental condition changes. (Of course there is no guarantee.)

In short, it is desirable to find a consistent optimum condition mainly from main effects.

(2) Additivity of Effects

In an experiment, if only interaction $A \times B$ is assigned because it is suspicious to be significant, and the results showed $A \times B$ is significant, it is dangerous to conclude the assignment was right. Because other interactions such as $A \times B \times C$ or $A \times B \times D$ which were not assigned could be significant. These unassigned interactions would confound with other main effects, and the conclusions on the main effects might be questionable.

If characteristics that are not additive are used, it is very often to have such results. Since various interactions confound with any column, there is always a worry that the selection of a truly significant factor might not be made.

To this point, the thought of trying a layout with some large interaction terms seems to be meaningless. There is a case when the assignment of interactions is valuable. That is when those interactions are confirmed to be small.

In order to find the optimum condition which is reproducible, it is most important to select the characteristic with additivity. In that case, the layout only with main effects is good enough. But in many cases, a perfect additivity does not exist, therefore, the use of mixed-type orthogonal arrays is recommended.

10. Discount Coefficient and the Meaning of Over Estimate

Consider a 2-level factor, A. Let the true values of A_1 and A_2 be m_1 and m_2 respectively. A_1- A_2 is the effect of A. To estimate the true effect, m_1- m_2 from n pieces of data each,

$$\overline{A}_1 - \overline{A}_2 \tag{1}$$

is calculated. Its error is given by

$$L = \overline{A}_1 - \overline{A}_2 - (m_1 - m_2) \tag{2}$$

It is desireable of course that L is equal to zero. According to the concept of the Loss Function in quality engineering, the loss due to error, L, is proportional to the square of the error.

The expectation of Equation (2) is given by

$$E(L^2) = E[\{(\overline{A}_1 - \overline{A}_2) - (m_1 - m_2)\}^2]$$
$$= (m_1 - m_2)^2 + \frac{2}{n}\sigma^2 \tag{3}$$

where n is the number of replications, and σ^2 is experiment error variance.

Let's consider if it is possible to reduce the squares of L. For example, the following is used to estimate m_1-m_2:

$$\beta(\overline{A}_1 - \overline{A}_2) \tag{4}$$

The error, L', is then

$$L' = \beta(\overline{A}_1 - \overline{A}_2) - (m_1 - m_2) \tag{5}$$

therefore,

$$E(L'^2) = E[\{\beta(\overline{A}_1 - \overline{A}_2) - (m_1 - m_2)\}^2]$$
$$= (1 - \beta)^2(m_1 - m_2)^2 + \beta^2 \frac{2}{n}\sigma^2 \tag{6}$$

to minimize Equation (6), let

$$\frac{\partial E(L'^2)}{\partial \beta} = 0 \tag{7}$$

β is given by

$$\beta = \frac{\frac{n}{2}(m_1 - m_2)^2}{\sigma^2 + \frac{n}{2}(m_1 - m_2)} \tag{8}$$

Both σ^2 and $(m_1$-$m_2)^2$ are unknown. The error variance, V_e, in the ANOVA table is used to estimate σ^2. Let the level totals of A_1 and A_2 be also denoted by A_1 and A_2 respectively. Variation of A is given by

$$S_A = \frac{(A_1 - A_2)^2}{2n} \tag{9}$$

the expectation of V_A is made by dividing S_A by its degrees of freedom, 1.

$$E(V_A) = \sigma^2 + \frac{n}{2}(m_1 - m_2)^2 \tag{10}$$

Therefore, V_A and V_e are used to substitute σ^2 and σ^2 + $(m_1$-$m_2)^2$ respectively.

$$\hat{\sigma}^2 = V_e \tag{11}$$

$$\sigma^2 + \frac{n}{2}(m_1 - m_2)^2 \approx V_A \tag{12}$$

Putting Equations (11) and (12) into Equation (8),

$$\beta = 1 - \frac{\sigma^2}{\sigma^2 + \frac{n}{2}(m_1 - m_2)^2} \approx 1 - \frac{1}{F_0} \tag{13}$$

where F is the variance ratio, given by

$$F_0 = \frac{V_A}{V_e} \tag{14}$$

Figure 1. Discount Coefficient

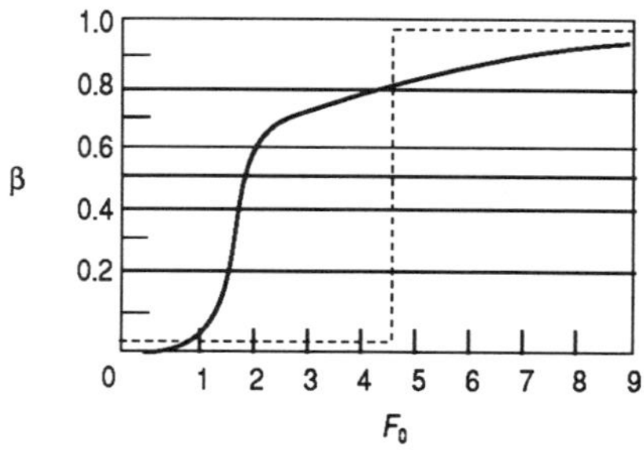

When the variance ratio is equal to ∞, β is equal to 1; it gets near to 0 when the variance ratio gets near to 1. When the variance ratio is smaller than 1, the differene m_1-m_2 is smaller than the case when the variance ratio is equal to 1, therefore, use 0 for β. As a conclusion, the following equation which minimized the mean square error of estimating m_1-m_2, the difference between the true values of A_1 and A_2, is used:

$$\text{Estimate of } m_1 - m_2 = \beta(\overline{A}_1 - \overline{A}_2) \tag{15}$$

where β is

$$\beta = \begin{cases} 0 & \text{When } F_0 = \dfrac{V_A}{V_e} \leq 1 \\ 1 - \dfrac{1}{F_0} & \text{When } F_0 = \dfrac{V_A}{V_e} > 1 \end{cases} \tag{16}$$

Let σ_{est}^2 denote the mean square error when discount coefficient, β, is used to estimate m_1-m_2. σ_{est}^2 is given by putting Equation (8) into Equation (6).

$$\sigma_{est}^2 = [1 - \frac{\sigma^2}{\sigma^2 + \frac{n}{2}(m_1 - m_2)^2}] \times \frac{2}{n}\sigma^2 \tag{17}$$

It becomes simpler by putting Equation (13)

$$\sigma_{est}{}^2 = \beta \times \frac{2}{n}\sigma^2 \tag{18}$$

Actually, the value of β varies. Letting its average by $\bar{\beta}$, Equation (18) becomes

$$\sigma_{est}{}^2 = \bar{\beta} \times \frac{2}{n}\sigma^2 \tag{19}$$

Figure 1 shows the value of discount coefficient β at different value of F_0. It is seen from the figure that when the variance ratio is equal to 2, the use of half of the difference obtained from the experiment is considered to be the estimate. This is so called "discount the story to a half." Even when the variance ratio is as high as 10 showing a high significance, it is better to multiply 0.9 to the observed difference. A practician must have the experience of applying the results obtained from the experiment and resulted in having a smaller effect than expected in most cases, rather than having a larger effect than expected. It shows that even if an experiment was conducted "conscientiously," the effect tends to be larger in many cases.

The same fact applies to the estimate of $\bar{A}_i - \bar{T}$. For example, the optimum condition was obtained from an experiment with 6 factors to be $A_2B_1C_3D_1E_2F_3$. If all factorial effects were added together, it would be an over-estimate.

$$\hat{\mu} = \bar{T} + (\bar{A}_2 - \bar{T}) + (\bar{B}_1 - \bar{T}) + (\bar{C}_3 - \bar{T}) + \cdots + (\bar{F}_3 - \bar{T}) \tag{20}$$

It has been clear that the use of a discount coefficient is the most appropriate way for estimation. To estimate the optimum condition,

$$\begin{cases} \beta_A = 1 - \dfrac{1}{F_A} & \text{where,} F_A = \dfrac{V_A}{V_e} \\ \beta_B = 1 - \dfrac{1}{F_B} & \text{where, } F_B = \dfrac{V_B}{V_e} \\ \cdots & \\ \beta_F = 1 - \dfrac{1}{F_F} & \text{where, } F_F = \dfrac{V_F}{V_e} \end{cases} \tag{21}$$

are calculated, then the following equation is used:

$$\hat{m} = \bar{T} + \beta_A(\bar{A}_2 - \bar{T}) + \beta_B(\bar{B}_1 - \bar{T}) + \cdots + \beta_F(\bar{F}_3 - \bar{T}) \tag{22}$$

However, this calculation is cumbersome. For convenience, only larger effects are used for estimation. It is similar to use 1 as β for larger effects and 0 for smaller effects as shown in the dotted line in figure 1.

It is the same situation when significant effects from the F test are used.

11. Equation for Confidence Interval, Effective Number of Replications

(1) Confidence Interval for a Factorial Effect

A significant factor from analysis of variance is to be estimated. The linear equation of data $y_1, y_2, \ldots, y_n$ is

$$\hat{\mu} = c_1 y_1 + c_2 y_2 + \cdots + c_n y_n$$

Its average is

$$\hat{\mu} = \frac{y_1 + y_2 + \cdots + y_n}{n}$$

$\hat{\mu}$ is a linear equation with $c_1 = c_2 = \cdots = c_n = \dfrac{1}{n}$

Letting r pieces of data at "i"th level be $y_{i1}, y_{i2}, \ldots, y_{ir}$, the estimate of the average of the second level of A, or $\overline{A}_2$, is

$$\overline{A}_2 = \frac{y_{21} + y_{22} + \cdots + y_{2r}}{r}$$

For the data excluding y_{2j} (j=1, 2, ..., r) are not used in the estimation, therefore, coefficient 0 is multiplied to those data. For those data of y_{2j} (j=1, 2, ..., r), coefficient $1/r$ is multiplied.

Let the true value of linear equation $\hat{\mu}$ be μ. We are interested in the difference between μ and the estimate $\hat{\mu}$. Letting the difference be L,

$$L = \hat{\mu} - \mu$$

Since L is a linear equation, its variation is given by

$$S_L = \frac{L^2}{\text{No. of units}} = \frac{(c_1 y_1 + c_2 y_2 + \cdots + c_n y_n - \mu)^2}{c_1^{\,2} + c_2^{\,2} + \cdots + c_n^{\,2}} = (\hat{\mu} - \mu)^2 n_e$$

where

$$n_e = \frac{1}{c_1^{\,2} + c_2^{\,2} + \cdots + c_n^{\,2}}$$

Let the error variance be V_e. S_L/V_e belongs to the F distribution with 1 degree of freedom for the numerator and with the degrees of freedom of the error variance (f_e) for the denominator. Therefore, the probability that S_L/V_e be smaller than $F^1_{f_e}(1-\alpha)$ is α.

Letting α=0.95%,

$$P_r\left[F = \frac{S_L}{V_e} \leq F^1_{fe}(0.05)\right] = 0.95$$

$$P_r\left[S_L \leq F^1_{fe}(0.05) \times V_e\right] = 0.95$$

$$P_r\left[(\hat{\mu} - \mu)^2 n_e \leq F^1_{fe}(0.05) \times V_e\right] = 0.95$$

$$P_r\left[\hat{\mu} - \sqrt{F^1_{fe}(0.05) \times \frac{V_e}{n_e}} \leq \mu \leq \hat{\mu} + \sqrt{F^1_{fe}(0.05) \times \frac{V_e}{n_e}}\right] = 0.95$$

the confidence interval at 95% level is

$$\mu = \hat{\mu} \pm \sqrt{F^1_{fe}(0.05) \times \frac{V_e}{n_e}}$$

n_e is the reciprocal of $(c_1{}^2 + c_2{}^2 + \cdots + c_n{}^2)$, the number of units of the linear equation, $\hat{\mu}$ called the **effective number of replications.**

Effective number of replications shows how many effective number of data points is used to estimate $\hat{\mu}$.

The confidence interval of the average $\hat{\mu}$ from n pieces of data, y_1, y_2, ..., y_n is given by the following equation:

$$\pm \sqrt{F^1_{fe}(0.05) \times \frac{V_e}{n}} \tag{1}$$

where

$$n_e = \frac{1}{c_1{}^2 + c_2{}^2 + \cdots + c_n{}^2} = \frac{1}{\left(\frac{1}{n}\right)^2 n} = n$$

(2) Confidence Interval for the Optimum Condition

The equation of the confidence interval for the optimum condition is also given by Equation (1). For example, a 4-level factor A and a 3-level factor B are assigned in a 2-way layout. Assume that both A and B are significant and the optimum condition is A_3B_4. The estimate of $\hat{\mu}$ for A_3B_4 is

$$\begin{aligned}\hat{\mu} &= \overline{T} + (\overline{A}_3 - \overline{T}) + (\overline{B}_4 - \overline{T}) \\ &= \overline{A}_3 + \overline{B}_4 - \overline{T} \end{aligned} \tag{2}$$

where $\overline{T}$ is the general mean.

$\hat{\mu}$ in Equation (2) is a linear equation where each of the 20 pieces of data is multiplied by the coefficients in Table 1.

Table 1. Coefficients of $\hat{\mu}$

	B_1	B_2	B_3	B_4
A_1	$-\frac{1}{20}$	$-\frac{1}{20}$	$-\frac{1}{20}$	$\frac{1}{5}-\frac{1}{20}$
A_2	$-\frac{1}{20}$	$-\frac{1}{20}$	$-\frac{1}{20}$	$\frac{1}{5}-\frac{1}{20}$
A_3	$\frac{1}{4}-\frac{1}{20}$	$\frac{1}{4}-\frac{1}{20}$	$\frac{1}{4}-\frac{1}{20}$	$\frac{1}{4}+\frac{1}{5}-\frac{1}{20}$
A_4	$-\frac{1}{20}$	$-\frac{1}{20}$	$-\frac{1}{20}$	$\frac{1}{5}-\frac{1}{20}$
A_5	$-\frac{1}{20}$	$-\frac{1}{20}$	$-\frac{1}{20}$	$\frac{1}{5}-\frac{1}{20}$

The total of the squares of coefficients is

$$\frac{1}{n_e} = (-\frac{1}{20})^2 \times 12 + \left(\frac{1}{4}-\frac{1}{20}\right)^2 \times 3 + \left(\frac{1}{5}-\frac{1}{10}\right)^2 \times 4 + \left(\frac{1}{4}+\frac{1}{5}-\frac{1}{20}\right)^2$$
$$= \frac{8}{20} \qquad (3)$$

The effective number of replications is equal to 2.5. That means if the estimate is made from one result of A_3B_4, the precision is equivalent to 1. But by using Equation (3), the precision is equivalent to 2.5 pieces of data and its confidence intervals are

$$\pm\sqrt{F^1_{fe}(0.05) \times \frac{V_e}{2.5}}$$

It is not necessary to use the coefficients in Table 1 to calculate the n_e in Equation (3). Instead, the following equation can be used. The equation is called Taguchi's Equation.

$$n_e = \frac{\text{Total number of experiments}}{\text{Total of degrees of freedom considered in the calculation of } \hat{\mu}}$$

This equation is applicable for 2-way layout, 3-way layout, Latin Square, or orthogonal array. In general, it is applicable for orthogonal type of experiments. In this example, the factorial effects considered in $\hat{\mu}$ are the general mean and main effects A and B. Its n_e is

$$n_e = \frac{20}{1(m)+4(A)+3(B)}$$
$$= \frac{20}{8} = 2.5$$

(3) Confidence Interval for Confirmatory Experiment

The correctness of significant factorial effects, or the repeatability of those effects must be checked through confirmatory experiment in principle. That is to confirm the actual result of the optimum condition agrees with the estimate.

Although the confirmed result does not completely agree with the estimate, it is important that the result is close to the estimate. Confidence intervals are used for this purpose.

Suppose the estimate from significant effects is calculated by the following equation:

$$\hat{\mu} = c_1y_1 + c_2y_2 + \cdots + c_ny_n$$

r pieces of data, $z_1, z_2, \ldots, z_r$ were collected in the confirmatory experiment. Letting the average of r pieces of data by $\bar{x}$,

$$\bar{x} = \frac{z_1 + z_2 + \cdots + z_r}{r}$$

The difference between $\hat{\mu}$ and $\bar{x}$ is a linear equation.

$$L = \hat{\mu} - \bar{x}$$

The variation of the above linear equation, S_L, is

$$S_L = \frac{(\hat{\mu} - \bar{x})^2}{\text{No. of units}} = \frac{(\hat{\mu} - \bar{x})^2}{c_1{}^2 + c_2{}^2 + \cdots + c_n{}^2 + (-\frac{1}{r})^2 r}$$

$$= \frac{(\hat{\mu} - \bar{x})^2}{\frac{1}{n_e} + \frac{1}{r}}$$

If the probability of S_L being smaller than F^1_{fe} is set at 0.95,

$$P_r\ [F = \frac{S_L}{V_e} \le F^1_{fe}\ (0.05)\ ] = 0.95$$

$$P_r\ [S_L \le F^1_{fe}\ (0.05) \times V_e] = 0.95$$

$$P_r \left[\frac{(\hat{\mu} - \bar{x})^2}{\frac{1}{n_e} + \frac{1}{r}} \le F^1_{fe}\ (0.05) \times V_e \right] = 0.95$$

$$P_r\ [(\hat{\mu} - \bar{x})^2 \le F^1_{fe}\ (0.05) \times V_e \times \left(\frac{1}{n_e} + \frac{1}{r} \right)] = 0.95$$

The confidence interval is given by

$$\bar{x} = \hat{\mu} \pm \sqrt{F^1_{fe}\ (0.05) \times V_e \times \left(\frac{1}{n_e} + \frac{1}{r} \right)}$$

Putting 1 to r in the above equation, the confidence interval of the confirmatory experiment from one data point is obtained.

$$x = \hat{\mu} \pm \sqrt{F_{fe}^{1}(0.05) \times V_e \times (\frac{1}{n_e} + 1)}$$

If the actual result is held in the confidence interval, the reproducibility of factorial effects and the error is recognized and the experiment is reliable.

The confidence interval from Equation (1) can be considered as the interval of the average from an infinite number of replications. The above equation gives the interval of individual data point.

The confidence intervals calculated from significant factorial effects for two conditions, the current condition and the optimum condition, are shown in Figure 1.

Since individual results are estimated with a fairly wide range, it is quite possible that the individual results from the optimum condition could be worse than the current condition. But if the averages from many data points are compared, it will show the optimum is better than current. Otherwise, there is no reproducibility and the analysis of variance is not reliable.

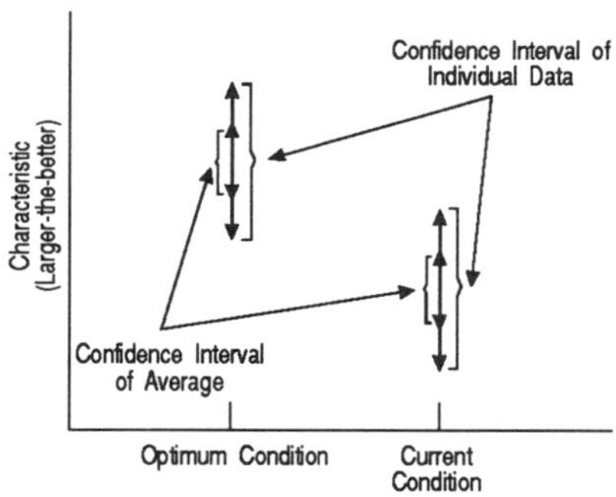

Figure 1. Confidence Interval

12. Polynomial Expansion and Chevyshev's Orthogonal Polynomial Equation

(1) The Least Square Method

When factors which cause variation are orthogonal, the effect of a factor will not influence the effects of other factors. To study the magnitude of causes of variation, it is effective to layout an experiment in such a way that the factors are orthogonal.

When the factors which affect the output are continuous variables, the output can be expressed by a polynomial equation of these factors. For example, temperature, A, is varied as A_1, A_2, ..., A_6, as shown in Figure 1, the output characteristic, y, may be expressed by the linear equation of A such as

$$y = a_0 + a_1 A \tag{1}$$

or

$$y = a_0' + a_1' A + a_2' A^2 \tag{2}$$

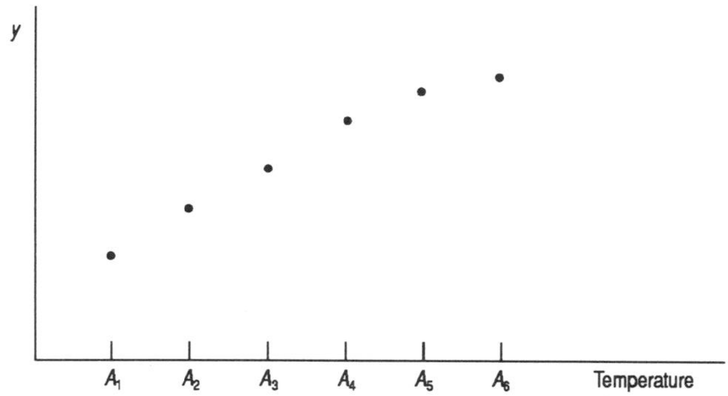

Figure 1. Temperature (A) and Characteristic (y)

The coefficients in Equations (1) and (2) can be determined by using the least squares method.

Let the equation to be determined by

$$y = a_0 + a_1 A + a_2 A^2 + a_3 A^3 + \cdots + a_k A^k \tag{3}$$

Letting

$$A = x_1, A^2 = x_2, \cdots, A^k = x_k$$

Equation (3) is rewritten as

$$y = a_0 + a_1 x_1 + a_2 x_2 + \cdots + a_k x_k \tag{4}$$

From n sets of observational data ($n \geq k+1$), coefficients $a_1, a_2, ..., a_k$ are determined to minimize the following equation:

$$S_e = \sum_{i=1}^{n} (y_i - a_1 x_{1i} - a_2 x_{2i} - \cdots - a_k x_{ki})^2 \tag{5}$$

This is the least square's solution. Such a least square solution is obtained by partially differentiating S_e in Equation (5) by $a_1, a_2, ..., a_k$ and putting it equal to zero to get

$$\begin{cases} S_{11}a_1 + S_{12}a_2 + \cdots + S_{1k}a_k = S_{1y} \\ S_{21}a_1 + S_{22}a_2 + \cdots + S_{2k}a_k = S_{2y} \\ \cdots \\ S_{k1}a_1 + S_{k2}a_2 + \cdots + S_{kk}a_k = S_{ky} \end{cases} \tag{6}$$

and solve the above normal equations. In the equations, S_{ij} is the covariance of x and y.

In normal equations, (6), if all values of S_{ij} for all combinations of i and j ($i \neq j$) are equal to zero, then it is very simple to determine $a_1, a_2, \ldots, a_k$ from the following equation:

$$a_i = \frac{S_{iy}}{S_{ij}} \tag{7}$$

(2) Orthogonal Polynomial Equation

If a modification were made on Equation (3), it would be possible to let S_{ij} be equal to zero for all combinations of i and j for the cases where i is not equal to j. The expansion called orthogonal polynomial equation is used to satisfy such a requirement. Especially if the intervals of A are equal,

$$y = m + a_1(A - \overline{A}) + a_2[(A - \overline{A})^2 - \frac{k^2 - 1}{12} h_A{}^2] + \cdots \tag{8}$$

where h_a is the interval of A and k is the number of levels. This is called Chevyshev's Orthogonal Polynomial Equation.

By using this expansion, a_1 and a_2 are calculated from Equation (7) as

$$a_1 = \frac{(A_1 - \overline{A})y_1 + (A_2 - \overline{A})y_2 + \cdots + (A_k - \overline{A})y_k}{(A_1 - \overline{A})^2 + (A_2 - \overline{A})^2 + \cdots + (A_k - \overline{A})^2} \tag{9}$$

$$a_2 = \frac{\left[(A_1 - \overline{A})^2 - \frac{k^2 - 1}{12} h_A{}^2\right] y_1 + \cdots + \left[(A_k - \overline{A})^2 - \frac{k^2 - 1}{12} h_A{}^2\right] y_k}{\left[(A_1 - \overline{A})^2 - \frac{k^2 - 1}{12} h_A{}^2\right]^2 + \cdots + [(A_k - \overline{A})^2 - \frac{k^2 - 1}{12} h_A{}^2]^2} \tag{10}$$

It looks to be complicated, but those coefficients such as

$$A_i - \overline{A}, \quad (A_i - \overline{A})^2 - \frac{k^2 - 1}{12} h_A{}^2$$

have been given as a table called The Table of Orthogonal Polynomial Equation. For example, when k=5, the coefficients are found to be

$$a_1 = \frac{-2y_1 - y_2 + 0y_3 + y_4 + 2y_5}{10 \times h_A} \tag{11}$$

$$a_2 = \frac{2y_1 - y_2 - 2y_3 - y_4 + 2y_5}{14 \times h_A{}^2} \tag{12}$$

In Equations (1) or (2), the terms are not orthogonal; therefore, simultaneous equations (6) must be solved in order to determine coefficients such as a_1, a_1' or a_2', etc. Today, computers are available to calculate those coefficients, and it is not very tedious as before. The merit of using orthogonal polynomial equation can be seen when analysis of variance is performed.

By the use of orthogonal polynomial equations, it is easy to decompose the total variation into components such as linear, quadratic, ..., and the kth power terms. From the evaluation of the significant terms or from the percent contribution, it is possible to judge which terms should be used in an equation.

The fact that factorial effects are orthogonal is essentially the same as saying that all terms are orthogonal. It can be expressed as

$$S_T = S_1 + S_2 + \cdots + S_{k-1} \tag{13}$$

where $S_1, S_2, ..., S_{k-1}$ are the terms in an orthogonal polynomial equation.

13. PERCENT CONTRIBUTION WHEN e_1 IS SIGNIFICANT AGAINST e_2

Consider the case in Table 1.

First, the percent contributions of A, B, C and D which are significant against the primary error, (e_1), denoted by ρ_A, ρ_B, ρ_C, ρ_D respectively, are given by

$$\begin{cases} \rho_A = \dfrac{S_A - (a-1) \times V_{e1}}{S_T} \times 100 \\ \rho_B = \dfrac{S_B - (b-1) \times V_{e1}}{S_T} \times 100 \\ \rho_C = \dfrac{S_C - (c-1) \times V_{e1}}{S_T} \times 100 \\ \rho_D = \dfrac{S_D - (d-1) \times V_{e1}}{S_T} \times 100 \end{cases}$$

Percent contribution of e_1 is

$$\rho_{e1} = \frac{S_{e1} + f^* V_{e1} - (f^* + f_{e1}) V_{e2}}{S_T} \times 100$$

where f^* is the total degrees of freedom of A, B, C and D which are significant against the primary error, or $(a\text{-}1) + (b\text{-}1) + (c\text{-}1) + (d\text{-}1)$.

Percent contribution of the secondary error is given by

$$\rho_{e2} = \frac{S_{e2} + (f^* + f_{e1})V_{e2}}{S_T} \times 100$$

Table 1. ANOVA Table

Factor	f	S	V	ρ(%)
A	a-1	S_A	V_A**	ρ_A
B	b-1	S_B	V_B**	ρ_B
C	c-1	S_C	V_C**	ρ_C
D	d-1	S_D	V_D**	ρ_D
E	e-1	S_E	V_E°	
F	f-1	S_F	V_F°	
(e_1)*	(f_{e1})	(S_{e1})	(V_{e1})*	ρ_{e1}
e_2	f_{e2}	S_{e2}	V_{e2}	ρ_{e2}
T	n-1	S_T		100.0

* (e_1) is the primary error variance by pooling e_1, E and F.

14. Square of Sensitivity and Decibel

In an experiment for nominal-the-best case where no signal factors are assigned, the following characteristic is used to find the effect on the mean:

$$S = 10 \log m^2$$
$$= 10 \log \frac{1}{r}(S_m - V_e) \qquad (1)$$

Some people are questioning if m could be simply used for the analysis. In the analysis of Equation (1), the following S/N Ratio must be analyzed too:

$$\eta = 10 \log \frac{\frac{1}{r}(S_m - V_e)}{V_e} \qquad (2)$$

The purpose is to find the condition with a small variablity against the mean, or a condition which is stable. In a parameter design, the level intervals of the inner factors are widely set for experiment. It must be wide since with narrow intervals, there would be not much difference in stability in many cases. With wide intervals, the mean becomes one or two digits larger sometimes, depending on the combination of factors. For such a case, the conclusions with a high reproducibility can be expected if a decibel value is used instead of raw data.

Although it is all right to use the means for the analysis if there is not too much difference between factor combinations, raw data are converted into decibel values for the analysis in parameter design. That is the same reason in the case of analyzing the sensitivity of nominal-the-best with a signal factor.

$$
\begin{aligned}
S &= 10 \log \beta^2 \\
&= 10 \log \frac{1}{r}(S_\beta - V_e)
\end{aligned} \tag{3}
$$

15. NOISE FACTOR LEVEL SETTING

In order to appropriately reflect the effect of noise factor K in an experiment, its levels are set as follows:

$$
\begin{cases}
K_1 = \text{average of } K - \sqrt{\dfrac{3}{2}}\ \sigma_k \\
K_2 = \text{average of } K \\
K_3 = \text{average of } K + \sqrt{\dfrac{3}{2}}\ \sigma_k
\end{cases} \tag{1}
$$

There should be no problem in setting the mean as the second level. The reason for using $\sqrt{3/2}\ \sigma_k$ as an interval is explained as follows.

Assume the noise factor varies at random to be $K_1, K_2, \ldots, K_n$ and outputs $y_1, y_2, \ldots, y_n$ were obtained. In other words, observations (k_i, y_i) $(i = 1, 2, \ldots, n)$ were obtained with respect to K and y. When y is expressed by the linear term of K,

$$
y = \bar{y} + \beta(K - \bar{K}) \tag{2}
$$

where

$$
\bar{y} = \frac{1}{n}(y_1 + y_2 + \cdots + y_n)
$$

$$
\beta = \frac{(K_1 - \bar{K})y_1 + \cdots + (K_n - \bar{K})y_n}{(K_1 - \bar{K})^2 + \cdots + (K_n - \bar{K})^2}
$$

$$
\bar{K} = \frac{1}{n}(K_1 + K_2 + \cdots + K_n)
$$

The variation of β, or S_β, is given by

$$
S_\beta = \frac{[(K_1 - \bar{K})y_1 + \cdots + (K_n - \bar{K})y_n]^2}{(K_1 - \bar{K})^2 + \cdots + (K_n - \bar{K})^2} \tag{3}
$$

The expectation of S_β is

$$
E(S_\beta) = \sigma^2 + [(K_1 - \bar{K})^2 + \cdots + (K_n - \bar{K})^2]\beta^2 = \sigma^2 + (n-1)\sigma_k{}^2\beta^2 \tag{4}
$$

where σ^2 is the experimental error variance.

If the levels of K are set at equal intervals instead of randomly,

$$\begin{cases} K_1 = \text{average} - h \\ K_2 = \text{average} \\ K_3 = \text{average} + h \end{cases} \tag{5}$$

If the numbers of repetitions of the levels are equal, the number of repetitions of each level, r_0, is equal to $n/3$.

The variation S_β is

$$S_\beta = \frac{(-y_1 + y_3)^2}{\frac{n}{3} \times 2} \tag{6}$$

where

y_1 is the total of $n/3$ pieces of data at the first level of K

y_3 is the total of $n/3$ pieces of data at the third level of K

The expectations, $E(S_\beta)$

$$E(S_\beta) = \sigma^2 + r_0 s h^2 \beta^2 = \sigma^2 + \frac{n}{3} \times 2 \times h^2 \beta^2 \tag{7}$$

the value of h which makes (4) equal to (7) is derived as follows.

$$\sigma^2 + \frac{n}{3} \times 2 \times h^2 \beta^2 = \sigma^2 + (n-1)\sigma_k{}^2 \beta^2 \tag{8}$$

$$h^2 = \frac{n-1}{n} \times \frac{3}{2} \sigma_k{}^2 \tag{9}$$

Letting

$$\frac{n-1}{n} \approx 1$$

h is given by

$$h^2 \approx \frac{3}{2} \sigma_k{}^2 \tag{10}$$

$$\therefore h \approx \sqrt{\frac{2}{3}}\ \sigma_k \tag{11}$$

When a 3-level factor is set as shown in Equation (1), the expectation of its linear effect is the same as the case when K varies randomly.

In the case of a 2-level factor, it is recommended to set as

$K_1 = \text{mean} - \sigma_k$

$K_2 = \text{mean} + \sigma_k$

16. Effective Number of Replications for Dummy Treatment

(1) Equation of the Effective Number of Replications, n_e, for Orthogonal Type Experiments

The n_e in the equation of the confidence intervals for a process average is called the effective number of replications.

$$\pm\sqrt{F \times V_e \times \frac{1}{n_e}} \tag{1}$$

The estimation of a process average is given by a linear equation with constant coefficients to observations $y_1, y_2, \ldots, y_n$.

$$\hat{\mu} = c_1 y_1 + c_2 y_2 + \cdots + c_n y_n \tag{2}$$

the effective number of replications of μ is given by the reciprocal of D, which is the number of units of the estimate, $\hat{\mu}$.

$$n_e = \frac{1}{D} = \frac{1}{c_1{}^2 + c_2{}^2 + \cdots + c_n{}^2} \tag{3}$$

When there are a few factorial effects, it would be tedious to calculate the reciprocal of the number of units after rewriting the estimate of a process average into a linear equation of observations. Instead, Taguchi's Equation may be used to calculate the effective number of replications for the case of orthogonal type experiment.

$$n_e = \frac{\text{Total number of experiments}}{\begin{array}{c}\text{Sum of the number of degrees of freedom}\\ \text{of the factorial effects that are considered in the}\\ \text{estimation of } f \text{ (including the general mean)}\end{array}} \tag{4}$$

When orthogonal arrays are used with treatment, the direct summing method or pseudo factor design are used, the numbers of replications for different levels of the same factor will be different. Therefore, the weight of individual observation which affects the estimate will be different.

That means the degrees of freedom of the factors in Equation (4) will be different if a factor has a different number of replications.

In an experiment using orthogonal array L_9 for example, A and C are significant. Since A is treated, the numbers of replications for A_1 and A_2 are different.

Table 1. Layout of Treatment

	A 1	B 2	C 3	D 4	Results
1	1	1	1	1	y_1
2	1	2	2	2	y_2
3	1	3	3	3	y_3
4	2	1	2	3	y_4
5	2	2	3	1	y_5
6	2	3	1	2	y_6
7	1′	1	3	2	y_7
8	1′	2	1	3	y_8
9	1′	3	2	1	y_9

The estimates of A_1B_3 and A_2B_3 are written as

$$\hat{\mu}(A_1B_3) = \overline{A}_1 + \overline{B}_3 - \overline{T}$$

$$= \frac{y_1 + y_2 + y_3 + y_7 + y_8 + y_9}{6} + \frac{y_3 + y_6 + y_9}{3} - \frac{y_1 + \cdots + y_9}{9} \tag{5}$$

$$\hat{\mu}(A_2B_3) = \overline{A}_2 + \overline{B}_3 - \overline{T}$$

$$= \frac{y_4 + y_5 + y_6}{3} + \frac{y_3 + y_6 + y_9}{3} - \frac{y_1 + \cdots + y_9}{9} \tag{6}$$

In Equations (5) and (6), the coefficients $c_1, c_2, \ldots, c_9$ to observations $y_1, y_2, \ldots, y_9$ are summarized in Table 2.

Table 2. Coefficients

Estimate / Coefficient	$\hat{\mu}(A_1B_3)$			$\hat{\mu}(A_2B_3)$		
c_1	$\frac{1}{6}$		$-\frac{1}{9}$			$-\frac{1}{9}$
c_2	$\frac{1}{6}$		$-\frac{1}{9}$			$-\frac{1}{9}$
c_3	$\frac{1}{6}$	$+\frac{1}{3}$	$-\frac{1}{9}$		$+\frac{1}{3}$	$-\frac{1}{9}$
c_4			$-\frac{1}{9}$	$\frac{1}{3}$		$-\frac{1}{9}$
c_5			$-\frac{1}{9}$	$\frac{1}{3}$		$-\frac{1}{9}$
c_6		$+\frac{1}{3}$	$-\frac{1}{9}$	$\frac{1}{3}$	$+\frac{1}{3}$	$-\frac{1}{9}$
c_7	$\frac{1}{6}$		$-\frac{1}{9}$			$-\frac{1}{9}$
c_8	$\frac{1}{6}$		$-\frac{1}{9}$			$-\frac{1}{9}$
c_9	$\frac{1}{6}$	$+\frac{1}{3}$	$-\frac{1}{9}$		$+\frac{1}{3}$	$-\frac{1}{9}$

The number of units for A_1B_3 and A_2B_3 are calculated as

$$D(A_1B_3) = \left(\frac{1}{6}-\frac{1}{9}\right)^2 \times 4 + \left(\frac{1}{6}+\frac{1}{3}-\frac{1}{9}\right)^2 \times 2 + \left(\frac{1}{3}-\frac{1}{9}\right)^2 \times 1 + \left(-\frac{1}{9}\right)^2 \times 2 = \frac{3.5}{9} \tag{7}$$

$$D(A_2B_3) = \left(\frac{1}{3}-\frac{1}{9}\right)^2 \times 4 + \left(\frac{1}{3}+\frac{1}{3}-\frac{1}{9}\right)^2 \times 1 + \left(\frac{1}{9}\right)^2 \times 4 = \frac{5}{9} \tag{8}$$

It shows that the numbers of units for $\hat{\mu}$ (A_1B_3) and $\hat{\mu}$ (A_2B_3) are different. To use Equation (4) to calculate the effective number of replications, 0.5 and 2 are used for A_1 and A_2 respectively, as the degrees of freedom of A. This is called the effective degrees of freedom. In the case of treatment, the effective degrees of freedom is calculated by the following equation:

$$\begin{array}{c}\text{Effective degrees}\\ \text{of freedom}\end{array} = \frac{\begin{array}{c}\text{No. of levels of}\\ \text{the column before dummy treated}\end{array}}{\text{No. of repetitions of dummy level}} - 1$$

17. Estimation for Idle Column Method

When a 3-level factor A is assigned using idle column method, the effects of A_1A_2 and A_2A_3 are separately calculated. Both effects become significant sometimes, and only one is significant at other times.

The following table is an example of the layout:

Example of Layout

	(1) 1	A $\cap$ 2 3	B $\cap$ 4 5	C 6	D 7	Results
1	1	1	1	1	1	y_1
2	1	1	2	2	2	y_2
3	1	2	1	2	2	y_3
4	1	2	2	1	1	y_4
5	2	2	1	1	2	y_5
6	2	2	3	2	1	y_6
7	2	3	1	2	1	y_7
8	2	3	3	1	2	y_8

Let the estimate of A_1 and A_2 from the 1st level of the idle column be $\overline{A}_1'$ and $\overline{A}_2'$, and the estimate of A_2 and A_3 from the 2nd level of the idle column be $\overline{A}_2''$ and $\overline{A}_3''$.

Similarly, let the estimate of B_1 and B_2 from the 1st level of the idle column be $\overline{B}_1'$ and $\overline{B}_2'$, and the estimate of B_1 and B_3 from the 2nd level of the idle column be $\overline{B}_1''$ and $\overline{B}_3'''$. $\overline{A}_1'$, $\overline{A}_2'$, ..., $\overline{B}_3''$ are given by

$$\begin{cases} \overline{A}_1' = \dfrac{y_1 + y_2}{2} \\ \overline{A}_2' = \dfrac{y_3 + y_4}{2} \\ \overline{A}_2'' = \dfrac{y_5 + y_6}{2} \\ \overline{A}_3'' = \dfrac{y_7 + y_8}{2} \\ \overline{B}_1' = \dfrac{y_1 + y_3}{2} \\ \overline{B}_2' = \dfrac{y_2 + y_4}{2} \\ \overline{B}_1'' = \dfrac{y_5 + y_7}{2} \\ \overline{B}_3'' = \dfrac{y_6 + y_8}{2} \end{cases} \tag{1}$$

(1) Estimation of A_1

(a) The Case When Both A_1A_2 and A_2A_3 Are Significant

The level of A experimented in both the 1st and the 2nd levels of the idle column (hereafter referred to idle 1 and idle 2 respectively) is A_2. This is called the contrast level. When both A_1A_2 and A_2A_3 are significant, A_1, A_2 and A_3 are estimated using the contrast level. To do this, an average is calculated from $\overline{A}_2'$ which is estimated from idle 1 and $\overline{A}_2''$ from idle 2 to be the contrast level. Based on this contrast level, $\overline{A}_1'$ and $\overline{A}_3''$ are corrected to obtain $\overline{A}_1$ and $\overline{A}_3$ respectively as shown below:

$$\overline{A}_1 = \overline{A}_1' - \frac{1}{2}(\overline{A}_2' - \overline{A}_2'') \tag{2}$$

$$\overline{A}_2 = \overline{A}_2' - \frac{1}{2}(\overline{A}_2' - \overline{A}_2'') \tag{3}$$

$$\overline{A}_3 = \overline{A}_3'' + \frac{1}{2}(\overline{A}_2' - \overline{A}_2'') \tag{4}$$

(b) The Case When Only A_1A_2 Is Significant

Since A_2A_3 is insignificant, $\overline{A}_2''$, and $\overline{A}_3''$ in idle 2 are both the estimate of A_2. An average is calculated to be the estimate of A_2. Since $\overline{A}_2'$ in idle 1 is also the estimate of A_2, the overall estimate of A_2, or $\overline{A}_2$, is calculated from idle 1 and idle 2. Based on this figure, $\overline{A}_1'$ is corrected to be $\overline{A}_1$. $\overline{A}_1$ and $\overline{A}_2$ are calculated as follows:

$$\overline{A}_1 = \overline{A}_1' - \frac{1}{2}\left(\overline{A}_2' - \frac{\overline{A}_2'' + \overline{A}_3''}{2}\right) = \overline{A}_1' - \frac{\overline{A}_2'}{2} + \frac{\overline{A}_2'' + \overline{A}_3''}{4} \tag{5}$$

$$\overline{A}_2 = \frac{1}{2}\left(A_2' + \frac{\overline{A}_2'' + \overline{A}_3''}{2}\right) = \frac{2\overline{A}_2' + \overline{A}_2'' + \overline{A}_3''}{4} \tag{6}$$

(c) The Case When Only A_2A_3 Is Significant

In the same way, $\overline{A}_2$ and $\overline{A}_3$ are calculated from

$$\overline{A}_2 = \frac{1}{2}\left(\frac{\overline{A}_1' + \overline{A}_2'}{2} + \overline{A}_2''\right)$$

$$= \frac{\overline{A}_1' + \overline{A}_2' + 2\overline{A}_2''}{4} \tag{7}$$

$$\overline{A}_3 = \overline{A}_3'' + \frac{1}{2}\left(\frac{\overline{A}_1' + \overline{A}_2'}{2} - \overline{A}_2''\right)$$

$$= \overline{A}_3'' + \frac{\overline{A}_1' + \overline{A}_2'}{4} - \frac{\overline{A}_2''}{2} \tag{8}$$

(2) Estimation of B

Unlike the case of A, the contrast level for B is B_1. However, the method is basically the same. The estimating equations are summarized in the following table:

Estimation of B_1, B_2 and B_3

Case / Level	Both B_1B_2 and B_1B_3 are Significant	Only B_1B_2 is Significant	Only B_1B_3 is Significant
$\overline{B}_1$	$\frac{\overline{B}_1' + \overline{B}_1''}{2}$	$\frac{2\overline{B}_1' + \overline{B}_1'' + \overline{B}_3''}{4}$	$\frac{\overline{B}_1' + \overline{B}_2' + 2\overline{B}_1''}{4}$
$\overline{B}_2$	$\overline{B}_2' - \frac{1}{2}(\overline{B}_1' - \overline{B}_1'')$	$\frac{4\overline{B}_2' - 2\overline{B}_1' + \overline{B}_1'' + \overline{B}_3''}{4}$	
$\overline{B}_3$	$\overline{B}_3'' + \frac{1}{2}(\overline{B}_1' - \overline{B}_1'')$		$\frac{\overline{B}_1' + \overline{B}_2' - 2\overline{B}_1'' + 4\overline{B}_3''}{4}$

18. Objective of Using Cumulated Data in Accumulation Analysis

(1) A Simple Example

There are two countermeasures to improve the appearance of a product. Using a current method before taking any countermeasures, appearance of the products were classified into good, normal and bad and it was found that 3 products were found in each class.

Using countermeasure *A*, the number of products in good, normal and bad class was 6, 3 and 0 respectively. For countermeasure *B*, they were 3, 6 and 0.

Let the 1st level be the case without taking countermeasures and the 2nd level is the case when a countermeasure was taken. Table 1 shows the results.

Table 1. Non-cumulative Frequency

	Good	Normal	Bad	Total		Good	Normal	Bad	Total
A_1	3	3	3	9	B_1	3	3	3	9
A_2	6	3	0	9	B_2	3	6	0	9

From those results, the effects of *A* and *B* are compared. In this case, the order of classes has technical significance: "good" is better than "normal" and "normal" is better than "bad." In other words, classes have an order. That is why cumulative frequency is taken and accumulation anaysis is recommended. Following is the case when cumulative frequency is not taken.

(2) Analysis Using Frequency Analysis

The effect of *A* is calculated without using cumulative frequency.

$$\begin{cases} W_1 = \dfrac{18^2}{9 \times (18-9)} = 4.0 \\[2ex] W_2 = \dfrac{18^2}{6 \times (18-6)} = 4.5 \\[2ex] W_3 = \dfrac{18^2}{3 \times (18-3)} = 7.2 \end{cases}$$

Table 2. Results of *A*

	Good	Normal	Bad	Total
A_1	3	3	3	9
A_2	6	3	0	9
Total	9	6	3	18

Table 3. ANOVA Table for *A* (Frequency Analysis)

Factor	f	S	V	F
A	2	5.6	2.8	1.85
e	32	48.4	1.51	
T	34	54.0		

Table 4. Results of *B*

	Good	Normal	Bad	Total
B_1	3	3	3	9
B_2	3	6	0	9
Total	6	9	3	18

Table 5. ANOVA Table for *B* (Frequency Analysis)

Factor	f	S	V	F
B	2	5.6	2.8	1.85
e	32	48.4	1.51	
T	34	54.0		

$$S_T = 18 \times 3 = 54 \qquad (f = 17 \times 2 = 34)$$

$$S_A = \frac{(3-6)^2 \times W_1 + (3-3)^2 \times W_2 + (3-0)^2 \times W_3}{18}$$

$$= 5.6 \qquad (f = 1 \times 2 = 2)$$

The effect of *B* is then calculated

$$\begin{cases} W_1 = \dfrac{18^2}{6 \times (18-6)} = 4.5 \\ W_2 = \dfrac{18^2}{9 \times (18-9)} = 4.0 \\ W_3 = \dfrac{18^2}{3 \times (18-3)} = 7.2 \end{cases}$$

$$S_T = 18 \times 3 = 54 \qquad (f = 17 \times 2 = 34)$$

$$S_B = \frac{(3-3)^2 \times W_1 + (3-6)^2 \times W_2 + (3-0)^2 \times W_3}{18}$$

$$= 5.6 \qquad (f{=}2)$$

$$S_e = 54 - 5.6 = 48.4 \qquad (f = 32)$$

From the results, the effects of A and B are evaluated as the same which does not make sense. When either countermeasure A or B was taken there was no occurrence in class "bad." But the number of occurrence of class "good" increased for A and class "normal" increased for B. It is appropriate therefore to conclude that A is a better countermeasure without talking the significance.

(3) Anlaysis Using Accumulation Analysis

The effect of A is calculated using cumulative frequency.

$$\begin{cases} W_1 = \dfrac{18^2}{9 \times (18-9)} = 4.0 \\ W_2 = \dfrac{18^2}{15 \times (18-15)} = 7.2 \end{cases}$$

$$S_T = 18 \times 2 = 36 \qquad (f = 17 \times 2 = 34)$$

$$S_A = \frac{(3-6)^2 \times W_1 + (6-9)^2 \times W_2}{18} = 5.6$$

$$S_e = 36 - 5.6 = 30.4$$

The effect of B is calculated.

$$\begin{cases} W_1 = \dfrac{18^2}{6 \times (18-6)} = 4.5 \\ W_2 = \dfrac{18^2}{15 \times (18-15)} = 7.2 \end{cases}$$

$$S_T = 18 \times 2 = 36 \qquad (f = 34)$$

$$S_B = \frac{(3-3)^2 \times W_1 + (6-9)^2 \times W_2}{18} = 3.6 \qquad (f{=}2)$$

$$S_e = 36 - 3.6 = 32.4 \qquad (f = 32)$$

From Tables 7 and 9, it is seen that the effect of A is evaluated to be larger.

By taking cumulative frequency, the evaluation of factorial effects matches with realistic feeling.

Table 6. Cumulative Frequency for A

	I G	II $G+N$	III $G+N+B$
A_1	3	6	9
A_2	6	9	9
Total	9	15	18

Table 7. ANOVA Table for A (Accumulation Analysis)

Factor	f	S	V	F
A	2	5.6	2.8	2.95
e	32	30.4	0.95	
T	34	36.0		

Table 8. Cumulative Frequency for B

	I	II	III
B_1	3	6	9
B_2	3	9	9
Total	6	15	18

Table 9. ANOVA Table for B (Accumulation Analysis)

Factor	f	S	V	F
B	2	3.6	1.8	1.78
e	32	32.4	1.01	
T	34	36.0		

Appendix B

APPENDIX TABLE 1: F Table (5%, 1%)

No. of Degrees of Freedom of Denominator, f_2	No. of Degrees of Freedom of Numerator, f_1										
	1	2	3	4	5	6	7	8	9	10	11
1	161 4,052	200 4,999	126 5,403	225 5,625	230 5,764	234 5,859	237 5,928	239 5,981	241 6,022	242 6,056	243 6,082
2	18.51 98.49	19.00 99.01	19.16 99.17	19.25 99.25	19.30 99.30	19.33 99.33	19.36 99.34	19.37 99.36	19.38 99.38	19.39 99.40	19.40 99.41
3	10.13 34.12	9.55 30.81	9.28 29.46	9.12 28.71	9.01 28.24	8.94 27.91	8.88 27.67	8.84 27.49	8.81 27.34	8.78 27.23	8.76 27.13
4	7.71 21.20	6.94 18.00	6.59 16.69	6.39 15.98	6.26 15.52	6.16 15.21	6.09 14.98	6.04 14.80	6.00 14.66	5.96 14.54	5.93 14.45
5	6.61 16.26	5.79 13.27	5.41 12.06	5.19 11.39	5.05 10.97	4.95 10.67	4.88 10.45	4.82 10.27	4.78 10.15	4.74 10.05	4.70 9.96
6	5.99 13.74	5.14 10.92	4.76 9.78	4.53 9.15	4.39 8.75	4.28 8.47	4.21 8.26	4.15 8.10	4.10 7.98	4.06 7.87	4.03 7.79
7	5.59 12.25	4.74 9.55	4.35 8.45	4.12 7.85	3.97 7.46	3.87 7.19	3.79 7.00	3.73 6.84	3.68 6.71	3.63 6.62	3.60 6.54
8	5.32 11.26	4.46 8.65	4.07 7.59	3.84 7.01	3.69 6.63	3.58 6.37	3.50 6.19	3.44 6.03	3.39 5.91	3.34 5.82	3.31 5.74
9	5.12 10.56	4.26 8.02	3.86 6.99	3.63 6.42	3.48 6.06	3.37 5.80	3.29 5.62	3.23 5.47	3.18 5.35	3.13 5.26	3.10 5.18
10	4.96 10.04	4.10 7.56	3.71 6.55	3.48 5.99	3.33 5.64	3.22 5.39	3.14 5.21	3.07 5.06	3.02 4.95	2.97 4.85	2.94 4.78
11	4.84 9.65	3.98 7.20	3.59 6.22	3.36 5.67	3.20 5.32	3.09 5.07	3.01 4.88	2.95 4.74	2.90 4.63	2.86 4.54	2.82 4.46
12	4.75 9.33	3.88 6.93	3.49 5.95	3.26 5.41	3.11 5.06	3.00 4.82	2.92 4.65	2.85 4.50	2.80 4.39	2.76 4.30	2.72 4.22
13	4.67 9.07	3.80 6.70	3.41 5.74	3.18 5.20	3.02 4.86	2.92 4.62	2.84 4.44	2.77 4.30	2.72 4.19	2.67 4.10	2.63 4.02
14	4.60 8.86	3.74 6.51	3.34 5.56	3.11 5.03	2.96 4.69	2.85 4.46	2.77 4.28	2.70 4.14	2.65 4.03	2.60 3.94	2.56 3.86
15	4.54 8.68	3.68 6.36	3.29 5.42	3.06 4.89	2.90 4.56	2.79 4.32	2.70 4.14	2.64 4.00	2.59 3.89	2.55 3.80	2.51 3.73
16	4.49 8.53	3.63 6.23	3.24 5.29	3.01 4.77	2.85 4.44	2.74 4.20	2.66 4.03	2.59 3.89	2.54 3.78	2.49 3.69	2.45 3.61
17	4.45 8.40	3.59 6.11	3.20 5.18	2.96 4.67	2.81 4.34	2.70 4.10	2.62 3.93	2.55 3.79	2.50 3.68	2.45 3.59	2.41 3.52
18	4.41 8.28	3.55 6.01	3.16 5.09	2.93 4.58	2.77 4.25	2.66 4.01	2.58 3.85	2.51 3.71	2.46 3.60	2.41 3.51	2.37 3.44
19	4.38 8.18	3.52 5.93	3.13 5.01	2.90 4.50	2.74 4.17	2.63 3.94	2.55 3.77	2.48 3.63	2.43 3.52	2.38 3.43	2.34 3.36
20	4.35 8.10	3.49 5.85	3.10 4.94	2.87 4.43	2.71 4.10	2.60 3.87	2.52 3.71	2.45 3.56	2.40 3.45	2.35 3.37	2.31 3.30

Genichi Taguchi (1972): Statistical Analysis, Maruzen

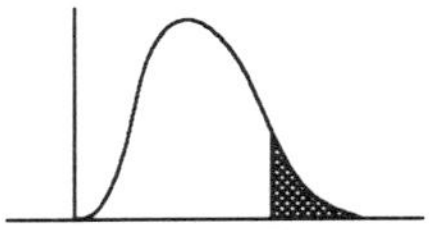

Upper figures are 5% point of F distribution.
Lower figures are 1% point of F distribution

No. of Degrees of Freedom of Numerator, f_1												
12	14	16	20	24	30	40	50	75	100	200	500	∞
244 6,106	245 6,142	246 6,169	248 6,208	249 6,234	250 6,258	251 6,286	252 6,302	253 6,323	253 6,334	254 6,352	2.54 6,361	254 6,366
19.41 99.42	19.42 99.43	19.43 99.44	19.44 99.45	19.45 99.46	19.46 99.47	19.47 99.48	19.47 99.48	19.48 99.49	19.49 99.49	19.49 99.49	19.50 99.50	19.50 99.50
8.74 27.05	8.71 26.92	8.69 26.83	8.66 26.69	8.64 26.60	8.62 26.50	8.60 26.41	8.58 26.30	8.57 26.27	8.56 26.23	8.54 26.18	8.54 26.14	8.53 26.12
5.91 14.37	5.87 14.24	5.84 14.15	5.80 14.02	5.77 13.93	5.74 13.83	5.71 13.74	5.70 13.69	5.68 13.61	5.66 13.57	5.65 13.52	5.64 13.48	5.63 13.46
4.68 9.89	4.64 9.77	4.60 9.68	4.56 9.55	4.53 9.47	4.50 9.38	4.46 9.29	4.44 9.24	4.42 9.17	4.40 9.13	4.38 9.07	4.37 9.04	4.36 9.02
4.00 7.72	3.96 7.60	3.92 7.52	3.87 7.39	3.84 7.31	3.81 7.23	3.77 7.14	3.75 7.09	3.72 7.02	3.71 6.99	3.69 6.94	3.68 6.90	3.67 6.88
3.57 6.47	3.52 6.35	3.49 6.27	3.44 6.15	3.41 6.07	3.38 5.98	3.34 5.90	3.32 5.85	3.29 5.78	3.28 5.75	3.25 5.70	3.24 5.67	3.23 5.65
3.28 5.67	3.23 5.56	3.20 5.48	3.15 5.36	3.12 5.28	3.08 5.20	3.05 5.11	3.03 5.06	3.00 5.00	2.98 4.96	2.96 4.91	2.94 4.88	2.93 4.86
3.07 5.11	3.02 5.00	2.98 4.92	2.93 4.80	2.90 4.73	2.86 4.64	2.82 4.56	2.80 4.51	2.77 4.45	2.76 4.41	2.73 4.36	2.72 4.33	2.71 4.31
2.91 4.71	2.86 4.60	2.82 4.52	2.77 4.41	2.74 4.33	2.70 4.25	2.67 4.17	2.64 4.12	2.61 4.05	2.59 4.01	2.56 3.96	2.55 3.93	2.54 3.91
2.79 4.40	2.74 4.29	2.70 4.21	2.65 4.10	2.61 4.02	2.57 3.94	2.53 3.86	2.50 3.80	2.47 3.74	2.45 3.70	2.42 3.66	2.41 3.62	2.40 3.60
2.69 4.16	2.64 4.05	2.60 3.98	2.54 3.86	2.50 3.78	2.46 3.70	2.42 3.61	2.40 3.56	2.36 3.49	2.35 3.46	2.32 3.41	2.31 3.38	2.30 3.36
2.60 3.96	2.55 3.85	2.51 3.78	2.46 3.67	2.42 3.59	2.38 3.51	2.34 3.42	2.32 3.37	2.28 3.30	2.26 3.27	2.24 3.21	2.22 3.18	2.21 3.16
2.53 3.80	2.48 3.70	2.44 3.62	2.39 3.51	2.35 3.43	2.31 3.34	2.27 3.26	2.24 3.21	2.21 3.14	2.19 3.11	2.16 3.06	2.14 3.02	2.13 3.00
2.48 3.67	2.43 3.56	2.39 3.48	2.33 3.36	2.29 3.29	2.25 3.20	2.21 3.12	2.18 3.07	2.15 3.00	2.12 2.97	2.10 2.92	2.08 2.89	2.07 2.87
2.42 3.55	2.37 3.45	2.33 3.37	2.28 3.26	2.24 3.18	2.20 3.10	2.16 3.01	2.13 2.96	2.09 2.89	2.07 2.86	2.04 2.80	2.02 2.77	2.01 2.75
2.38 3.45	2.33 3.35	2.29 3.27	2.23 3.16	2.19 3.08	2.15 3.00	2.11 2.92	2.08 2.86	2.04 2.79	2.02 2.76	1.99 2.70	1.97 2.67	1.96 2.65
2.34 3.37	2.29 3.27	2.25 3.19	2.19 3.07	2.15 3.00	2.11 2.91	2.07 2.83	2.04 2.78	2.00 2.71	1.98 2.68	1.95 2.62	1.93 2.59	1.92 2.57
2.31 3.30	2.26 3.19	2.21 3.12	2.15 3.00	2.11 2.92	2.07 2.84	2.02 2.76	2.00 2.70	1.96 2.63	1.94 2.60	1.91 2.54	1.90 2.51	1.88 2.49
2.28 3.23	2.23 3.13	2.18 3.05	2.12 2.94	2.08 2.86	2.04 2.77	1.99 2.69	1.96 2.63	1.92 2.56	1.90 2.53	1.87 2.47	1.85 2.44	1.84 2.42

Appendix Table 1 (cont.)

No. of Degrees of Freedom of Denominator, f_2	No. of Degrees of Freedom of Numerator, f_1										
	1	2	3	4	5	6	7	8	9	10	11
21	4.32	3.47	3.07	2.84	2.68	2.57	2.49	2.42	2.37	2.32	2.28
	8.02	5.78	4.87	4.37	4.04	3.81	3.65	3.51	3.40	3.31	3.24
22	4.30	3.44	3.05	2.82	2.66	2.55	2.47	2.40	2.35	2.30	2.26
	7.94	5.72	4.82	4.31	3.99	3.76	3.59	3.45	3.35	3.26	3.18
23	4.28	3.42	3.03	2.80	2.64	2.53	2.45	2.38	2.32	2.28	2.24
	7.88	5.66	4.76	4.26	3.94	3.71	3.54	3.41	3.30	3.21	3.14
24	4.26	3.40	3.01	2.78	2.62	2.51	2.43	2.36	2.30	2.26	2.22
	7.82	5.61	4.72	4.22	3.90	3.67	3.50	3.36	3.25	3.17	3.09
25	4.24	3.38	2.99	2.76	2.60	2.49	2.41	2.34	2.28	2.24	2.20
	7.77	5.57	4.68	4.18	3.86	3.63	3.46	3.32	3.21	3.13	3.05
26	4.22	3.37	2.89	2.74	2.59	2.47	2.39	2.32	2.27	2.22	2.18
	7.72	5.53	4.64	4.14	3.82	3.59	3.42	3.29	3.17	3.09	3.02
27	4.21	3.35	2.96	2.73	2.57	2.46	2.37	2.30	2.25	2.20	2.16
	7.68	5.49	4.60	4.11	3.79	3.56	3.39	3.26	3.14	3.06	2.98
28	4.20	3.34	2.95	2.71	2.56	2.44	2.36	2.29	2.24	2.19	2.15
	7.64	5.45	4.57	4.07	3.76	3.53	3.36	3.23	3.11	3.03	2.95
29	4.18	3.33	2.93	2.70	2.54	2.43	2.35	2.28	2.22	2.18	2.14
	7.60	5.52	4.54	4.04	3.73	3.50	3.33	3.20	3.03	3.00	2.92
30	4.17	3.32	2.92	2.69	2.53	2.42	2.34	2.27	2.21	2.16	2.12
	7.56	5.39	4.51	4.02	3.70	3.47	3.30	3.17	3.06	2.98	2.90
32	4.15	3.30	2.90	2.67	2.51	2.40	2.32	2.25	2.19	2.14	2.10
	7.50	5.34	4.46	3.97	3.66	3.42	3.25	3.12	3.01	2.94	2.86
34	4.13	3.28	2.88	2.65	2.49	2.38	2.30	2.23	2.17	2.12	2.08
	7.44	5.29	4.42	3.93	3.61	3.38	3.21	3.08	2.97	2.89	2.82
36	4.11	3.26	2.86	2.63	2.48	2.36	2.28	2.21	2.15	2.10	2.06
	7.39	5.25	4.38	3.89	3.58	3.35	3.18	3.04	2.94	2.86	2.78
38	4.10	3.25	2.85	2.62	2.46	2.35	2.26	2.19	2.14	2.09	2.05
	7.35	5.21	4.34	3.86	3.54	3.32	3.15	3.02	2.91	2.82	2.75
40	4.08	3.23	2.84	2.61	2.45	2.34	2.25	2.18	2.12	2.07	2.04
	7.31	5.18	4.31	3.83	3.51	3.29	3.12	2.99	2.88	2.80	2.73
42	4.07	3.22	2.83	2.59	2.44	2.32	2.24	2.17	2.11	2.06	2.02
	7.27	5.15	4.29	3.80	3.49	3.26	3.10	2.96	2.86	2.77	2.70
44	4.06	3.21	2.82	2.58	2.43	2.31	2.23	2.16	2.10	2.05	2.01
	7.24	5.12	4.26	3.78	3.46	3.24	3.07	2.94	2.84	2.75	2.68
46	4.05	3.20	2.81	2.57	2.42	2.30	2.22	2.15	2.09	2.04	2.00
	7.21	5.10	4.24	3.76	3.44	3.22	3.05	2.92	2.82	2.73	2.66
48	4.04	3.19	2.80	2.56	2.41	2.30	2.21	2.14	2.08	2.03	1.99
	7.19	5.08	4.22	3.74	3.42	3.20	3.04	2.90	2.80	2.71	2.64
50	4.03	3.18	2.79	2.56	2.40	2.29	2.20	2.13	2.07	2.02	1.98
	7.17	5.06	4.20	3.72	3.41	3.18	3.02	2.88	2.78	2.72	2.62
55	4.02	3.17	2.78	2.54	2.38	2.27	2.18	2.11	2.05	2.00	1.97
	7.12	5.01	4.16	3.68	3.37	3.15	2.98	2.85	2.75	2.66	2.59
60	4.00	3.15	2.76	2.52	2.37	2.25	2.17	2.10	2.04	1.99	1.95
	7.08	4.98	4.13	3.65	3.34	3.12	2.95	2.82	2.72	2.63	2.56
65	3.99	3.14	2.75	2.51	2.36	2.24	2.15	2.08	2.02	1.98	1.94
	7.04	4.95	4.10	3.62	3.31	3.09	2.93	2.79	2.70	2.61	2.54
70	3.98	3.13	2.74	2.50	2.35	2.32	2.14	2.07	2.01	1.97	1.93
	7.01	4.92	4.08	3.60	3.29	3.07	2.91	2.77	2.67	2.59	2.51

No. of Degrees of Freedom of Numerator, f_1												
12	14	16	20	24	30	40	50	75	100	200	500	∞
2.25	2.20	2.15	2.09	2.05	2.00	1.96	1.93	1.89	1.87	1.84	1.82	1.81
3.17	3.07	2.99	2.88	2.80	2.72	2.63	2.58	2.51	2.47	2.42	2.38	2.36
2.23	2.18	2.13	2.07	2.03	1.98	1.93	1.91	1.87	1.84	1.81	1.80	1.78
3.12	3.02	2.94	2.83	2.75	2.67	2.58	2.53	2.46	2.42	2.37	2.33	2.31
2.20	2.14	2.10	2.04	2.00	1.96	1.91	1.88	1.84	1.82	1.79	1.77	1.76
3.07	2.97	2.89	2.78	2.70	2.62	2.53	2.48	2.41	2.37	2.32	2.28	2.26
2.18	2.13	2.09	2.02	1.98	1.94	1.89	1.86	1.82	1.80	1.76	1.74	1.73
3.03	2.93	2.85	2.74	2.66	2.58	2.49	2.44	2.36	2.33	2.27	2.23	2.21
2.16	2.11	2.06	2.00	1.96	1.92	1.87	1.84	1.80	1.77	1.74	1.72	1.71
2.99	2.89	2.81	2.70	2.62	2.54	2.45	2.40	2.32	2.29	2.23	2.19	2.17
2.15	2.10	2.05	1.99	1.95	1.90	1.85	1.82	1.78	1.76	1.72	1.70	1.69
2.96	2.86	2.77	2.66	2.58	2.50	2.41	2.36	2.28	2.25	2.19	2.15	2.13
2.13	2.08	2.03	1.97	1.93	1.88	1.84	1.80	1.76	1.74	1.71	1.68	1.67
2.93	2.83	2.74	2.63	2.55	2.47	2.38	2.33	2.25	2.21	2.16	2.12	2.10
2.12	2.06	2.02	1.96	1.91	1.87	1.81	1.78	1.75	1.72	1.69	1.67	1.65
2.90	2.80	2.71	2.60	2.52	2.44	2.35	2.30	2.22	2.18	2.13	2.09	2.06
2.10	2.05	2.00	1.94	1.90	1.85	1.80	1.77	1.73	1.71	1.68	1.65	1.64
2.87	2.77	2.68	2.57	2.49	2.41	2.32	2.27	2.19	2.15	2.10	2.06	2.03
2.09	2.04	1.99	1.93	1.89	1.84	1.79	1.76	1.72	1.69	1.66	1.64	1.62
2.84	2.74	2.66	2.55	2.47	2.38	2.29	2.24	2.16	2.13	2.07	2.03	2.01
2.07	2.02	1.97	1.91	1.86	1.82	1.76	1.74	1.69	1.67	1.64	1.61	1.59
2.80	2.70	2.62	2.51	2.42	2.34	2.25	2.20	2.12	2.08	2.02	1.98	1.96
2.05	2.00	1.95	1.89	1.84	1.80	1.74	1.71	1.67	1.64	1.61	1.59	1.57
2.76	2.66	2.58	2.47	2.38	2.30	2.21	2.15	2.08	2.04	1.98	1.94	1.91
2.03	1.98	1.93	1.87	1.82	1.78	1.72	1.69	1.65	1.62	1.59	1.56	1.53
2.72	2.62	2.54	2.43	2.35	2.26	2.17	2.12	2.04	2.00	1.94	1.90	1.87
2.02	1.96	1.92	1.85	1.80	1.76	1.71	1.67	1.63	1.60	1.57	1.54	1.55
2.69	2.59	2.51	2.40	2.32	2.22	2.14	2.08	2.00	1.97	1.90	1.86	1.84
2.00	1.95	1.90	1.84	1.79	1.74	1.69	1.66	1.61	1.59	1.55	1.53	1.51
2.66	2.56	2.49	2.37	2.29	2.20	2.11	2.05	1.97	1.94	1.88	1.84	1.81
1.99	1.94	1.89	1.82	1.78	1.73	1.68	1.64	1.60	1.57	1.54	1.51	1.49
2.64	2.54	2.46	2.35	2.26	2.17	2.08	2.02	1.94	1.91	1.85	1.80	1.78
1.98	1.92	1.88	1.81	1.76	1.72	1.66	1.63	1.58	1.56	1.52	1.50	1.48
2.62	2.52	2.44	2.32	2.24	2.15	2.06	2.00	1.92	1.88	1.82	1.78	1.75
1.97	1.91	1.87	1.80	1.75	1.71	1.65	1.62	1.57	1.54	1.51	1.48	1.46
2.60	2.50	2.42	2.30	2.22	2.13	2.04	1.98	1.90	1.86	1.80	1.76	1.72
1.96	1.90	1.86	1.79	1.74	1.70	1.64	1.61	1.56	1.53	1.50	1.47	1.45
2.58	2.48	2.40	2.28	2.20	2.11	2.02	1.96	1.88	1.84	1.78	1.73	1.70
1.95	1.90	1.85	1.78	1.74	1.69	1.63	1.60	1.55	1.52	1.48	1.46	1.44
2.56	2.46	2.39	2.26	2.18	2.10	2.00	1.94	1.86	1.82	1.76	1.71	1.68
1.93	1.88	1.83	1.76	1.72	1.67	1.61	1.58	1.52	1.50	1.46	1.43	1.41
2.53	2.43	2.35	2.23	2.15	2.06	1.96	1.90	1.82	1.78	1.71	1.66	1.64
1.92	1.86	1.81	1.75	1.70	1.65	1.59	1.56	1.50	1.48	1.44	1.41	1.39
2.50	2.40	2.32	2.20	2.12	2.03	1.93	1.87	1.79	1.74	1.68	1.63	1.60
1.90	1.85	1.80	1.73	1.68	1.63	1.57	1.54	1.49	1.46	1.42	1.39	1.37
2.47	2.37	2.30	2.18	2.09	2.00	1.90	1.84	1.76	1.71	1.64	1.60	1.56
1.89	1.84	1.79	1.72	1.67	1.62	1.56	1.53	1.47	1.45	1.40	1.37	1.35
2.45	2.35	2.28	2.15	2.07	1.98	1.88	1.82	1.74	1.69	1.62	1.56	1.53

Appendix Table 1 (cont.)

No. of Degrees of Freedom of Denominator, f_2	No. of Degrees of Freedom of Numerator, f_1										
	1	2	3	4	5	6	7	8	9	10	11
80	3.96	3.11	2.72	2.48	2.33	2.21	2.12	2.05	1.99	1.95	1.91
	6.96	4.88	4.04	3.56	3.25	3.04	2.87	2.74	2.64	2.55	2.48
100	3.94	3.09	2.70	2.46	2.30	2.19	2.10	2.03	1.97	1.92	1.88
	6.90	4.82	3.98	3.51	3.20	2.99	2.82	2.69	2.59	2.51	2.43
125	3.92	3.07	2.68	2.44	2.29	2.17	2.08	2.01	1.95	1.90	1.86
	6.84	4.78	3.94	3.47	3.17	2.95	2.79	2.65	2.56	2.47	2.40
150	3.91	3.06	2.67	2.43	2.27	2.16	2.07	2.00	1.94	1.89	1.85
	6.81	4.75	3.91	3.44	3.13	2.92	2.76	2.62	2.53	2.44	2.37
200	3.89	3.04	2.65	2.41	2.26	2.14	2.05	1.98	1.92	1.87	1.83
	6.76	4.71	3.88	3.41	3.11	2.90	2.73	2.60	2.50	2.41	2.34
400	3.86	3.02	2.62	2.39	2.23	2.12	2.03	1.96	1.90	1.85	1.81
	6.70	4.66	3.83	3.36	3.06	2.85	2.69	2.55	2.46	2.37	2.29
1000	3.85	3.00	2.61	2.38	2.22	2.10	2.02	1.95	1.89	1.84	1.80
	6.66	4.62	3.80	3.34	3.04	2.82	2.66	2.53	2.43	2.34	2.26
∞	3.84	2.99	2.60	2.37	2.21	2.09	2.01	1.94	1.88	1.83	1.79
	6.64	4.60	3.78	3.32	3.02	2.80	2.64	2.51	2.41	2.32	2.24

No. of Degrees of Freedom of Numerator, f_1												
12	14	16	20	24	30	40	50	75	100	200	500	∞
1.88 2.41	1.82 2.32	1.77 2.24	1.70 2.11	1.65 2.03	1.60 1.94	1.54 1.84	1.51 1.78	1.45 1.70	1.42 1.65	1.38 1.57	1.35 1.52	1.32 1.49
1.85 2.36	1.79 2.26	1.75 2.19	1.68 2.06	1.63 1.98	1.57 1.89	1.51 1.79	1.48 1.73	1.42 1.64	1.39 1.59	1.34 1.51	1.30 1.46	1.28 1.43
1.83 2.33	1.77 2.23	1.72 2.15	1.65 2.03	1.60 1.94	1.55 1.85	1.49 1.75	1.45 1.68	1.39 1.59	1.36 1.54	1.31 1.46	1.27 1.40	1.25 1.37
1.82 2.30	1.76 2.20	1.71 2.12	1.64 2.00	1.59 1.91	1.54 1.83	1.47 1.72	1.44 1.66	1.37 1.56	1.34 1.51	1.29 1.43	1.25 1.37	1.22 1.33
1.80 2.28	1.74 2.17	1.69 2.09	1.62 1.97	1.57 1.88	1.52 1.79	1.45 1.69	1.42 1.62	1.35 1.53	1.32 1.48	1.26 1.39	1.22 1.33	1.19 1.28
1.78 2.23	1.72 2.12	1.67 2.04	1.60 1.92	1.54 1.84	1.49 1.74	1.42 1.64	1.38 1.57	1.32 1.47	1.28 1.42	1.22 1.32	1.16 1.24	1.13 1.19
1.76 2.20	1.70 2.09	1.65 2.01	1.58 1.89	1.53 1.81	1.47 1.71	1.41 1.61	1.36 1.54	1.30 1.44	1.26 1.38	1.19 1.28	1.13 1.19	1.08 1.11
1.75 2.18	1.69 2.07	1.64 1.99	1.57 1.87	1.52 1.79	1.46 1.69	1.40 1.59	1.35 1.52	1.28 1.41	1.24 1.36	1.17 1.25	1.11 1.15	1.00 1.00

APPENDIX TABLE 2: ORTHOGONAL POLYNOMIAL EQUATION

Orthogonal Polynomial Equation for Equal Interval

b_1: Linear term

b_2: Quadratic term

$\vdots \quad \vdots$

$$b_i = \frac{W_1 A_1 + \cdots + W_k A_k}{r \lambda S h^i}$$

$$n_e \text{ of } b_i = r \cdot s \cdot h^{2i}$$

$$S_{bi} = \frac{(W_1 A_1 + \cdots + W_k A_k)^2}{r \cdot \lambda^2 S}$$

No. of Levels / Coefficients	k=2	k=3		k=4			k=5			
	b_1	b_1	b_2	b_1	b_2	b_3	b_1	b_2	b_3	b_4
W_1	-1	-1	1	-3	1	-1	-2	2	-1	1
W_2	1	0	-2	-1	-1	3	-1	-1	2	-4
W_3		1	1	1	-1	-3	0	-2	0	6
W_4				3	1	1	1	-1	-2	-4
W_5							2	2	1	1
$\lambda^2 S$	2	2	6	20	4	20	10	14	10	70
λS	1	2	2	10	4	6	10	14	12	24
S	$\frac{1}{2}$	2	$\frac{2}{3}$	5	4	$\frac{9}{5}$	10	14	$\frac{72}{5}$	$\frac{288}{35}$
λ	2	1	3	2	1	$\frac{10}{3}$	1	1	$\frac{5}{6}$	$\frac{35}{12}$

	k=6					k=7				
	b_1	b_2	b_3	b_4	b_5	b_1	b_2	b_3	b_4	b_5
W_1	-5	5	-5	1	-1	-3	5	-1	3	-1
W_2	-3	-1	7	-3	5	-2	0	1	-7	4
W_3	-1	-4	4	2	-10	-1	-3	1	1	-5
W_4	1	-4	-4	2	10	0	-4	0	6	0
W_5	3	-1	-7	-3	-5	1	-3	-1	1	5
W_6	5	5	5	1	1	2	0	-1	-7	-4
W_7						3	5	1	3	1
$\lambda^2 S$	70	84	180	28	252	28	84	6	154	84
λS	35	56	108	48	120	28	84	36	264	240
S	$\frac{35}{2}$	$\frac{112}{3}$	$\frac{324}{5}$	$\frac{576}{7}$	$\frac{400}{7}$	28	84	216	$\frac{3,168}{7}$	$\frac{4,800}{7}$
λ	2	$\frac{3}{2}$	$\frac{5}{3}$	$\frac{7}{12}$	$\frac{21}{10}$	1	1	$\frac{1}{6}$	$\frac{7}{12}$	$\frac{7}{20}$

Genichi Taguchi (1972): Statistical Analysis, Maruzen

Appendix Table 2 (cont.)

No. of Levels / Coefficients	k=8					k=9				
	b_1	b_2	b_3	b_4	b_5	b_1	b_2	b_3	b_4	b_5
W_1	-7	7	-7	7	-7	-4	28	-14	14	-4
W_2	-5	1	5	-13	23	-3	7	7	-21	11
W_3	-3	-3	7	-3	-17	-2	-8	13	-11	-4
W_4	-1	-5	3	9	-15	-1	-17	9	9	-9
W_5	1	-5	-3	9	15	0	-20	0	18	0
W_6	3	-3	-7	-3	17	1	-17	-9	9	9
W_7	5	1	-5	-13	-23	2	-8	-13	-11	4
W_8	7	7	7	7	7	3	7	-7	-21	-11
W_9						4	28	14	14	4
$\lambda^2 S$	168	168	264	616	2,184	60	2,772	990	2,002	468
λS	84	168	396	1,056	3,120	60	924	1,188	3,432	3,120
S	42	168	594	$\frac{12,672}{7}$	$\frac{31,200}{7}$	60	308	$\frac{7,128}{5}$	$\frac{41,184}{7}$	20,800
λ	2	1	$\frac{2}{3}$	$\frac{7}{12}$	$\frac{7}{10}$	1	3	$\frac{5}{6}$	$\frac{7}{12}$	$\frac{3}{20}$

	k=10					k=11				
	b_1	b_2	b_3	b_4	b_5	b_1	b_2	b_3	b_4	b_5
W_1	-9	6	-42	18	-6	-5	15	-30	6	-3
W_2	-7	2	14	-22	14	-4	6	6	-6	6
W_3	-5	-1	35	-17	-1	-3	-1	22	-6	1
W_4	-3	-3	31	3	-11	-2	-6	23	-1	-4
W_5	-1	-4	12	18	-6	-1	-9	14	4	-4
W_6	1	-4	-12	18	6	0	-10	0	6	0
W_7	3	-3	-31	3	11	1	-9	-14	4	4
W_8	5	-1	-35	-17	1	2	-6	-23	-1	4
W_9	7	2	-14	-22	-14	3	-1	-22	-6	-1
W_{10}	9	6	42	18	6	4	6	-6	-6	-6
W_{11}						5	15	30	6	3
$\lambda^2 S$	330	132	8,580	2,860	780	110	858	4,290	286	156
λS	165	264	5,148	6,864	7,800	110	858	5,148	3,432	6,240
S	$\frac{165}{2}$	528	$\frac{15,444}{5}$	$\frac{82,368}{5}$	78,000	110	858	$\frac{30,888}{5}$	41,184	249,600
λ	2	$\frac{1}{2}$	$\frac{5}{3}$	$\frac{5}{12}$	$\frac{1}{10}$	1	1	$\frac{5}{6}$	$\frac{1}{12}$	$\frac{1}{40}$

APPENDIX TABLE 3: OMEGA CONVERSION TABLE

p (%)	dB	p (%)	dB	p (%)	dB	p (%)	dB	p (%)	dB	p (%)	dB
0.0	∞*	5.0	-12.787	10.0	-9.541	15.0	-7.532	20.0	-6.020	25.0	-4.770
0.1	-29.995	5.1	-12.696	10.1	-9.493	15.1	-7.498	20.1	-5.993	25.1	-4.747
0.2	-26.980	5.2	-12.607	10.2	-9.446	15.2	-7.465	20.2	-5.966	25.2	-4.724
0.3	-25.215	5.3	-12.520	10.3	-9.399	15.3	-7.431	20.3	-5.939	25.3	-4.701
0.4	-23.961	5.4	-12.434	10.4	-9.352	15.4	-7.397	20.4	-5.912	25.4	-4.678
0.5	-22.988	5.5	-12.350	10.5	-9.305	15.5	-7.364	20.5	-5.885	25.5	-4.655
0.6	-22.191	5.6	-12.267	10.6	-9.259	15.6	-7.331	20.6	-5.859	25.6	-4.632
0.7	-21.518	5.7	-12.185	10.7	-9.214	15.7	-7.298	20.7	-5.832	25.7	-4.610
0.8	-20.933	5.8	-12.105	10.8	-9.168	15.8	-7.266	20.8	-5.806	25.8	-4.587
0.9	-20.417	5.9	-12.026	10.9	-9.124	15.9	-7.233	20.9	-5.779	25.9	-4.564
1.0	-19.955	6.0	-11.949	11.0	-9.079	16.0	-7.201	21.0	-5.753	26.0	-4.542
1.1	-19.537	6.1	-11.872	11.1	-9.035	16.1	-7.168	21.1	-5.727	26.1	-4.519
1.2	-19.155	6.2	-11.797	11.2	-8.991	16.2	-7.136	21.2	-5.701	26.2	-4.497
1.3	-18.803	6.3	-11.723	11.3	-8.947	16.3	-7.104	21.3	-5.675	26.3	-4.474
1.4	-18.476	6.4	-11.650	11.4	-8.904	16.4	-7.073	21.4	-5.649	26.4	-4.452
1.5	-18.172	6.5	-11.578	11.5	-8.861	16.5	-7.041	21.5	-5.623	26.5	-4.429
1.6	-17.888	6.6	-11.507	11.6	-8.819	16.6	-7.010	21.6	-5.598	26.6	-4.407
1.7	-17.620	6.7	-11.437	11.7	-8.777	16.7	-6.978	21.7	-5.572	26.7	-4.385
1.8	-17.367	6.8	-11.368	11.8	-8.735	16.8	-6.947	21.8	-5.547	26.8	-4.363
1.9	-17.128	6.9	-11.300	11.9	-8.693	16.9	-6.916	21.9	-5.521	26.9	-4.341
2.0	-16.901	7.0	-11.233	12.0	-8.652	17.0	-6.885	22.0	-5.496	27.0	-4.319
2.1	-16.685	7.1	-11.167	12.1	-8.611	17.1	-6.855	22.1	-5.470	27.1	-4.297
2.2	-16.478	7.2	-11.101	12.2	-8.570	17.2	-6.824	22.2	-5.445	27.2	-4.275
2.3	-16.281	7.3	-11.037	12.3	-8.530	17.3	-6.794	22.3	-5.420	27.3	-4.253
2.4	-16.091	7.4	-10.973	12.4	-8.490	17.4	-6.763	22.4	-5.395	27.4	-4.231
2.5	-15.910	7.5	-10.910	12.5	-8.450	17.5	-6.733	22.5	-5.370	27.5	-4.209
2.6	-15.735	7.6	-10.848	12.6	-8.410	17.6	-6.703	22.6	-5.345	27.6	-4.187
2.7	-15.566	7.7	-10.786	12.7	-8.371	17.7	-6.673	22.7	-5.321	27.7	-4.166
2.8	-15.404	7.8	-10.725	12.8	-8.332	17.8	-6.644	22.8	-5.296	27.8	-4.144
2.9	-15.247	7.9	-10.665	12.9	-8.293	17.9	-6.614	22.9	-5.271	27.9	-4.122
3.0	-15.096	8.0	-10.606	13.0	-8.255	18.0	-6.584	23.0	-5.247	28.0	-4.101
3.1	-14.949	8.1	-10.547	13.1	-8.216	18.1	-6.555	23.1	-5.222	28.1	-4.079
3.2	-14.806	8.2	-10.489	13.2	-8.178	18.2	-6.526	23.2	-5.198	28.2	-4.058
3.3	-14.668	8.3	-10.432	13.3	-8.141	18.3	-6.497	23.3	-5.173	28.3	-4.036
3.4	-14.534	8.4	-10.375	13.4	-8.103	18.4	-6.468	23.4	-5.149	28.4	-4.015
3.5	-14.404	8.5	-10.319	13.5	-8.066	18.5	-6.439	23.5	-5.125	28.5	-3.994
3.6	-14.277	8.6	-10.263	13.6	-8.029	18.6	-6.410	23.6	-5.101	28.6	-3.972
3.7	-14.153	8.7	-10.209	13.7	-7.992	18.7	-6.381	23.7	-5.077	28.7	-3.951
3.8	-14.033	8.8	-10.154	13.8	-7.955	18.8	-6.353	23.8	-5.053	28.8	-3.930
3.9	-13.916	8.9	-10.100	13.9	-7.919	18.9	-6.325	23.9	-5.029	28.9	-3.909
4.0	-13.801	9.0	-10.047	14.0	-7.883	19.0	-6.296	24.0	-5.005	29.0	-3.888
4.1	-13.689	9.1	- 9.994	14.1	-7.847	19.1	-6.268	24.1	-4.981	29.1	-3.867
4.2	-13.580	9.2	- 9.942	14.2	-7.811	19.2	-6.240	24.2	-4.958	29.2	-3.846
4.3	-13.473	9.3	- 9.890	14.3	-7.775	19.3	-6.212	24.3	-4.934	29.3	-3.825
4.4	-13.369	9.4	- 9.839	14.4	-7.740	19.4	-6.184	24.4	-4.910	29.4	-3.804
4.5	-13.267	9.5	-9.788	14.5	-7.705	19.5	-6.157	24.5	-4.887	29.5	-3.783
4.6	-13.167	9.6	-9.738	14.6	-7.670	19.6	-6.129	24.6	-4.863	29.6	-3.762
4.7	-13.069	9.7	-9.688	14.7	-7.635	19.7	-6.101	24.7	-4.840	29.7	-3.741
4.8	-12.973	9.8	-9.639	14.8	-7.601	19.8	-6.074	24.8	-4.817	29.8	-3.720
4.9	-12.879	9.9	-9.590	14.9	-7.566	19.9	-6.047	24.9	-4.793	29.9	-3.699

Seiso Konishi, et al (1978): S/N Ratio Manual, Japanese Standards Association

Appendix Table 3 (cont.)

p (%)	dB	p (%)	dB	p (%)	dB	p (%)	dB	p (%)	dB	p (%)	dB
30.0	-3.679	35.0	-2.687	40.0	-1.760	45.0	-0.871	50.0	0.000	55.0	0.872
30.1	-3.658	35.1	-2.668	40.1	-1.742	45.1	-0.853	50.1	0.017	55.1	0.889
30.2	-3.637	35.2	-2.649	40.2	-1.724	45.2	-0.835	50.2	0.035	55.2	0.907
30.3	-3.617	35.3	-2.630	40.3	-1.706	45.3	-0.818	50.3	0.052	55.3	0.924
30.4	-3.596	35.4	-2.611	40.4	-1.688	45.4	-0.800	50.4	0.069	55.4	0.942
30.5	-3.576	35.5	-2.592	40.5	-1.670	45.5	-0.783	50.5	0.087	55.5	0.959
30.6	-3.555	35.6	-2.573	40.6	-1.652	45.6	-0.765	50.6	0.104	55.6	0.977
30.7	-3.535	35.7	-2.554	40.7	-1.634	45.7	-0.748	50.7	0.122	55.7	0.995
30.8	-3.515	35.8	-2.536	40.8	-1.616	45.8	-0.730	50.8	0.139	55.8	1.012
30.9	-3.494	35.9	-2.517	40.9	-1.598	45.9	-0.713	50.9	0.156	55.9	1.030
31.0	-3.474	36.0	-2.498	41.0	-1.580	46.0	-0.695	51.0	0.174	56.0	1.047
31.1	-3.454	36.1	-2.479	41.1	-1.562	46.1	-0.678	51.1	0.191	56.1	1.065
31.2	-3.433	36.2	-2.460	41.2	-1.544	46.2	-0.660	51.2	0.209	56.2	1.083
31.3	-3.413	36.3	-2.441	41.3	-1.526	46.3	-0.643	51.3	0.226	56.3	1.100
31.4	-3.393	36.4	-2.423	41.4	-1.508	46.4	-0.625	51.4	0.243	56.4	1.118
31.5	-3.373	36.5	-2.404	41.5	-1.490	46.5	-0.608	51.5	0.261	56.5	1.136
31.6	-3.353	36.6	-2.385	41.6	-1.472	46.6	-0.591	51.6	0.278	56.6	1.153
31.7	-3.333	36.7	-2.366	41.7	-1.454	46.7	-0.573	51.7	0.295	56.7	1.171
31.8	-3.313	36.8	-2.348	41.8	-1.436	46.8	-0.556	51.8	0.313	56.8	1.189
31.9	-3.293	36.9	-2.329	41.9	-1.419	46.9	-0.538	51.9	0.330	56.9	1.206
32.0	-3.273	37.0	-2.310	42.0	-1.401	47.0	-0.521	52.0	0.348	57.0	1.224
32.1	-3.253	37.1	-2.292	42.1	-1.383	47.1	-0.503	52.1	0.365	57.1	1.242
32.2	-3.233	37.2	-2.273	42.2	-1.365	47.2	-0.486	52.2	0.382	57.2	1.260
32.3	-3.213	37.3	-2.255	42.3	-1.347	47.3	-0.468	52.3	0.400	57.3	1.277
32.4	-3.193	37.4	-2.236	42.4	-1.330	47.4	-0.451	52.4	0.417	57.4	1.295
32.5	-3.173	37.5	-2.217	42.5	-1.312	47.5	-0.434	52.5	0.435	57.5	1.313
32.6	-3.153	37.6	-2.199	42.6	-1.294	47.6	-0.416	52.6	0.452	57.6	1.331
32.7	-3.134	37.7	-2.180	42.7	-1.276	47.7	-0.399	52.7	0.469	57.7	1.348
32.8	-3.114	37.8	-2.162	42.8	-1.259	47.8	-0.381	52.8	0.487	57.8	1.366
32.9	-3.094	37.9	-2.144	42.9	-1.241	47.9	-0.364	52.9	0.504	57.9	1.384
33.0	-3.075	38.0	-2.125	43.0	-1.223	48.0	-0.347	53.0	0.522	58.0	1.402
33.1	-3.055	38.1	-2.107	43.1	-1.205	48.1	-0.329	53.1	0.539	58.1	1.420
33.2	-3.035	38.2	-2.088	43.2	-1.188	48.2	-0.312	53.2	0.557	58.2	1.437
33.3	-3.016	38.3	-2.070	43.3	-1.170	48.3	-0.294	53.3	0.574	58.3	1.455
33.4	-2.996	38.4	-2.051	43.4	-1.152	48.4	-0.277	53.4	0.592	58.4	1.473
33.5	-2.977	38.5	-2.033	43.5	-1.135	48.5	-0.260	53.5	0.609	58.5	1.491
33.6	-2.957	38.6	-2.015	43.6	-1.117	48.6	-0.242	53.6	0.626	58.6	1.509
33.7	-2.938	38.7	-1.996	43.7	-1.099	48.7	-0.225	53.7	0.644	58.7	1.527
33.8	-2.918	38.8	-1.978	43.8	-1.082	48.8	-0.208	53.8	0.661	58.8	1.545
33.9	-2.899	38.9	-1.960	43.9	-1.064	48.9	-0.190	53.9	0.679	58.9	1.563
34.0	-2.880	39.0	-1.942	44.0	-1.046	49.0	-0.173	54.0	0.696	59.0	1.581
34.1	-2.860	39.1	-1.923	44.1	-1.029	49.1	-0.155	54.1	0.714	59.1	1.599
34.2	-2.841	39.2	-1.905	44.2	-1.011	49.2	-0.138	54.2	0.731	59.2	1.617
34.3	-2.822	39.3	-1.887	44.3	-0.994	49.3	-0.121	54.3	0.749	59.3	1.635
34.4	-2.802	39.4	-1.869	44.4	-0.976	49.4	-0.103	54.4	0.766	59.4	1.653
34.5	-2.783	39.5	-1.851	44.5	-0.958	49.5	-0.086	54.5	0.784	59.5	1.671
34.6	-2.764	39.6	-1.832	44.6	-0.941	49.6	-0.068	54.6	0.801	59.6	1.689
34.7	-2.745	39.7	-1.814	44.7	-0.923	49.7	-0.051	54.7	0.819	59.7	1.707
34.8	-2.726	39.8	-1.796	44.8	-0.906	49.8	-0.034	54.8	0.836	59.8	1.725
34.9	-2.707	39.9	-1.778	44.9	-0.888	49.9	-0.016	54.9	0.854	59.9	1.743

Appendix Table 3 (cont.)

p (%)	dB	p (%)	dB	p (%)	dB	p (%)	dB	p (%)	dB	p (%)	dB
60.0	1.761	65.0	2.688	70.0	3.680	75.0	4.771	80.0	6.021	85.0	7.533
60.1	1.779	65.1	2.708	70.1	3.700	75.1	4.794	80.1	6.048	85.1	7.567
60.2	1.797	65.2	2.727	70.2	3.721	75.2	4.818	80.2	6.075	85.2	7.602
60.3	1.815	65.3	2.746	70.3	3.742	75.3	4.841	80.3	6.102	85.3	7.636
60.4	1.833	65.4	2.765	70.4	3.763	75.4	4.864	80.4	6.130	85.4	7.671
60.5	1.852	65.5	2.784	70.5	3.784	75.5	4.888	80.5	6.158	85.5	7.706
60.6	1.870	65.6	2.803	70.6	3.805	75.6	4.911	80.6	6.185	85.6	7.741
60.7	1.888	65.7	2.823	70.7	3.826	75.7	4.935	80.7	6.213	85.7	7.776
60.8	1.906	65.8	2.842	70.8	3.847	75.8	4.959	80.8	6.241	85.8	7.812
60.9	1.924	65.9	2.861	70.9	3.868	75.9	4.982	80.9	6.269	85.9	7.848
61.0	1.943	66.0	2.881	71.0	3.889	76.0	5.006	81.0	6.297	86.0	7.884
61.1	1.961	66.1	2.900	71.1	3.910	76.1	5.030	81.1	6.326	86.1	7.920
61.2	1.979	66.2	2.919	71.2	3.931	76.2	5.054	81.2	6.354	86.2	7.956
61.3	1.997	66.3	2.939	71.3	3.952	76.3	5.078	81.3	6.382	86.3	7.993
61.4	2.016	66.4	2.958	71.4	3.973	76.4	5.102	81.4	6.411	86.4	8.030
61.5	2.034	66.5	2.978	71.5	3.995	76.5	5.126	81.5	6.440	86.5	8.067
61.6	2.052	66.6	2.997	71.6	4.016	76.6	5.150	81.6	6.469	86.6	8.104
61.7	2.071	66.7	3.017	71.7	4.037	76.7	5.174	81.7	6.498	86.7	8.142
61.8	3.089	66.8	3.036	71.8	4.059	76.8	5.199	81.8	6.527	86.8	8.179
61.9	2.108	66.9	3.056	71.9	4.080	76.9	5.223	81.9	6.556	86.9	8.217
62.0	2.126	67.0	3.076	72.0	4.102	77.0	5.248	82.0	6.585	87.0	8.256
62.1	2.145	67.1	3.095	72.1	4.123	77.1	5.272	82.1	6.615	87.1	8.294
62.2	2.163	67.2	3.115	72.2	4.145	77.2	5.297	82.2	6.645	87.2	8.333
62.3	2.181	67.3	3.135	72.3	4.167	77.3	5.322	82.3	6.674	87.3	8.372
62.4	2.200	67.4	3.154	72.4	4.188	77.4	5.346	82.4	6.704	87.4	8.411
62.5	2.218	67.5	3.174	72.5	4.210	77.5	5.371	82.5	6.734	87.5	8.451
62.6	2.237	67.6	3.194	72.6	4.232	77.6	5.396	82.6	6.764	87.6	8.491
62.7	2.256	67.7	3.214	72.7	4.254	77.7	5.421	82.7	6.795	87.7	8.531
62.8	2.274	67.8	3.234	72.8	4.276	77.8	5.446	82.8	6.825	87.8	8.571
62.9	2.293	67.9	3.254	72.9	4.298	77.9	5.471	82.9	6.856	87.9	8.612
63.0	2.311	68.0	3.274	73.0	4.320	78.0	5.497	83.0	6.886	88.0	8.653
63.1	2.330	68.1	3.294	73.1	4.342	78.1	5.522	83.1	6.917	88.1	8.694
63.2	2.349	68.2	3.314	73.2	4.364	78.2	5.548	83.2	6.948	88.2	8.736
63.3	2.367	68.3	3.334	73.3	4.386	78.3	5.573	83.3	6.979	88.3	8.778
63.4	2.386	68.4	3.354	73.4	4.408	78.4	5.599	83.4	7.011	88.4	8.820
63.5	2.405	68.5	3.374	73.5	4.430	78.5	5.624	83.5	7.042	88.5	8.862
63.6	2.424	68.6	3.394	73.6	4.453	78.6	5.650	83.6	7.074	88.6	8.905
63.7	2.442	68.7	3.414	73.7	4.475	78.7	5.676	83.7	7.105	88.7	8.948
63.8	2.461	68.8	3.434	73.8	4.498	78.8	5.702	83.8	7.137	88.8	8.992
63.9	2.480	68.9	3.455	73.9	4.520	78.9	5.728	83.9	7.169	88.9	9.036
64.0	2.499	69.0	3.475	74.0	4.543	79.0	5.754	84.0	7.202	89.0	9.080
64.1	2.518	69.1	3.495	74.1	4.565	79.1	5.780	84.1	7.234	89.1	9.125
64.2	2.537	69.2	3.516	74.2	4.588	79.2	5.807	84.2	7.267	89.2	9.169
64.3	2.555	69.3	3.536	74.3	4.611	79.3	5.833	84.3	7.299	89.3	9.215
64.4	2.574	69.4	3.556	74.4	4.633	79.4	5.860	84.4	7.332	89.4	9.260
64.5	2.593	69.5	3.577	74.5	4.656	79.5	5.886	84.5	7.365	89.5	9.306
64.6	2.612	69.6	3.597	74.6	4.679	79.6	5.913	84.6	7.398	89.6	9.353
64.7	2.631	69.7	3.618	74.7	4.702	79.7	5.940	84.7	7.432	89.7	9.400
64.8	2.650	69.8	3.638	74.8	4.725	79.8	5.967	84.8	7.466	89.8	9.447
64.9	2.669	69.9	3.659	74.9	4.748	79.9	5.994	84.9	7.499	89.9	9.494

Appendix Table 3 (cont.)

p (%)	dB	p (%)	dB	p (%)	dB	p (%)	dB	p (%)	dB	p (%)	dB
90.0	9.542	92.0	10.607	94.0	11.950	96.0	13.802	98.0	16.902	100.0	∞*
90.1	9.591	92.1	10.666	94.1	12.027	96.1	13.917	98.1	17.129		
90.2	9.640	92.2	10.726	94.2	12.106	96.2	14.034	98.2	17.368		
90.3	9.689	92.3	10.787	94.3	12.186	96.3	14.154	98.3	17.621		
90.4	9.739	92.4	10.849	94.4	12.268	96.4	14.278	98.4	17.889		
90.5	9.789	92.5	10.911	94.5	12.351	96.5	14.405	98.5	18.173		
90.6	9.840	92.6	10.974	94.6	12.435	96.6	14.535	98.6	19.477		
90.7	9.891	92.7	11.038	94.7	12.521	96.7	14.669	98.7	18.804		
90.8	9.943	92.8	11.102	94.8	12.608	96.8	14.807	98.8	19.156		
90.9	9.995	92.9	11.168	94.9	12.697	96.9	14.950	98.9	19.538		
91.0	10.048	93.0	11.234	95.0	12.788	97.0	15.097	99.0	19.956		
91.1	10.111	93.1	11.301	95.1	12.880	97.1	15.248	99.1	20.418		
91.2	10.155	93.2	11.369	95.2	12.974	97.2	15.405	99.2	20.934		
91.3	10.210	93.3	11.438	95.3	13.070	97.3	15.567	99.3	21.519		
91.4	10.264	93.4	11.508	95.4	13.168	97.4	15.736	99.4	22.192		
91.5	10.320	93.5	11.579	95.5	13.268	97.5	15.911	99.5	22.989		
91.6	10.376	93.6	11.651	95.6	13.370	97.6	16.092	99.6	23.962		
91.7	10.433	93.7	11.724	95.7	13.474	97.7	16.282	99.7	25.216		
91.8	10.490	93.8	11.798	95.8	13.581	97.8	16.479	99.8	26.981		
91.9	10.548	93.9	11.873	95.9	13.690	97.9	16.686	99.9	29.996		

APPENDIX TABLE 4: ORTHOGONAL ARRAYS AND LINEAR GRAPHS

[Note]

1. "No" shows experimental trial number. "Column" shows the column number in the orthogonal array.
2. Interaction tables are used to obain the interaction between two columns.
3. The following symbols are used to show the group number associated with that particular column in the orthogonal array.

Group	Symbol	Only For L_{32} (2^{31})
Group 1	○	Group 1 and 2
Group 2	◎	Group 3
Group 3	⊙	Group 4
Group 4	●	Group 5

L_8 (2^7)

No. \ Column	1	2	3	4	5	6	7
1	1	1	1	1	1	1	1
2	1	1	1	2	2	2	2
3	1	2	2	1	1	2	2
4	1	2	2	2	2	1	1
5	2	1	2	1	2	1	2
6	2	1	2	2	1	2	1
7	2	2	1	1	2	2	1
8	2	2	1	2	1	1	2
Components	a	b	a b	c	a c	b c	a b c
	Group 1	Group 2		Group 3			

L_4 (2^3)

No. \ Column	1	2	3
1	1	1	1
2	1	2	2
3	2	1	2
4	2	2	1
Components	a	b	a b
	Group 1	Group 2	

(1)

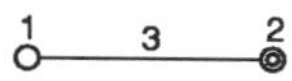

Interactions Between Two Columns

Column \ Column	1	2	3	4	5	6	7
	(1)	3	2	5	4	7	6
		(2)	1	6	7	4	5
			(3)	7	6	5	4
				(4)	1	2	3
					(5)	3	2
						(6)	1
							(7)

(1)

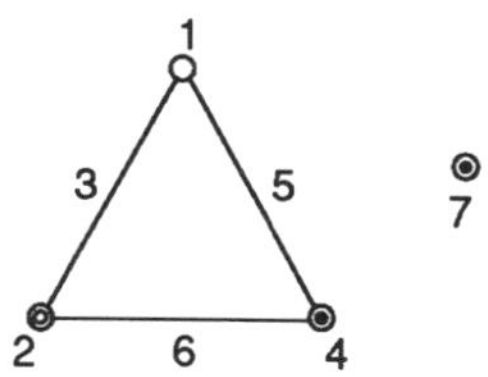

(2)

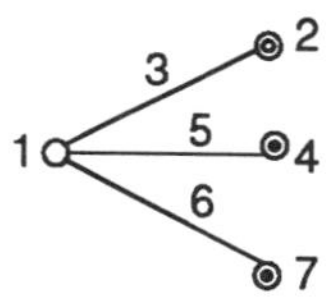

Genichi Taguchi (1961): *Orthogonal Arrays and Linear Graphs*, Maruzen. Jiro Yamaguchi, et. al. (1962): *Quality Control Handbook*, Japanese Standards Association.

Appendix Table 4 (cont.)

L_{16} (2^{15})

No. \ Column	1	2	3	4	5	6	7	8	9	10	11	12	13	14	15
1	1	1	1	1	1	1	1	1	1	1	1	1	1	1	1
2	1	1	1	1	1	1	1	2	2	2	2	2	2	2	2
3	1	1	1	2	2	2	2	1	1	1	1	2	2	2	2
4	1	1	1	2	2	2	2	2	2	2	2	1	1	1	1
5	1	2	2	1	1	2	2	1	1	2	2	1	1	2	2
6	1	2	2	1	1	2	2	2	2	1	1	2	2	1	1
7	1	2	2	2	2	1	1	1	1	2	2	2	2	1	1
8	1	2	2	2	2	1	1	2	2	1	1	1	1	2	2
9	2	1	2	1	2	1	2	1	2	1	2	1	2	1	2
10	2	1	2	1	2	1	2	2	1	2	1	2	1	2	1
11	2	1	2	2	1	2	1	1	2	1	2	2	1	2	1
12	2	1	2	2	1	2	1	2	1	2	1	1	2	1	2
13	2	2	1	1	2	2	1	1	2	2	1	1	2	2	1
14	2	2	1	1	2	2	1	2	1	1	2	2	1	1	2
15	2	2	1	2	1	1	2	1	2	2	1	2	1	1	2
16	2	2	1	2	1	1	2	2	1	1	2	1	2	2	1
Components	a	b	a b	c	a c	b c	a b c	d	a d	b d	a b d	c d	a c d	b c d	a b c d
	Group 1	Group 2		Group 3				Group 4							

(1)

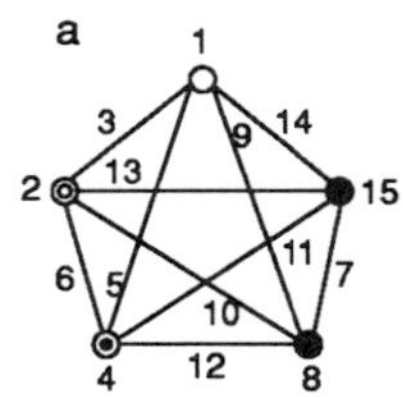

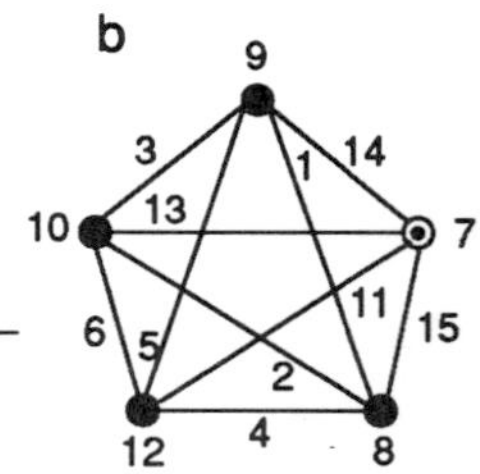

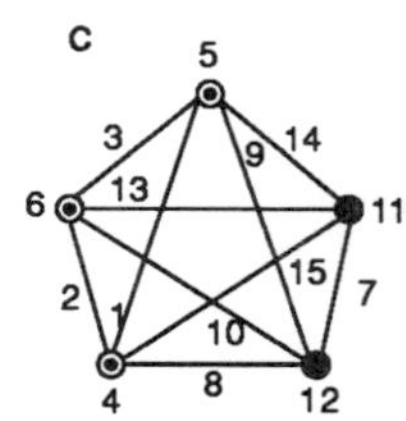

Interactions Between Two Columns

Column \ Column	1	2	3	4	5	6	7	8	9	10	11	12	13	14	15
	(1)	3	2	5	4	7	6	9	8	11	10	13	12	15	14
		(2)	1	6	7	4	5	10	11	8	9	14	15	12	13
			(3)	7	6	5	4	11	10	9	8	15	14	13	12
				(4)	1	2	3	12	13	14	15	8	9	10	11
					(5)	3	2	13	12	15	14	9	8	11	10
						(6)	1	14	15	12	13	10	11	8	9
							(7)	15	14	13	12	11	10	9	8
								(8)	1	2	3	4	5	6	7
									(9)	3	2	5	4	7	6
										(10)	1	6	7	4	5
											(11)	7	6	5	4
												(12)	1	2	3
													(13)	3	2
														(14)	1

Appendix Table 4 (cont.)

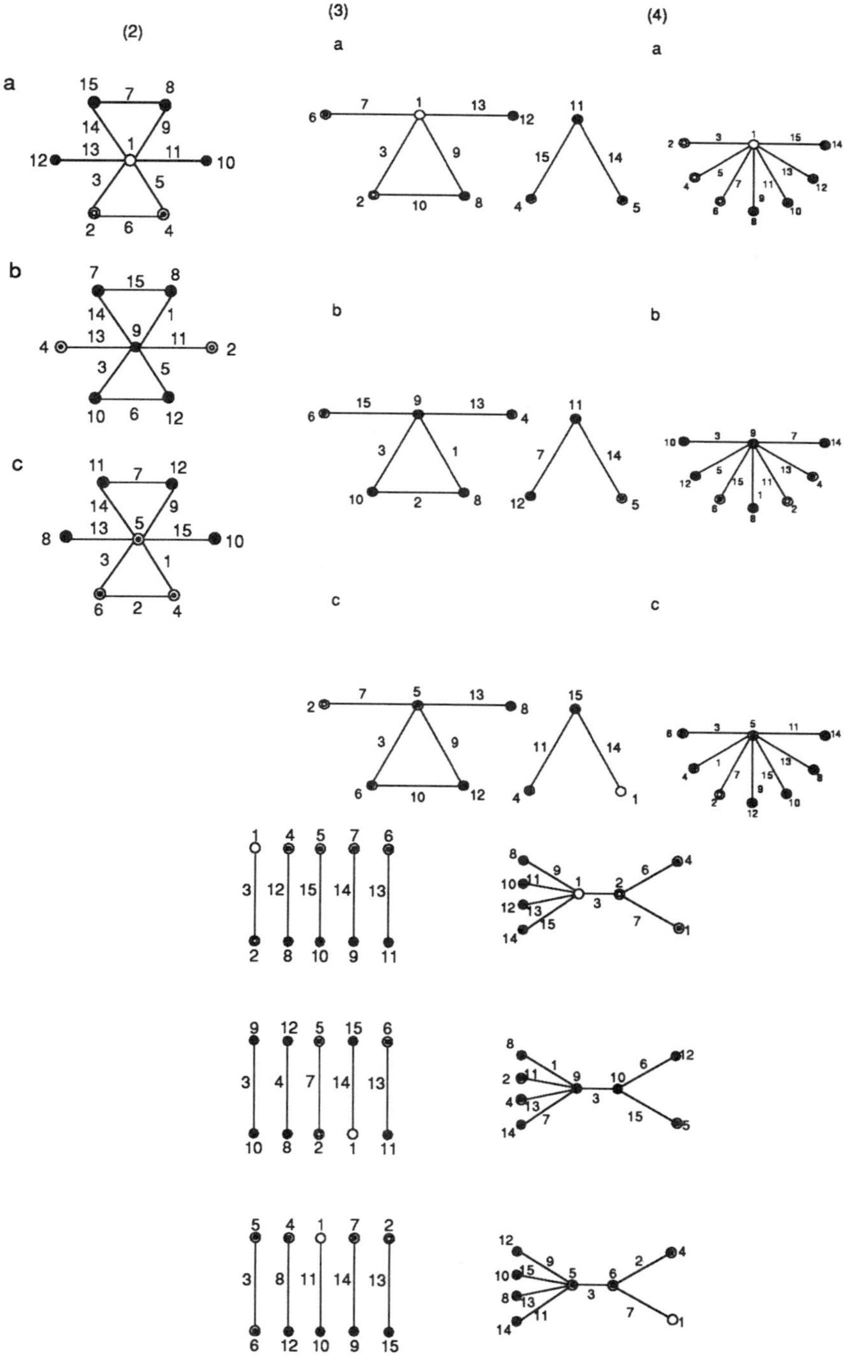

Appendix Table 4 (cont.)

$L_{32}(2^{31})$

No. \ Col. No.	1	2	3	4	5	6	7	8	9	10	11	12	13	14	15	16	17	18	19	20	21	22	23	24	25	26	27	28	29	30	31
1	1	1	1	1	1	1	1	1	1	1	1	1	1	1	1	1	1	1	1	1	1	1	1	1	1	1	1	1	1	1	1
2	1	1	1	1	1	1	1	1	1	1	1	1	1	1	1	2	2	2	2	2	2	2	2	2	2	2	2	2	2	2	2
3	1	1	1	1	1	1	1	2	2	2	2	2	2	2	2	1	1	1	1	1	1	1	1	2	2	2	2	2	2	2	2
4	1	1	1	1	1	1	1	2	2	2	2	2	2	2	2	2	2	2	2	2	2	2	2	1	1	1	1	1	1	1	1
5	1	1	1	2	2	2	2	1	1	1	1	2	2	2	2	1	1	1	1	2	2	2	2	1	1	1	1	2	2	2	2
6	1	1	1	2	2	2	2	1	1	1	1	2	2	2	2	2	2	2	2	1	1	1	1	2	2	2	2	1	1	1	1
7	1	1	1	2	2	2	2	2	2	2	2	1	1	1	1	1	1	1	1	2	2	2	2	2	2	2	2	1	1	1	1
8	1	1	1	2	2	2	2	2	2	2	2	1	1	1	1	2	2	2	2	1	1	1	1	1	1	1	1	2	2	2	2
9	1	2	2	1	1	2	2	1	1	2	2	1	1	2	2	1	1	2	2	1	1	2	2	1	1	2	2	1	1	2	2
10	1	2	2	1	1	2	2	1	1	2	2	1	1	2	2	2	2	1	1	2	2	1	1	2	2	1	1	2	2	1	1
11	1	2	2	1	1	2	2	2	2	1	1	2	2	1	1	1	1	2	2	1	1	2	2	2	2	1	1	2	2	1	1
12	1	2	2	1	1	2	2	2	2	1	1	2	2	1	1	2	2	1	1	2	2	1	1	1	1	2	2	1	1	2	2
13	1	2	2	2	2	1	1	1	1	2	2	2	2	1	1	1	1	2	2	2	2	1	1	1	1	2	2	2	2	1	1
14	1	2	2	2	2	1	1	1	1	2	2	2	2	1	1	2	2	1	1	1	1	2	2	2	2	1	1	1	1	2	2
15	1	2	2	2	2	1	1	2	2	1	1	1	1	2	2	1	1	2	2	2	2	1	1	2	2	1	1	1	1	2	2
16	1	2	2	2	2	1	1	2	2	1	1	1	1	2	2	2	2	1	1	1	1	2	2	1	1	2	2	2	2	1	1
17	2	1	2	1	2	1	2	1	2	1	2	1	2	1	1	2	1	2	1	2	1	2	1	2	1	2	1	2	1	2	1
18	2	1	2	1	2	1	2	1	2	1	2	1	2	1	2	1	2	1	2	1	2	1	2	1	2	1	2	1	2	1	2
19	2	1	2	1	2	1	2	2	1	2	1	2	1	2	1	1	2	1	2	1	2	1	2	2	1	2	1	2	1	2	1
20	2	1	2	1	2	1	2	2	1	2	1	2	1	2	1	2	1	2	1	2	1	2	1	1	2	1	2	1	2	1	2
21	2	1	2	2	1	2	1	1	2	1	2	2	1	2	1	1	2	1	2	2	1	2	1	1	2	1	2	2	1	2	1
22	2	1	2	2	1	2	1	1	2	1	2	2	1	2	1	2	1	2	1	1	2	1	2	2	1	2	1	1	2	1	2
23	2	1	2	2	1	2	1	2	1	2	1	1	2	1	2	1	2	1	2	2	1	2	1	2	1	2	1	1	2	1	2
24	2	1	2	2	1	2	1	2	1	2	1	1	2	1	2	2	1	2	1	1	2	1	2	1	2	1	2	2	1	2	1
25	2	2	1	1	2	2	1	1	2	2	1	1	2	2	1	1	2	2	1	1	2	2	1	1	2	2	1	1	2	2	1
26	2	2	1	1	2	2	1	1	2	2	1	1	2	2	1	2	1	1	2	2	1	1	2	2	1	1	2	2	1	1	2
27	2	2	1	1	2	2	1	2	1	1	2	2	1	1	2	1	2	2	1	1	2	2	1	2	1	1	2	2	1	1	2
28	2	2	1	1	2	2	1	2	1	1	2	2	1	1	2	2	1	1	2	2	1	1	2	1	2	2	1	1	2	2	1
29	2	2	1	2	1	1	2	1	2	2	1	2	1	1	2	1	2	2	1	2	1	1	2	1	2	2	1	2	1	1	2
30	2	2	1	2	1	1	2	1	2	2	1	2	1	1	2	2	1	1	2	1	2	2	1	2	1	1	2	1	2	2	1
31	2	2	1	2	1	1	2	2	1	1	2	1	2	2	1	1	2	2	1	2	1	1	2	2	1	1	2	1	2	2	1
32	2	2	1	2	1	1	2	2	1	1	2	1	2	2	1	2	1	1	2	1	2	2	1	1	2	2	1	2	1	1	2
Components	a	b	a b	c	a c	b c	a b c	d	a d	b d	a b d	c d	a c d	b c d	a b c d	e	a e	b e	a b e	c e	a c e	b c e	a b c e	d e	a d e	b d e	a b d e	c d e	a c d e	b c d e	a b c d e
Group	1	2		3				4								5															

Appendix Table 4 (cont.)

Col.	1	2	3	4	5	6	7	8	9	10	11	12	13	14	15	16	17	18	19	20	21	22	23	24	25	26	27	28	29	30	31
(1)		3	2	5	4	7	6	9	8	11	10	13	12	15	14	17	16	19	18	21	20	23	22	25	24	27	26	29	28	31	30
(2)			1	6	7	4	5	10	11	8	9	14	15	12	13	18	19	16	17	22	23	20	21	26	27	24	25	30	31	28	29
(3)				7	6	5	4	11	10	9	8	15	14	13	12	19	18	17	16	23	22	21	20	27	26	25	24	31	30	29	28
(4)					1	2	3	12	13	14	15	8	9	10	11	20	21	22	23	16	17	18	19	28	29	30	31	24	25	26	27
(5)						3	2	13	12	15	14	9	8	11	10	21	20	23	22	17	16	19	18	29	28	31	30	25	24	27	26
(6)							1	14	15	12	13	10	11	8	9	22	23	20	21	18	19	16	17	30	31	28	29	26	27	24	25
(7)								15	14	13	12	11	10	9	8	23	22	21	20	19	18	17	16	31	30	29	28	27	26	25	24
(8)									1	2	3	4	5	6	7	24	25	26	27	28	29	30	31	16	17	18	19	20	21	22	23
(9)										3	2	5	4	7	6	25	24	27	26	29	28	31	30	17	16	19	18	21	20	23	22
(10)											1	6	7	4	5	26	27	24	25	30	31	28	29	18	19	16	17	22	23	20	21
(11)												7	6	5	4	27	26	25	24	31	30	29	28	19	18	17	16	23	22	21	20
(12)													1	2	3	28	29	30	31	24	25	26	27	20	21	22	23	16	17	18	19
(13)														3	2	29	28	31	30	25	24	27	26	21	20	23	22	17	16	19	18
(14)															1	30	31	28	29	26	27	24	25	22	23	20	21	18	19	16	17
(15)																31	30	29	28	27	26	25	24	23	22	21	20	19	18	17	16
(16)																	1	2	3	4	5	6	7	8	9	10	11	12	13	14	15
(17)																		3	2	5	4	7	6	9	8	11	10	13	12	15	14
(18)																			1	6	7	4	5	10	11	8	9	14	15	12	13
(19)																				7	6	5	4	11	10	9	8	15	14	13	12
(20)																					1	2	3	12	13	14	15	8	9	10	11
(21)																						3	2	13	12	15	14	9	8	11	10
(22)																							1	14	15	12	13	10	11	8	9
(23)																								15	14	13	12	11	10	9	8
(24)																									1	2	3	4	5	6	7
(25)																										3	2	5	4	7	6
(26)																											1	6	7	4	5
(27)																												7	6	5	4
(28)																													1	2	3
(29)																														3	2
(30)																															1

Appendix Table 4 (cont.)

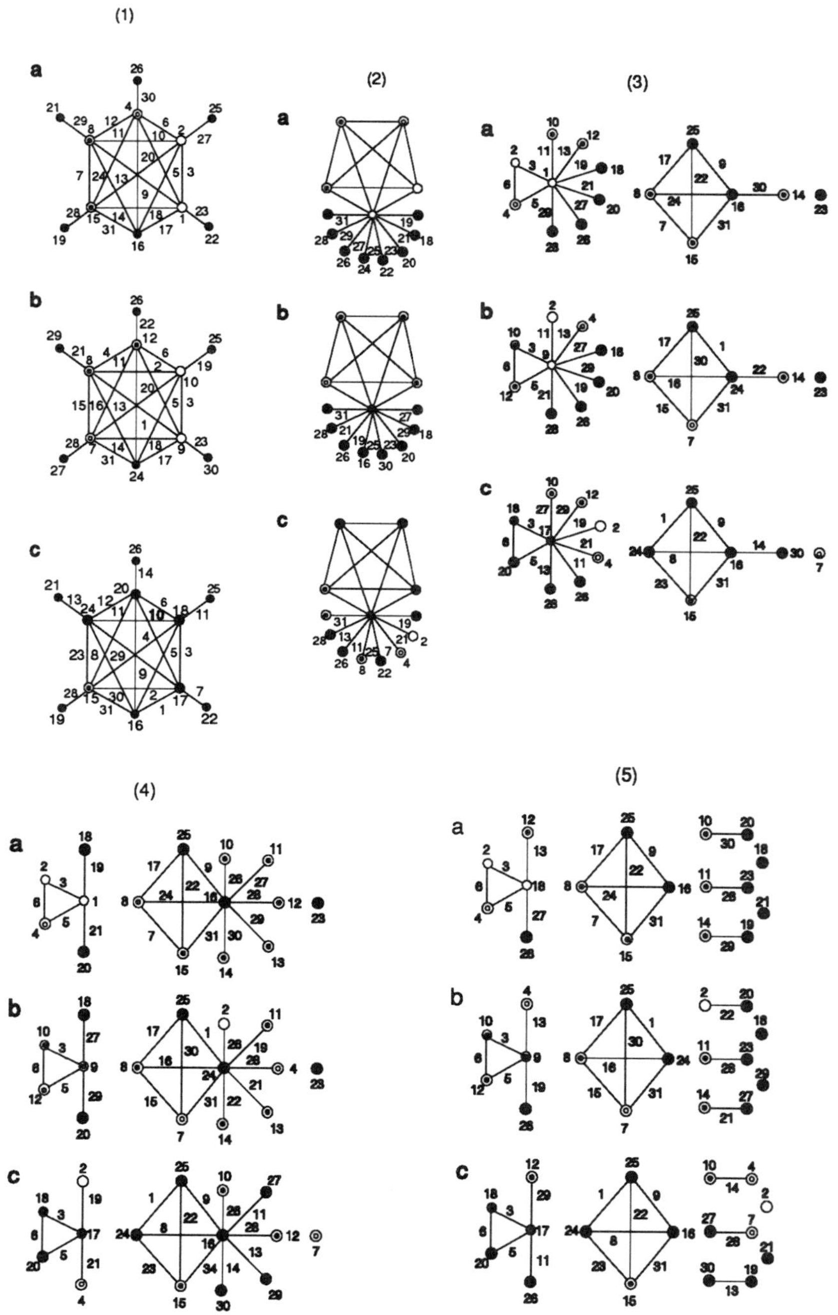

Appendix Table 4 (cont.)

(6)

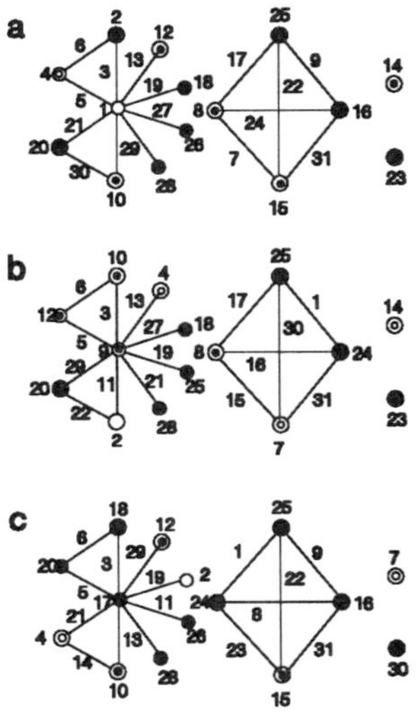

(7)

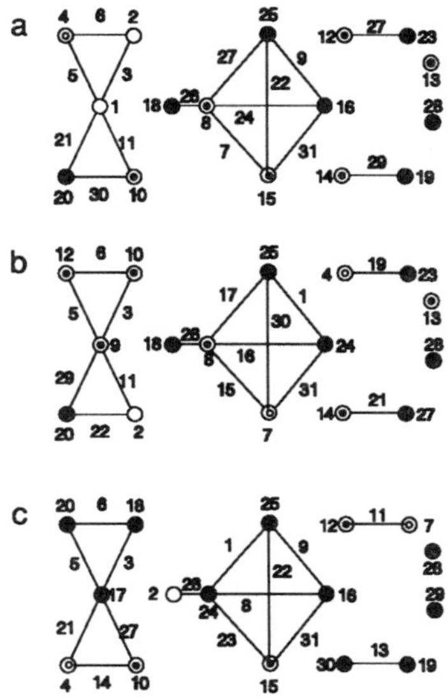

(8)

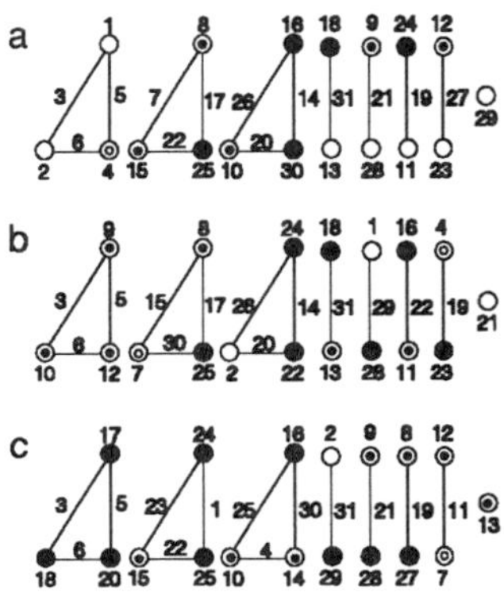

(9)

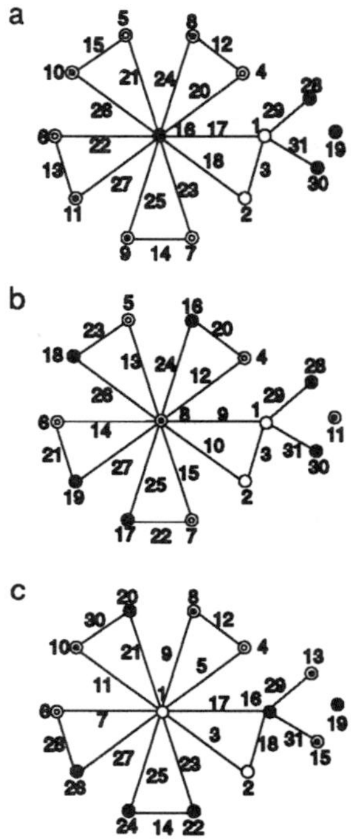

Appendix Table 4 (cont.)

(11)

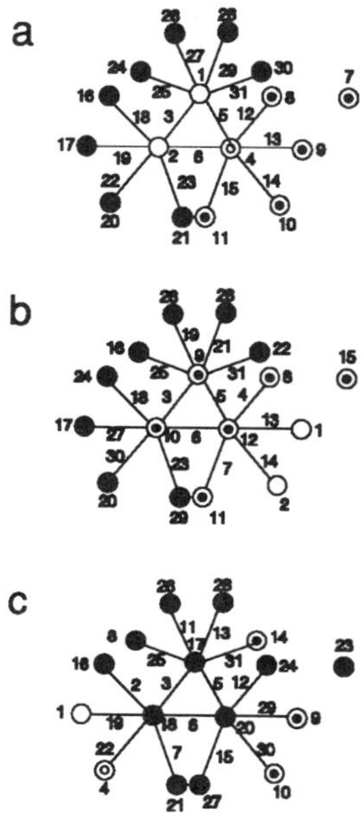

(10)

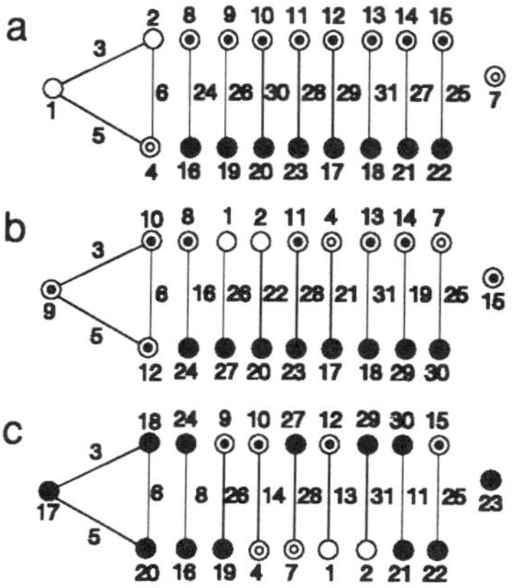

(12)

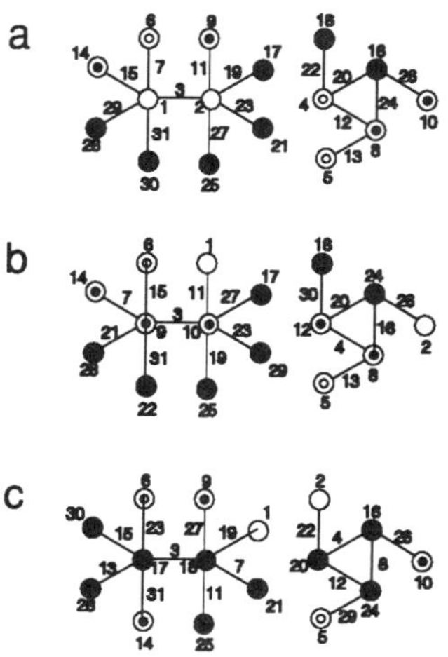

(13)

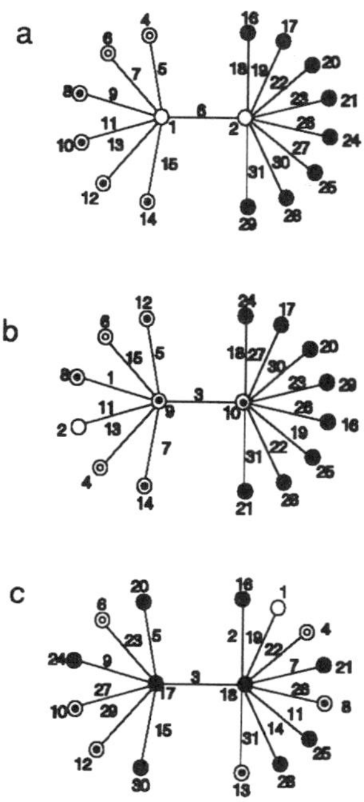

Appendix Table 4 (cont.)

L_9 (3^4)

No. \ Column	1	2	3	4
1	1	1	1	1
2	1	2	2	2
3	1	3	3	3
4	2	1	2	3
5	2	2	3	1
6	2	3	1	2
7	3	1	3	2
8	3	2	1	3
9	3	3	2	1
Components	a	b	a b	a b^2
	Group 1	Group 2		

(1)

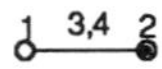

Appendix Table 4 (cont.)

L_{27} (3^{13})

No. \ Column	1	2	3	4	5	6	7	8	9	10	11	12	13
1	1	1	1	1	1	1	1	1	1	1	1	1	1
2	1	1	1	1	2	2	2	2	2	2	2	2	2
3	1	1	1	1	3	3	3	3	3	3	3	3	3
4	1	2	2	2	1	1	1	2	2	2	3	3	3
5	1	2	2	2	2	2	2	3	3	3	1	1	1
6	1	2	2	2	3	3	3	1	1	1	2	2	2
7	1	3	3	3	1	1	1	3	3	3	2	2	2
8	1	3	3	3	2	2	2	1	1	1	3	3	3
9	1	3	3	3	3	3	3	2	2	2	1	1	1
10	2	1	2	3	1	2	3	1	2	3	1	2	3
11	2	1	2	3	2	3	1	2	3	1	2	3	1
12	2	1	2	3	3	1	2	3	1	2	3	1	2
13	2	2	3	1	1	2	3	2	3	1	3	1	2
14	2	2	3	1	2	3	1	3	1	2	1	2	3
15	2	2	3	1	3	1	2	1	2	3	2	3	1
16	2	3	1	2	1	2	3	3	1	2	2	3	1
17	2	3	1	2	2	3	1	1	2	3	3	1	2
18	2	3	1	2	3	1	2	2	3	1	1	2	3
19	3	1	3	2	1	3	2	1	3	2	1	3	2
20	3	1	3	2	2	1	3	2	1	3	2	1	3
21	3	1	3	2	3	2	1	3	2	1	3	2	1
22	3	2	1	3	1	3	2	2	1	3	3	2	1
23	3	2	1	3	2	1	3	3	2	1	1	3	2
24	3	2	1	3	3	2	1	1	3	2	2	1	3
25	3	3	2	1	1	3	2	3	2	1	2	1	3
26	3	3	2	1	2	1	3	1	3	2	3	2	1
27	3	3	2	1	3	2	1	2	1	3	1	3	2
Components	a	b	a b	a b^2	c	a c	a c^2	b c	a b c	a b^2 c^2	b c^2	a b^2 c	a b c^2
	Group 1	Group 2			Group 3								

Appendix Table 4 (cont.)

2-Column Interaction Table for (L_{27} (3^{13}))

Column \ Column	1	2	3	4	5	6	7	8	9	10	11	12	13
	(1)	3 4	2 4	2 3	6 7	5 7	5 6	9 10	8 10	8 9	12 13	11 13	11 12
		(2)	1 4	1 3	8 11	9 12	10 13	5 11	6 12	7 13	5 8	6 9	7 10
			(3)	1 2	9 13	10 11	8 12	7 12	5 13	6 11	6 10	7 8	5 9
				(4)	10 12	8 13	9 11	6 13	7 11	5 12	7 9	5 10	6 8
					(5)	1 7	1 6	2 11	3 13	4 12	2 8	4 10	3 9
						(6)	1 5	4 13	2 12	3 11	3 10	2 9	4 8
							(7)	3 12	4 11	2 13	4 9	3 8	2 10
								(8)	1 10	1 9	2 5	3 7	4 6
									(9)	1 8	4 7	2 6	3 5
										(10)	3 6	4 5	2 7
											(11)	1 13	1 12
												(12)	1 11

(1)

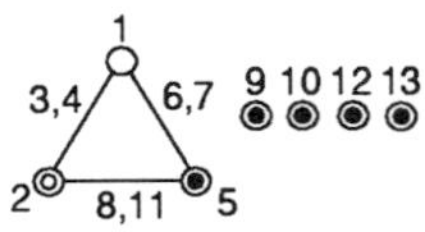

(2)

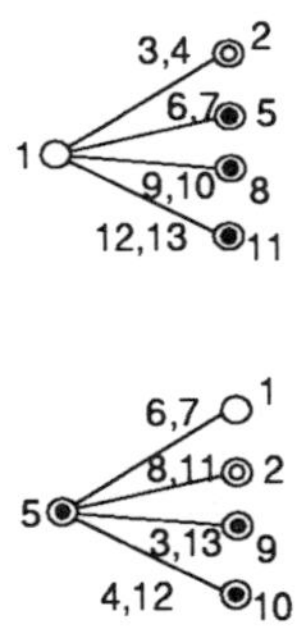

Appendix Table 4 (cont.)

$L_{12}(2^{11})$

No. \ Column	1	2	3	4	5	6	7	8	9	10	11
1	1	1	1	1	1	1	1	1	1	1	1
2	1	1	1	1	1	2	2	2	2	2	2
3	1	1	2	2	2	1	1	1	2	2	2
4	1	2	1	2	2	1	2	2	1	1	2
5	1	2	2	1	2	2	1	2	1	2	1
6	1	2	2	2	1	2	2	1	2	1	1
7	2	1	2	2	1	1	2	2	1	2	1
8	2	1	2	1	2	2	2	1	1	1	2
9	2	1	1	2	2	2	1	2	2	1	1
10	2	2	2	1	1	1	1	2	2	1	2
11	2	2	1	2	1	2	1	1	1	2	2
12	2	2	1	1	2	1	2	1	2	2	1

The interaction between two columns confounds little by little to the other nine columns. In order to determine interactions, it is necessary to use sequential analysis. Do not use this array for the experiments which require interactions.

$L_{18}(2^1 \times 3^7)$

No. \ Column	1	2	3	4	5	6	7	8
1	1	1	1	1	1	1	1	1
2	1	1	2	2	2	2	2	2
3	1	1	3	3	3	3	3	3
4	1	2	1	1	2	2	3	3
5	1	2	2	2	3	3	1	1
6	1	2	3	3	1	1	2	2
7	1	3	1	2	1	3	2	3
8	1	3	2	3	2	1	3	1
9	1	3	3	1	3	2	1	2
10	2	1	1	3	3	2	2	1
11	2	1	2	1	1	3	3	2
12	2	1	3	2	2	1	1	3
13	2	2	1	2	3	1	3	2
14	2	2	2	3	1	2	1	3
15	2	2	3	1	2	3	2	1
16	2	3	1	3	2	3	1	2
17	2	3	2	1	3	1	2	3
18	2	3	3	2	1	2	3	1
	Group 1	Group 2	Group 3					

(1)

(Interaction can be found without sacrificing columns, by using the two-way layout of columns 1 and 2.)
The interactions between three-level columns are partially confounded with the remaining three-level columns.

Appendix Table 4 (cont.)

(a) L_{36} ($2^{11} \times 3^{12}$)

Column / No.	1	2	3	4	5	6	7	8	9	10	11	12	13	14	15	16	17	18	19	20	21	22	23	1′	2′	3′	4′
1	1	1	1	1	1	1	1	1	1	1	1	1	1	1	1	1	1	1	1	1	1	1	1	1	1	1	1
2	1	1	1	1	1	1	1	1	1	1	1	2	2	2	2	2	2	2	2	2	2	2	2	1	1	1	1
3	1	1	1	1	1	1	1	1	1	1	1	3	3	3	3	3	3	3	3	3	3	3	3	1	1	1	1
4	1	1	1	1	1	2	2	2	2	2	2	1	1	1	1	2	2	2	2	3	3	3	3	1	2	2	1
5	1	1	1	1	1	2	2	2	2	2	2	2	2	2	2	3	3	3	3	1	1	1	1	1	2	2	1
6	1	1	1	1	1	2	2	2	2	2	2	3	3	3	3	1	1	1	1	2	2	2	2	1	2	2	1
7	1	1	2	2	2	1	1	1	2	2	2	1	1	2	3	1	2	3	3	1	2	2	3	2	1	2	1
8	1	1	2	2	2	1	1	1	2	2	2	2	2	3	1	2	3	1	1	2	3	3	1	2	1	2	1
9	1	1	2	2	2	1	1	1	2	2	2	3	3	1	2	3	1	2	2	3	1	1	2	2	1	2	1
10	1	2	1	2	2	1	2	2	1	1	2	1	1	3	2	1	3	2	3	2	1	3	2	2	2	1	1
11	1	2	1	2	2	1	2	2	1	1	2	2	2	1	3	2	1	3	1	3	2	1	3	2	2	1	1
12	1	2	1	2	2	1	2	2	1	1	2	3	3	2	1	3	2	1	2	1	3	2	1	2	2	1	1
13	1	2	2	1	2	2	1	2	1	2	1	1	2	3	1	3	2	1	3	3	2	1	2	1	1	1	2
14	1	2	2	1	2	2	1	2	1	2	1	2	3	1	2	1	3	2	1	1	3	2	3	1	1	1	2
15	1	2	2	1	2	2	1	2	1	2	1	3	1	2	3	2	1	3	2	2	1	3	1	1	1	1	2
16	1	2	2	2	1	2	2	1	2	1	1	1	2	3	2	1	1	3	2	3	3	2	1	1	2	2	2
17	1	2	2	2	1	2	2	1	2	1	1	2	3	1	3	2	2	1	3	1	1	3	2	1	2	2	2
18	1	2	2	2	1	2	2	1	2	1	1	3	1	2	1	3	3	2	1	2	2	1	3	1	2	2	2
19	2	1	2	2	1	1	2	2	1	2	1	1	2	1	3	3	3	1	2	2	1	2	3	2	1	2	2
20	2	1	2	2	1	1	2	2	1	2	1	2	3	2	1	1	1	2	3	3	2	3	1	2	1	2	2
21	2	1	2	2	1	1	2	2	1	2	1	3	1	3	2	2	2	3	1	1	3	1	2	2	1	2	2
22	2	1	2	1	2	2	2	1	1	1	2	1	2	2	3	3	1	2	1	1	3	3	2	2	2	1	2
23	2	1	2	1	2	2	2	1	1	1	2	2	3	3	1	1	2	3	2	2	1	1	3	2	2	1	2
24	2	1	2	1	2	2	2	1	1	1	2	3	1	1	2	2	3	1	3	3	2	2	1	2	2	1	2
25	2	1	1	2	2	2	1	2	2	1	1	1	3	2	1	2	3	3	1	3	1	2	2	1	1	1	3
26	2	1	1	2	2	2	1	2	2	1	1	2	1	3	2	3	1	1	2	1	2	3	3	1	1	1	3
27	2	1	1	2	2	2	1	2	2	1	1	3	2	1	3	1	2	2	3	2	3	1	1	1	1	1	3
28	2	2	2	1	1	1	1	2	2	1	2	1	3	2	2	2	1	1	3	2	3	1	3	1	2	2	3
29	2	2	2	1	1	1	1	2	2	1	2	2	1	3	3	3	2	2	1	3	1	2	1	1	2	2	3
30	2	2	2	1	1	1	1	2	2	1	2	3	2	1	1	1	3	3	2	1	2	3	2	1	2	2	3
31	2	2	1	2	1	2	1	1	1	2	2	1	3	3	3	2	3	2	2	1	2	1	1	2	1	2	3
32	2	2	1	2	1	2	1	1	1	2	2	2	1	1	1	3	1	3	3	2	3	2	2	2	1	2	3
33	2	2	1	2	1	2	1	1	1	2	2	3	2	2	2	1	2	1	1	3	1	3	3	2	1	2	3
34	2	2	1	1	2	1	2	1	2	2	1	1	3	1	2	3	2	3	1	2	2	3	1	2	2	1	3
35	2	2	1	1	2	1	2	1	2	2	1	2	1	2	3	1	3	1	2	3	3	1	2	2	2	1	3
36	2	2	1	1	2	1	2	1	2	2	1	3	2	3	1	2	1	2	3	1	1	2	3	2	2	1	3
Group	1	2										3															

Notes:

A. Substitute 1′, 2′, 3′ and 4′ for columns 1,2, ..., 11 to obtain L_{36} ($2^3 \times 3^{13}$).

B. In L_{36} ($2^{11} \times 3^{12}$), interactions are not orthogonal to other columns. Avoid assigning such interactions.

C. The assignment type is shown here only for L_{36} ($2^3 \times 3^{13}$).

Appendix Table 4 (cont)

(2) L_{36} (3^{13})

No. \ Column	1	2	3	4	5	6	7	8	9	10	11	12	13
1	1	1	1	1	1	1	1	1	1	1	1	1	1
2	2	2	2	2	2	2	2	2	2	2	2	2	1
3	3	3	3	3	3	3	3	3	3	3	3	3	1
4	1	1	1	1	2	2	2	2	3	3	3	3	1
5	2	2	2	2	3	3	3	3	1	1	1	1	1
6	3	3	3	3	1	1	1	1	2	2	2	2	1
7	1	1	2	3	1	2	3	3	1	2	2	3	1
8	2	2	3	1	2	3	1	1	2	3	3	1	1
9	3	3	1	2	3	1	2	2	3	1	1	2	1
10	1	1	3	2	1	3	2	3	2	1	3	2	1
11	2	2	1	3	2	1	3	1	3	2	1	3	1
12	3	3	2	1	3	2	1	2	1	3	2	1	1
13	1	2	3	1	3	2	1	3	3	2	1	2	2
14	2	3	1	2	1	3	2	1	1	3	2	3	2
15	3	1	2	3	2	1	3	2	2	1	3	1	2
16	1	2	3	2	1	1	3	2	3	3	2	1	2
17	2	3	1	3	2	2	1	3	1	1	3	2	2
18	3	1	2	1	3	3	2	1	2	2	1	3	2
19	1	2	1	3	3	3	1	2	2	1	2	3	2
20	2	3	2	1	1	1	2	3	3	2	3	1	2
21	3	1	3	2	2	2	3	1	1	3	1	2	2
22	1	2	2	3	3	1	2	1	1	3	3	2	2
23	2	3	3	1	1	2	3	2	2	1	1	3	2
24	3	1	1	2	2	3	1	3	3	2	2	1	2
25	1	3	2	1	2	3	3	1	3	1	2	2	3
26	2	1	3	2	3	1	1	2	1	2	3	3	3
27	3	2	1	3	1	2	2	3	2	3	1	1	3
28	1	3	2	2	2	1	1	3	2	3	1	3	3
29	2	1	3	3	3	2	2	1	3	1	2	1	3
30	3	2	1	1	1	3	3	2	1	2	3	2	3
31	1	3	3	3	2	3	2	2	1	2	1	1	3
32	2	1	1	1	3	1	3	3	2	3	2	2	3
33	3	2	2	2	1	2	1	1	3	1	3	3	3
34	1	3	1	2	3	2	3	1	2	2	3	1	3
35	2	1	2	3	1	3	1	2	3	3	1	2	3
36	3	2	3	1	2	1	2	3	1	1	2	3	3

INDEX

E

F

G

H

I

L

M

N

O

P

Q

R

S

T

U

V